Lecture Notes in Chemistry

Edited by G. Berthier, M
K. Fukui, H. Hartmann, H
W. Kutzelnigg, K. Rueden E. Scrocco, W. Zeil

10

Jens Ulstrup

Charge Transfer Processes in Condensed Media

Springer-Verlag
Berlin Heidelberg New York 1979

Author
Jens Ulstrup
The Technical University of Denmark
Chemistry Department A
Building 207
DK-2800 Lyngby

ISBN 3-540-09520-9 Springer-Verlag Berlin Heidelberg New York
ISBN 0-387-09520-9 Springer-Verlag New York Heidelberg Berlin

Printing and binding: Beltz Offsetdruck, Hemsbach/Bergstr.
2152/3140-543210

PREFACE

Molecules in liquid and solid media are exposed to strong interaction forces from the surrounding medium. The formulation of a comprehensive theory of chemical processes in condensed media is consequently an elaborate task involving concepts from several areas of the natural sciences. Within the last two and a half decades very notable results towards the formulation of a 'unified' quantum mechanical theory of such processes have in fact been achieved, and by the variety of physical, chemical, and biological processes which can be suitably covered by this framework, the new theory represents an adequate alternative to the transition state theory.

The present work has a two-fold purpose. Firstly, to provide a reasonably organized exposition of some basic aspects of these developments. This part emphasizes the fundamental similarities between chemical and other kinds of radiationless processes and includes the derivation of the most important rate expressions without resorting to involved mathematical techniques. The second major purpose is to illustrate the 'unified' character of the rate theory by analysis of a considerable amount of experimental data from both 'conventional' kinetics and from such untraditional areas as low-temperature, strongly exothermic, and biological processes. Particular attention is here given to those systems for which a classical description is inadequate, and which provide a diagnostic distinction between several alternative theoretical approaches.

Over the years I have had the great priviledge of enjoying a most inspiring collaboration with Professor R.R.Dogonadze and Drs.A.M.Kuznetsov, Yu.I.Kharkats, A.A.Kornyshev, E.D.German, and M.A.Vorotyntsev at the Institute of Electrochemistry of the Academy of Sciences of the USSR. I am extremely grateful to these people and to the directors of the institute, the late Academician A.N.Frumkin and Professor V.E.Kazarinov for the warm hospitality which I have enjoyed during several visits to the institute. I am also very grateful to Professor J.Jortner, Tel-Aviv University for our long collaboration and for the stimuli which initiated the present work.

I am much indebted to Professor H.Gerischer, Fritz-Haber-Institut der Max-Planck-Gesellschaft, Berlin, for hospitality over a six months visit during which part of the present work was completed, and to the following people for collaboration and many good talks over the years: Drs.P.Bindra and K.Doblhofer at the Fritz-Haber-Institut, Dr.W.Schmickler, the University of Bonn, Dr.P.P.Schmidt, University of Rochester, Michigan, and Dr.P.E.Sorensen, Miss N.Brüniche-Olsen, Mrs.A.M.Kjaer, and Mr.N.C.Sondergard, Chemistry Department A.

I would like to acknowledge encouragement and support from Professors N.Hofman-Bang and F.Woldbye, Chemistry Department A, and financial support from the Danish Ministry of Education, Statens Naturvidenskabelige Forskningsrad, and Otto Monsteds Fond.

I would finally like to thank Mrs.B.Rasmussen who prepared the manuscript very conscientiously using a text editing programme from the University of Waterloo. It was installed at the Technical University Computer Centre by Dr. Per Trinderup, whom I thank for a modification for this particular purpose.

Lyngby, April 1979

Jens Ulstrup

Table of Contents

1 INTRODUCTION

1.1 Nature of Elementary Chemical Processes

Electron and atom group transfer from a solvated donor to a solvated acceptor molecule, or between a solvated molecule and a macroscopic solid body such as an electrode or a membrane, have attracted attention since the days of Grotthuss, Davy, and Faraday(1). Such processes are of crucial importance in chemical, biological, and physical processes, which are conveniently viewed as individual or a series of consecutive reaction steps, the molecular nature of which involves electron or atom group transfer. Thus, in chemical and biological systems in liquid or solid media the elementary reaction steps most frequently consist of a synchronous bond break and bond formation in atom group transfer, or the transfer of an electron without accompanying bond breaks. Elementary chemical reaction steps are moreover conceptually closely related to such physical processes as radiative and radiationless electronic transitions in large isolated molecules(2,3) and in impurity centres in solid matrices(3,4), to defect diffusion in solids(5), and to nuclear tunnelling phenomena as manifested by 'abnormal' heat capacity effects(6-8) and vibrational level splitting(9,10).

The concept of elementary processes as a convenient notation for the simplest steps into which a composite process can be meaningfully split was invoked early in attempts to understand the microscopic nature of liquid state processes. Thus, in 1924, i.e. almost immediately after the introduction of Bronsted's acid base concept, Bronsted and Pedersen rationalized acid and base catalyzed reactions in terms of a general mechanism involving proton transfer(11), and in 1931 Gurney formulated the first quantum mechanical theory of electron transfer processes based on quantum mechanical electron tunnelling(12). However, the detailed molecular nature of elementary chemical processes in condensed phases, such as the dynamic role of the medium, the effect of high-frequency intramolecular modes, and the coupling between electronic and nuclear motion has only been studied relatively recently. This was associated with both the introduction of new relaxation and flow techniques for the study of fast reactions and with the formulation of new semiclassical and quantum mechanical theories of elementary chemical processes in condensed media.

The transfer of an electron from a solvated donor to a solvated acceptor in a solid or liquid medium represents the simplest conceivable chemical reaction. Such reactions can also be studied directly, since certain classes of simple chemical reactions can be regarded as possessing the character of 'isolated' elementary reactions. The most thoroughly studied representative of this class of processes is outer sphere electron transfer between transition metal complexes(13-16). The latter can often exist as stable entities in several oxidation states, and they can therefore also be conveniently followed experimentally. A well known example of a homonuclear outer sphere electron transfer reaction is

$$MnO_4^{2-} + MnO_4^{-} \rightarrow MnO_4^{-} + MnO_4^{2-} \qquad (1.1)$$

which can be followed by isotope substitution or by NMR line broadening of the central manganese atom. During this process the coordination spheres remain intact, and the overall process is controlled by electron transfer. An example of an outer sphere heteronuclear electron transfer reaction is

$$[Fe(CN)_6]^{4-} + [IrCl_6]^{3-} \rightarrow [Fe(CN)_6]^{3-} + [IrCl_6]^{4-} \qquad (1.2)$$

which is thus accompanied by a net chemical change. More involved examples of elementary reactions are found in the following processes

$$[(NH_3)_5CoCl]^{2+} + [Cr(H_2O)_6]^{2+} + 5H_2O \rightarrow$$

$$[Co(H_2O)_6]^{2+} + [ClCr(H_2O)_5]^{2+} + 5NH_3 \qquad (1.3)$$

or

$$-\overset{|}{C}H-C = O \rightarrow -\overset{|}{C} = C-OH \qquad (1.4)$$

The former reaction proceeds by an inner sphere mechanism via an intermediate binuclear complex of the form(13,16)

$$[(NH_3)_5Co-Cl-Cr(H_2O)_5]^{4+} \qquad (1.5)$$

The actual charge transfer in eq.(1.3) is thus both preceded and succeeded by ligand substitution, i.e. by other elementary steps. Moreover, the molecular nature of the elementary charge transfer itself may be viewed in several ways(16,17). For example, the chemical conversion may proceed by direct electron transfer from chromium(II) to cobalt(III) in which case the bridge ligand only brings the two centres together. The electron transfer may also proceed via an intermediate real or virtual electronic state using orbitals localized on the bridge group, or the reaction may be viewed as an atom group transfer(18).

The proton transfer from carbon to oxygen in the keto-enol conversion (eq.(1.4)) also involves a succession of elementary steps, i.e. proton transfers. The latter may be consecutive, i.e. independent elementary steps, or in some way coupled, i.e. the system has not relaxed after a given proton transfer before the next one occurs(19,20). We should notice that due to the long geometric distance between the carbon donor and the oxygen acceptor atoms a direct proton transfer, i.e. not involving 'proton mediating molecules' is excluded.

When chemical processes are viewed as being composed of consecutive steps, complete relaxation of all nuclear modes between each step is implicitly assumed. Each intermediate state can thus be represented by a potential energy minimum on a many-dimensional potential energy surface, and each step proceeds independently of the previous ones and statistically averaged over each new set of 'initial' energy values. This condition can be expressed as

$$\tau_r \ll [W_{fi}(E)]^{-1} \tag{1.6}$$

where τ_r is the average relaxation time of all system modes, and $W_{fi}(E)$ the reaction probability for a given value of the total energy E.

However, 'memory effects', from previous steps may operate in the sense that the nuclear modes do not or only partially relax before the subsequent step occurs. Eq.(1.6) is then no longer valid. Successive elementary steps proceed 'synchronously' or in a 'concerted' fashion (cf. the inner sphere electron transfer reaction and the keto-enol conversion), and all steps coupled in this way are conveniently viewed as a single, more involved elementary process. Synchronous electron transfer in which an electron is transferred from a donor to an intermediate state 'in concert' with an electron transfer from an intermediate state to the acceptor was invoked early in the theory of antiferromagnetic coupling(21,22) and electronic conduction in transition metal oxides(23). The concept was introduced as a possible mechanism for inner sphere electron transfer by Taube(24) and will also appear in our subsequent discussions of different types of chemical processes.

Most contemporary theoretical work on elementary condensed phase processes has focused on the calculation of the reaction probability of the process at infinite reactant dilution and for a fixed relative orientation of the reactants and products. For many applications of the theory, e.g. analysis of relations between rates and certain important parameters such as the free energy of reaction or the temperature, consideration only of the elementary act suffices, other effects being assumed constant or unimportant. However, the validity of several implications of these assumptions, should be checked. We thus notice

(A) Consideration solely of an elementary chemical process requires that the velocity of the process is determined by the particular electron or atom group transfer. We are thus interested in processes proceeding in the kinetic regime and ignore diffusion, unless the latter itself represents the elementary process. This requires in particular that the following condition is met

$$\tau_d \ll W_{fi}^{-1} \tag{1.7}$$

where τ_d is the average time during which the reactants are located sufficiently close to eachother (the diffusion time or the location time in a solvent cavity), and W_{fi} the statistically averaged reaction probability per unit time (W_{fi}^{-1} is then the average time between individual reaction acts). We shall thus consider 'slow' reactions only, and the statistical equilibrium between the reactants and the medium heat bath is therefore only disturbed locally on a molecular scale due to thermal fluctuations of the parameters of the field of interaction between the reactants and the medium.

(B) In most formulations of the theory of the rate of elementary chemical liquid state processes the reaction probability, $W_{fi}(\vec{R})$, is calculated for a given relative orientation of the reactants, characterized by a vector $\vec{R}$. $W_{fi}(\vec{R})$ is subsequently averaged with respect to $\vec{R}$ by means of the quantum statistical distribution function $\Phi(\vec{R})$ which expresses the probability that the configuration $\vec{R}$ is achieved(25). The procedure may have to be modified for other cases where the reactant and product distribution does not correspond to equilibrium. For example, for solid-state electron transfer processes between donor-acceptor couples which are randomly distributed with respect to energy and electron transfer distance, averaging with respect to these two parameters gives a rate constant which depends exponentially on the reciprocal of the fourth root of the temperature, instead of the Arrhenius-like dependence corresponding to an equilibrium distribution. In the classical limit of reactant motion $\Phi(\bar{R})$ takes the form

$$\Phi(\vec{R}) = Z_{\vec{R}}^{-1} \exp[-G(\vec{R})/k_B T] \qquad (1.8)$$

where $G(\vec{R})$ includes all free energy contributions necessary to achieve the configuration $\vec{R}$ from the infinitely separated reactants, e.g. coulomb interaction between charged reactants, energies required to break hydrogen bonds or for a partial desolvation of reactant species, and the effects of interaction with all other ions in the reaction zone. Moreover, k_B is Boltzmann's constant, T the absolute temperature, and $Z_{\vec{R}}$ the partition function, i.e.

$$Z_{\vec{R}} = \int \exp[-G(\vec{R})/k_B T]d\vec{R} \qquad (1.9)$$

By this 'quasistatic' averaging procedure the rate constant, k_r, takes the form

$$k_r = \int_{\vec{R}_{min}}^{\infty} \Phi(\vec{R}) W_{fi}(\vec{R}) d\vec{R} \qquad (1.10)$$

where the integration extends over all space outside the reactants, the volume of which is characterized by the quantity $\vec{R}_{min}$.

$\Phi(\vec{R})$ is a rapidly decreasing function of $\vec{R}$. On the other hand, the strongest dependence of $W_{fi}(\vec{R})$ on $\vec{R}$ is in the factors referring to an electronic transmission coefficient or to nuclear tunnelling through a potential barrier. Thus, if $W_{fi}(\vec{R})$ takes the approximate form

$$W_{fi}(\vec{R}) \approx A(\vec{R}) \exp(-E_A^{app}/k_BT) \tag{1.11}$$

where E_A^{app} ($\approx -d\ln W_{fi}(\vec{R})/d(1/k_BT)$) is an apparent activation energy which depends relatively slowly on $\vec{R}$, eq.(1.10) can be approximated by

$$k_r \approx \exp(-E_A^{app}/k_BT) \int_{\vec{R}_{min}}^{\infty} \Phi(\vec{R})A(\vec{R})d\vec{R} \approx$$

$$\exp(-E_A^{app}/k_BT)\Phi(\vec{R}^*)A(\vec{R}^*)\Delta\vec{R} = Z_{\vec{R}}^{-1}A(\vec{R}^*)\Delta\vec{R} \tag{1.12}$$

$$\exp -[G(\vec{R}^*)+E_A^{app}]/k_BT$$

$\vec{R}^*$ is the value of $\vec{R}$ for which the integral of eq.(1.10) or (1.12) is maximum, and $\Delta\vec{R}$ the extension of the 'effective' reaction zone.

Within this simplified description the role of the relative orientation of the reactants is thus reduced to a calculation of $W_{fi}(\vec{R})$ for fixed $\vec{R}$ and a subsequent averaging with respect to these coordinates. Ideologically this approach is in line with the assumption of prevalence of equilibrium in the reacting system and with the slow motion of the reactants as a whole compared to other nuclear modes. Conceptually it is equivalent to an adiabatic approximation in which the slow subsystem is that of the relative motion of the ions, and the solvent fluctuations and the intramolecular modes constitute the fast subsystem. However, even though eqs.(1.10) and (1.12) are adequate in most cases of electron and light atom group transfer we should notice reservations of two kinds: (a) We shall see in later sections that the expression for W_{fi} generally is much more complicated than implied by eq.(1.11). In particular, over sufficiently wide intervals of the free energy of reaction and the temperature, vibrationally excited states of the intramolecular modes of the reactants contribute to the overall reaction. In these cases $A(\vec{R}^*)$ displays a more complicated dependence on both these parameters and on $\vec{R}^*$, which itself becomes a function of the free energy change and the temperature. (b) The coordinates R really exert a dynamic role in the reaction which is not revealed by the 'static' averaging procedure above. For the reaction to proceed, distortion from equilibrium values must occur in a way analogous to the role of other classical modes in the system(26). Moreover, the motion along $\vec{R}$ can be of both classical and quantum (tunnelling) nature depending on the nature of the interaction between the reactants. For example, if the repulsion between donor and acceptor at small distances does not rise sufficiently sharply within given (small) $\vec{R}$-intervals the most favourable reaction path may involve tunnelling of the reactant molecules as a whole, at least for small molecules, through a classically forbidden region and from a position which has been reached by an otherwise classical motion. This finally implies that the nature (i.e. quantum or classical) of the relative motion of reactants and products in the transition region may change during the process.

(C) The study of elementary chemical rate processes involves essentially a consideration of two aspects, viz. the elementary act itself and a statistical aspect which deals with the spacial distribution of reactants and products. The latter is determined by a distribution function which incorporates the interaction not only between the reactant and product molecules but also between these molecules and all other species in the medium, i.e. the ionic sphere. This effect is commonly incorporated by Debye-Hückel corrections for homogeneous processes(27,28), or by a suitable model for the double layer structure in electrochemical processes(26,29). However, the ionic atmosphere also exerts a dynamic role in the reaction and is subject to reorganization during the process(26,28). In contrast to most reported ionic strength or double layer corrections of homogeneous and heterogeneous charge transfer processes, a nonequilibrium distribution of the surrounding ionic sphere must therefore be invoked - in a way which is analogous to the role of medium field fluctuations. Correspondingly, if equilibrium distribution functions are applied the appropriate ionic atmosphere in the transition state would be one equivalent to a partial charge transfer (i.e. an ionic distribution intermediate between the equilibrium distributions in the initial and final states).

The reservations (A)-(C) are seen to refer to the application of quantum mechanical rate theories rather than to the general formalism. They can be relaxed when sufficiently good representations of the actual reaction models are available. With this in mind we shall now proceed to an outline of the general theoretical framework for elementary rate processes in condensed phases, to its conceptual and formal relation to other molecular and condensed phase processes, and to a discussion of several recent extensions of its range of application. At first we notice, however, that throughout the last couples of decades the study of simple electron and atom group transfer reactions has occupied a prominent place in inorganic chemistry and electrochemistry. This is associated primarily with the fact that simple one-electron inorganic redox processes can be followed conveniently experimentally and also constitute a sufficiently simple category of chemical processes that they can be subject to comprehensive theoretical treatment which can be expected to agree well with experimental data. Many experiments have thus been analyzed in terms of the semiclassical formulations of electron transfer theory of Marcus and designed with the aim of testing this theory(30); this interplay between theory and experiment has certainly contributed immensely to an understanding of the nature of simple liquid state chemical processes. We shall discuss a few of these investigations in later sections, and others may be found in several previous reviews on both experimental and theoretical aspects of homogeneous(13-17,25,30-33) and heterogeneous(25,30,34-36) redox processes. However, although the importance of this work cannot be underestimated we should notice that much of this apparent agreement between theory and experiment can in fact be understood on the basis of very general assumptions. Experimental verification of the more subtle points of the elaborate quantum theory of condensed phase reactions - such as the effect of high-frequency quantum modes or nuclear tunnelling - has generally only very recently been achieved. On the other hand, several of these experiments, to which we shall return in the following, suggest the line of approach towards a test of the fundamental results of this theory.

1.2 Development of Theories for Elementary Chemical Processes

The earlier attempts towards a theoretical estimate of rate constants for homogeneous and heterogeneous electron and atom group transfer processes based on electron tunnelling and atom group transfer either by classical passage of an activation barrier or by nuclear tunnelling through the same, stationary barrier have been comprehensively reviewed in particular by Bell(19), Marcus(30), and by Bockris and his associates(35,36). We shall therefore only discuss a few features of particular interest in the light of later development. Thus, the first quantitative formulation of an electron transfer theory was that of Gurney in 1931(12). Gurney viewed the electron transfer between a metallic electrode and a depolariser ion close to the electrode surface as proceeding by electron tunnelling between a metal electrode level and a donor/acceptor level of the electrolyte in a way analogous to thermal electron emission. For the electron transfer to proceed as a radiationless process this requires that the energies of the donor and acceptor levels coincide. This is possible provided that (for the cathodic process) the ionization potential plus the solvation energy exceeds the work function of the metal (equal to the energy of the electron at the Fermi level). If this condition is not met at the equilibrium potential the process may proceed if the cathodic potential of the electrode with respect to the solution is increased, this leading to a shift of all the metal electronic levels. The electrons in the metal were assumed to be distributed according to a Fermi-Dirac distribution law, the electrolyte levels distributed by a Boltzmann law due to the strong coupling to a continuous manifold of vibrational and rotational medium modes, and the overall expression for the current density subsequently obtained by integration over the continuous energy spectrum.

As pointed out particularly by Bockris and his associates(35,36), Gurney's theory anticipated many fundamental results of much more recent work. Gurney's work thus included, although often implicitly: (a) the first clear understanding of the relation between current density and overvoltage for electrochemical processes; (b) introduction of the Franck Condon principle in the theoretical description of chemical processes; (c) the effect of strong coupling between electron and nuclear motion; (d) the effect of the continuous nature of the electronic energy spectrum of the metal electrode, and (e) a rationalization of the experimental data for the electrochemical hydrogen evolution reaction available at that time, in particular those relating to the current density dependence on electrode potential and temperature. Thus, the 'symmetry factor' (the Bronsted or Tafel coefficient) which determines the relation between the current density and the overpotential is given the first theoretical basis by the theory of Gurney.

On the other hand, seen in retrospect the theory of Gurney and later developments of this formalism display several shortcomings. Thus:(1) the proton is assumed to move classically. Quantum (tunnelling) corrections have, however, been introduced in later work by Conway(37,38), Bockris and Matthews(39) and by Christov(40), and we shall return to this question in later sections; (2) both the dynamic effect of the medium and the nature of the coupling between the electrons and the medium are not specified in an explicit manner. For example, the relation between coupling parameters and the medium properties does not enter the theory; (3) the electronic transmission coefficient of the rate constant is given effectively by the tunnelling probability. Although a tunnel factor formally similar to the Gamov factor can also be identified in present day multiphonon theories of electron and atom group transfer, its relation to the apparent transmission coefficient of the overall reaction is more involved. We shall return to this question also.

At the time of its appearence the theory of Gurney received little response. This was perhaps due to the novelty of applying quantum mechanics in the field of electrochemistry but also to the fact that in the context of the electrochemical hydrogen evolution reaction, the final state, i.e. the one prevailing after the proton discharge, was represented as a free hydrogen atom and a water molecule. It is known now that the final state of the electrochemical proton discharge should instead be represented as an adsorbed hydrogen atom, and it was shown first by Butler(41) that inclusion of this effect would lead to better values of the activation energy. Possibly as a consequence of this for the next two decades the attempts towards a theoretical formulation of elementary chemical processes took a different direction. These formulations were based almost entirely on adaptions of transition state theory, and work which is ideologically extensions of Gurney's only appeared in the 1960's. Thus, Gerischer(42) more explicitly considered both the role of the fluctuations of the solvation sheaths of the donor and acceptor molecules in simple electrochemical redox processes and the electronic spectrum of the substrate electrode. As a result he derived a rate expression valid for both metal and semiconductor electrodes. In the former electrons around the Fermi level provide the dominating contributions to the overall rate, whereas electrons at the top of the valence band and at the bottom of the conduction band contribute most for electron transfer via the valence and conduction band, respectively (hole and electron transfer). Bockris and Matthews(39) investigated several possible mechanisms of the proton discharge step of the electrochemical hydrogen evolution reaction and reached the conclusion that continuous stretching of the O-H bond of the depolariser H_3O^+ ion followed by electron tunnelling at the intersection point of a two-dimensional potential energy surface spanned by proton stretching coordinates constitutes the most plausible mechanism. We shall discuss this conclusion in later sections dealing with electrochemical processes and the theory of atom group transfer processes.

Some of the first attempts to describe homogeneous electron transfer processes, which appeared in the 1950's, were also based on electron tunnelling through a barrier of a simple rectangular or triangular form and to a considerable extent similar to the theory of Gurney. Thus, Marcus, Zwolinskij, and Eyring(43) viewed the electron transfer rate constant as a product of an electron tunnelling factor and an activation factor corresponding to a free energy of activation consisting of a coulomb repulsion term and a reorganization term. The latter was estimated by fitting to experimental data, and the electron transfer distance determined by maximizing the total rate expression with respect to this distance. Sacher and Laidler(44) also adopted this approach in essentials but included several modifications such as the variation of the solvent dielectric constant with the electric field of the ions.

Attempts to rationalize condensed phase elementary processes, and homogeneous and heterogeneous proton transfer reactions in particular, within the framework of transition state theory(45) can be traced back to the work of Horiuti and Polanyi(46), and Bell(19,47). It should also be recalled that the analogy between proton transfer in homogeneous processes and the electrochemical hydrogen evolution reaction was noted by Frumkin already in

1932(48), and that potential energy surfaces spanned by those nuclear coordinates which are reorganized during the reaction were also introduced in Gurney's theory. In a one-dimensional representation corresponding for example to motion of the proton from donor to acceptor the reaction is thus generally pictured as in fig.(1.1). In the initial state the energy has a minimum near some equilibrium value r_{io} (e.g. corresponding to the equilibrium

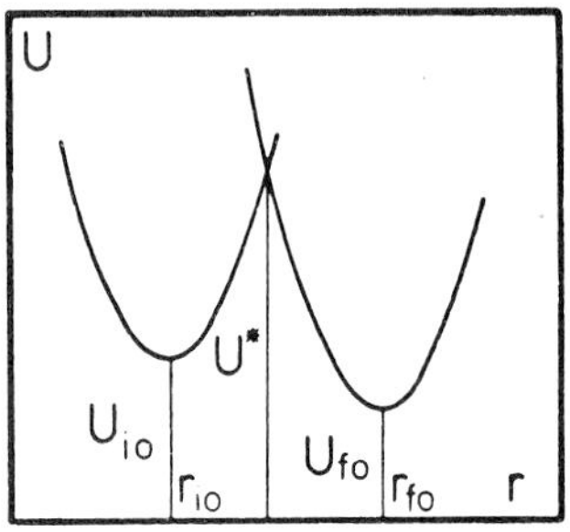

Fig (1.1)

position in the H_3O^+ ion) and in the final state near some different value r_{fo} (e.g. corresponding to an adsorbed hydrogen atom). In order to get from r_{io} to r_{fo} the system (the proton) must perform a thermally activated passage over the intervening potential energy barrier corresponding to the activation energy $U^* - U_{io}$. By means of this simplified approach a large amount of experimental data can in fact be explained. For example, a change of the overpotential for the electrochemical hydrogen evolution reaction or the difference in ΔpK values for the donor and acceptor in homogeneous processes shifts the relative position of the initial and final state surfaces vertically without distorting their form or horizontal position. This can explain the variation of rate or current density with ΔpK or overpotential, respectively. For approximately thermoneutral processes and potential curves of approximately the same (small) curvature, a plot of the logarithm of the rate constant against the heat of reaction over a wide interval of the latter is thus a straight line of slope 0.5, whereas this slope approaches the values zero and unity for strongly exothermic and endothermic processes, respectively. Similar considerations can explain the electrochemical rate dependence on the adsorption energy of the hydrogen atom on the metal, whereas the form of the potential surfaces can rationalize differences in the reaction patterns of substrate molecules or electrodes with different donor and acceptor atoms.

In the models of Bell and of Horiuti and Polanyi and in several later models(35,36) the coordinate spanning the potential surface was that of the proton. The bond between the proton and the donor fragment is thus continuously stretched until bonding with the acceptor fragment and electronic reorganization can occur. However, the fundamental conclusions concerning the rate dependence on the thermodynamic and structural properties of the donor and acceptor fragments are not associated with this particular choice of mechanism. Similar conclusions would be reached for other interpretations of the reaction coordinate r and for models of electron transfer as well. When diagnostic informatiom is to be extracted from experimental data it is therefore essential to focus on other, qualitative differences between the various theories and/or models.

We shall postpone a consideration of several important features of the theories briefly outlined so far to a later section dealing with the quantum mechanical formulation of atom group transfer theory and rather proceed to alternative approaches to a formulation of the rate theory, i.e. the theories of Marcus(49) and the earlier formulation of the theory of Levich, Dogonadze, and Kuznetsov(50-52). The concepts of the former bears some resemblance to the transition state theory. On the other hand, the latter represents the first quantum mechanical formulation of rate theory going beyond the static tunnelling concept and is formally closely analogous to the formulation of radiative and radiationless electronic transitions of trapped 'impurities' in crystalline and disordered solids. However, the concrete model for the reacting system first applied, i.e. that of hard structureless reactant and product charge distributions embedded in a continuous structureless dielectric medium, is common to both theories. Hence, the resulting rate expressions, although differing by the mathematical derivation, and physically, by the nature of the electronic coupling inducing the reaction, are also similar.

We recall at first that although the influence of the solvent was in principle taken into account in the earlier theories, its role - when explicitly stated - is usually static, rather than dynamic(27). Thus, the solvent may affect the geometry of the collision complex (for example, the electron or proton transfer distance, and the relative vertical position of the initial and final state surfaces corresponding to different solvation of reactants and products). The solvation of the activated state is furthermore generally expected to differ from that of both the reactants and products. On the other hand, the solvation of both the reactants, products, and the activated complex are assumed to correspond to equilibrium with the appropriate charge distribution of the molecular species, in other words, the reorientation of the solvent spheres is assumed to follow the motion of the electrons and intramolecular nuclear modes. Reorientation of solvent molecules is, however, a slow process compared to most intramolecular nuclear modes, and in particular compared to the electronic motion of the electron to be transferred in redox processes. The solvation of the activated complex in reactions involving electron transfer and intramolecular reorganization can therefore not be taken as the equilibrium value. In contrast, the solvent exerts a dynamic effect on the process analogous to other nuclear modes of the system, and it must therefore be viewed not solely as a medium and a heat bath but also as a part of the reacting system.

We can illustrate this role of the dynamic solvent further with reference to fig.(1.2). Suppose we consider a simple electron transfer reaction between structureless ions embedded in a dielectric continuum. The electron donor (D) and acceptor (A) levels are

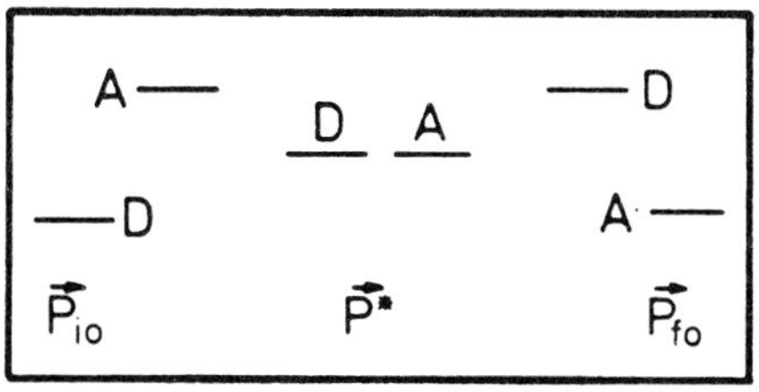

Fig (1.2)

represented by horizontal lines, and initially the corresponding energies differ strongly due to different strong coupling of the electronic charges of the donor and acceptor molecule to the dielectric medium. The initial state thus corresponds to an equilibrium value of the polarization vector $\vec{P} = \vec{P}_{io}$ (the induced dipole moment per unit volume). In this state the reaction cannot proceed without absorption or emission of electromagnetic radiation. Thus, due to the much lower speed of nuclear motion compared to electronic motion, stated by the Franck Condon principle, the electronic energy gap cannot be compensated by a change of the kinetic energy of the nuclei. On the other hand, due to the strong coupling between the electronic levels and the solvent modes, thermal fluctuations in the latter, i.e. thermal fluctuations in the polarization vector, also induce thermal fluctuations in the relative positions of the electronic energy levels. For some particular value of the polarization vector $\vec{P} = \vec{P}^*$, the levels coincide, and only at this value can the electron transfer occur as a radiationless process. Whether it actually does occur depends on the ratio between the characteristic times of the electron transfer itself and the time during which the polarization possesses such values that the donor and acceptor levels practically coincide. If the ratio is large, the polarization fluctuations must pass the value $\vec{P} = \vec{P}^*$ many times before the reaction succeeds, and the overall probability of reaction depends explicitly on the electronic coupling between the donor and acceptor levels (the nonadiabatic limit). On the other hand, if the ratio is small the reaction proceeds provided that the polarization reaches the value $\vec{P}^*$, and calculation of the reaction probability is essentially a calculation of the probability of the polarization value $\vec{P}^*$.

In the context of homogeneous electron transfer reactions the importance of these Franck Condon restrictions was first recognized by Libby(53), although in a somewhat different form. Thus, Libby viewed the electron transfer as occurring at the initial state equilibrium nuclear configuration which relaxes to its final state value after the transfer. The nonequilibrium polarization of the dielectric continuum was also considered by Weiss(54) in the context of thermal electron transfer processes, and by Platzmann and Franck(55) for optical transitions, but first incorporated in a quantitative theory of electron transfer by Marcus(30,49). However, before we discuss this theory we should recall that already prior to the work of Libby, Weiss, and Marcus a polarization concept closely related to the one above had been developed in general dielectric continuum theory(56) and for electrons trapped in polar crystals (the large polaron) in particular(57). Since the theoretical formulation of electronic processes in the latter systems also initiated the quantum mechanical approach to the theory of elementary chemical processes, we shall digress slightly for a consideration of these phenomena.

An electron in a stationary periodic field of the nuclei of a solid is characterized by delocalized wave functions and a set of continuous energy bands. On the other hand, if interaction with the lattice motion is taken into account, localized states with discrete energy levels may also be formed(57,58). Thus, the electric field of an excess electron in a polar medium, such as an alkali halide crystal or a solvent consisting of dipolar molecules, induces a polarization of the surrounding medium due to the strong interaction with the latter. The polarization involves a shift of the lattice nuclei from their equilibrium positions, but since the nuclei cannot follow the fast motion of the local electron, the distorted lattice in return constitutes a potential well for the electron, maintained stationary by the field of the latter. The trapped electron and its surrounding 'self-consistent' polarization field is called a polaron(51) and possesses in several ways particle-like properties. For example, the electron, together with the surrounding polarization can migrate through the medium, and it has an effective mass, which usually differs from the mass of a 'free' electron. Energetically it is localized in the band gap of the medium at a level which is determined by the degree of coupling to the nuclear modes.

Polaron mobility is of importance as a mechanism of charge transport in semiconductors, and the polaron serves also as a model for both solvated electrons and colour centres in crystals. With a view to the nonequilibrium polarization prevailing in electron transfer systems we shall consider the nature of the induced polarization by describing a process by which this polarization can be reversibly formed. Provided that the radius of the trapped electron is larger than the lattice constant (large polaron), details of the lattice structure can be ignored, and the energy of the system calculated by macroscopic electrostatic continuum theory. If this condition is not met (small polaron), structural details must also be considered, and this is commonly done by the introduction of potential energy surfaces spanned by the appropriate coordinates(58). For the sake of simplicity we shall adopt the former assumption at present. Thus, the total polarization, $\vec{P}_t$, consists of two components

$$\vec{P}_t = \vec{P}_{ir} + \vec{P}_e \tag{1.13}$$

of which the infrared polarization, $\vec{P}_{ir}$, is the response caused by (slow) atomic or reorientational motion, and the electronic (optical) component, $\vec{P}_e$, corresponds to (fast) electronic polarization of each medium molecule. However, in a uniform medium the 'external' electrons respond instantaneously to the field of the excess electron, and this polarization component is therefore implicit in the periodic field of the medium in the absence of trapping. The particular ionic configuration of the polaron may then be caused by an initial charging process which is so slow that both components of eq.(1.13) respond. The corresponding final value of the field causing the polarization, $\vec{D}$, is given by the equation

$$\vec{P}_t = \frac{\varepsilon_s - 1}{4\pi\varepsilon_s} \vec{D} \tag{1.14}$$

where ε_s is the static dielectric constant of the medium. If the field is subsequently rapidly switched off, the infrared component maintains its value, whereas the contribution

$$\vec{P}_e = \frac{\varepsilon_o - 1}{4\pi\varepsilon_o} \vec{D} \tag{1.15}$$

vanishes. ε_o is here the dielectric constant of the electronic polarization ($\varepsilon_o = n^2$, where n is the refractive index). The polarization of importance is thus

$$\vec{P}_{ir} = \frac{c}{4\pi} \vec{D} \quad ; \quad c = \frac{1}{\varepsilon_o} - \frac{1}{\varepsilon_s} \tag{1.16}$$

where $\vec{P}$ also contains the electronic polarization induced by the stationary shift of the nuclear positions. The potential free energy of the lattice, U_p, is correspondingly

$$U_p = \int dV \left\{ \int_0^{\vec{D}} \vec{D}\, d\vec{P}_t + \int_{\vec{D}}^{0} \vec{D}\, d\vec{P}_e \right\} = \frac{c}{8\pi} \int \vec{D}^2 dV \qquad (1.17)$$

where the integration with respect to V includes the whole volume. It is more convenient, however, to express U_p by the polarization which corresponds to the actual nuclear configuration rather than by the, so far arbitrary $\vec{D}$. Thus, in view of eq.(1.16)

$$U_p = \frac{2\pi}{c} \int \vec{P}_{ir}^2 dV \qquad (1.18)$$

A total characterization of the polarized medium, i.e. its Hamiltonian function, H, is obtained by adding the kinetic energy of the trapped electron, the kinetic energy of the nuclei, and the potential energy of interaction between the electron and the nuclear polarization. If we ignore effects of frequency and space dispersion of the medium nuclear modes(chapter 2) and only consider a single nuclear vibration frequency ω, H thus takes the form

$$H = \frac{P_e}{2\mu} + \frac{2\pi}{c} \int \left\{ \vec{P}_{ir}^2 + \omega^{-2} \dot{\vec{P}}_{ir}^2 \right\} dV - \int \vec{P}_{ir} \vec{D}_e dV \qquad (1.19)$$

p_e is the momentum of the trapped electron and μ its effective mass (by introducing the effective mass we have incorporated the effect of the periodic lattice). $\dot{\vec{P}}_{ir}$ is the rate of change of $\vec{P}_{ir}$ and is a measure of the kinetic energy associated with the dynamic polarization (see further chapter 2). Finally, $\vec{D}_e$ is the field created by the excess electron. If the electronic wave function of the latter is $\psi(\vec{r})$, then, provided that we ignore anisotropy and nonuniformity effects, $\vec{D}_e(\vec{r})$ at the point $\vec{r}$ is

$$\vec{D}_e(\vec{r}) = -e \int |\psi(\vec{r})|^2 \frac{\vec{r} - \vec{r}'}{|\vec{r} - \vec{r}'|^3} d\vec{r}' \qquad (1.20)$$

corresponding to a charge density $e|\psi(\vec{r})|^2$ (e is the charge of the electron).

We shall now exploit the concept of the inertial polarization outlined for the polaron problem to the electron transfer system with particular reference to the theory of Marcus(49). We shall thus consider hard structureless ions embedded in a structureless 'simple' dielectric continuum. According to the theory of Marcus the reaction sequence is then viewed in the following way

$$D + A \underset{k_{-1}}{\overset{k_1}{\rightleftarrows}} (DA) \tag{1.21}$$

$$(DA) \underset{k_{-2}}{\overset{k_2}{\rightleftarrows}} (D^+A^-) \tag{1.22}$$

$$(D^+A^-) \xrightarrow{k_3} D^+ + A^- \tag{1.23}$$

Thus, initially a solvated donor, D, approaches a solvated acceptor, A, to form an encounter complex, (DA), the equilibrium geometry of which is determined by the (coulomb) interaction between D and A as a whole and by the interaction with the dielectric continuum. Anticipating the discussion of chapter 3 we notice that the interaction between the reactants may also cause a distortion of the electronic levels compared with the isolated reactants but no electronic reorganization, i.e. no redistribution of electronic charge among levels characterized by different electronic quantum numbers. After the electron transfer the products are located in a similar encounter state before they diffuse apart. In this state the polarization is different from that of the reactants due to the different charge distribution of the molecules. The solvent molecules are for example more rigidly orientated towards a highly charged species.

As noted previously, fluctuations in the medium polarization resulting in some value $\vec{P}^*$ intermediate between those of the equilibrium values of the initial and final encounter states shifts the energy values of the latter in such a way that they become energetically identical. For this value of the polarization the system is characterized by two electronic wave functions, φ and φ^*, corresponding to a localization of the electron largely on the donor or acceptor, respectively. The two states may conveniently be denoted as the reactants' and products' activated complex, respectively. The real wave function would be some linear combination of φ and φ^*, but according to the Franck Condon principle the important thing is that the total energies of the hypothetical systems corresponding to φ and φ^* in the activated state are identical. This emphasizes the nonequilibrium character of the polarization state, since the equilibrium values would depend strongly on the different charge distributions associated with φ and φ^*.

φ and φ^* thus refer to a particular atomic configuration, and the electron transfer step itself would correspond to a transition from φ to φ^* at this configuration. Moreover, within the simple dielectric continuum model the inertial polarization contribution, $\vec{P}_{ir}$, which is a measure of the nuclear configuration, must remain constant during the electron transfer in the activated complex, i.e. independent of the instantaneous charge distribution, whereas the electronic component follows the field of the transferring electron. Just as for the polaron problem, we shall thus be interested in the free energy required for a reversible formation of the state characterized by the polarization $\vec{P}_{ir}$. In contrast to the polaron problem, however, this is no longer an equilibrium state.

An infinite manifold of pairs of states φ and φ^* satisfy the energy restriction given. In calculating the free energy of formation of the activated state, we are interested in finding the most probable of these configurations. This is done by a variational calculus minimizing the expression for the free energy of formation of the activated state from the reactants in the encounter complex, ΔG^*, subject to the restriction that the two states, φ and φ^* have the same total free energy. Thus,

$$\delta \Delta G^* = 0 \qquad (1.24)$$

subject to

$$\delta \Delta G_o = 0 \qquad (1.25)$$

where ΔG_o is the standard free energy of reaction when the reactants and products are located in their encounter complexes. ΔG_o differs from the 'experimental standard free energy in the appropriate medium by the work terms, w_r and w_p required to bring the reactants and products, respectively, from an infinite separation to their location in the encounter complexes.

In the theory of Marcus the important part of the electron transfer problem is the calculation of the most probable intermediate state of coinciding donor and acceptor levels. Once this state is reached the coupling between the donor and acceptor states is assumed to be strong enough that the electron is transferred with a probability of unity, but sufficiently weak that any resonance interaction energy between the levels has no significant effect on the activation energy. For details of the minimization procedure implied by eqs.(1.24) and (1.25) reference can be made both to Marcus' original derivation(49) and to a simplified and illustrative procedure also derived by Marcus but more easily available in a report by Schmidt(59). Exploiting these results we give here the expression for the bimolecular electron transfer rate constant

$$k_r \approx k_2 = Z\exp\left\{-[w_r + (\Delta G_o + w_p - w_r + \lambda_o)^2/4\lambda_o]k_BT\right\} \qquad (1.26)$$

Z is a collision number for the bimolecular collision of uncharged reactants (usually taken as 10^{11} $dm^3 mol^{-1} s^{-1}$) estimated from the loss of translational degrees of freedom when the collision complex is formed from the separated reactants. This quantity could be estimated from some quenching process provided that the latter would be effective in each collision. λ_0 is the free energy of solvent reorganization, i.e. the free energy required to change the inertial polarization from the equilibrium value for the reactants to that of the products in the encounter complex. For a structureless dielectric it takes the general form

$$\lambda_o = \frac{c}{8\pi} \int (\vec{D}_f - \vec{D}_i)^2 dV \qquad (1.27)$$

where $\bar{D}_i$ and $\bar{D}_f$ are the induction vectors of the reactants' and products' charge distributions respectively, and the parameter c was defined earlier. Eqs.(1.27) and (1.17) are thus a manifest of the analogy between this 'Franck Condon barrier' to electron transfer and the free energy of polarization in the polaron problem (eq.(1.17)). For the particular case of spherical reactants of radii a_1 and a_2 and an effective electron transfer distance R (usually the sum of the radii at close contact) Marcus derived the following expression for λ_0

$$\lambda_o = \frac{c}{8\pi}(ne)^2\left(\frac{1}{2a_1} + \frac{1}{2a_2} - \frac{1}{R}\right) \qquad (1.28)$$

where n is the number of electrons transferred in an elementary step.

The following features of eq.(1.26) should now be noted:

(a) The solvent dependence is reflected in all four quantities of the activation free energy expression. However, while w_r, w_p, and ΔG_0 are equilibrium quantities, λ_0 represents an 'intrinsic' activation energy relating to a nonequilibrium solvent configuration associated with the presence of a fast electronic and a slow nuclear subsystem.

(b) The expression bears a formal similarity to the rate expression of absolute rate theory(45). However, this apparent similarity should not conceal the fact that the role of the solvent is viewed in fundamentally different ways in the two theories. In the transition state theory the role of the solvent is ascribed solely to different equilibrium solvation in the initial and transition states. In contrast, as noted in (a), according to the theory of Marcus the electron transfer is induced by deviation from equilibrium (polarization fluctuations), and the concept of a transition state has a different meaning.

(c) The dependence of the activation free energy on the free energy of reaction is quadratic and gives a maximum rate for $\Delta G_0 = -\lambda_0$. As shown later, this reflects the linear response of the medium to the field of the ions, i.e. the assumed proportionality between $\vec{P}$ and $\vec{D}$, and does not refer to any harmonic motion of individual molecules. (Strictly speaking a linear response to the repolarization only, since both the initial and final state polarization may be nonlinear). A quadratic dependence is therefore not restricted to a polar medium but would also be manifested in other kinds of fields than electric, for example in a pressure-density dependence. Accordingly, the quadratic free energy relationship represents a very unrestrictive model, as long as intramolecular modes are ignored.

Marcus' later development of the semiclassical formulation has provided a more general theory for both homogeneous and heterogeneous electron transfer processes, in which a number of previously ignored effects were included, in particular reorganization of intramolecular modes (in the classical limit). However, due to its conceptual relative simplicity and apparently close relationship to the transition state theory, the earlier form of the Marcus theory has so far maintained by far the greatest appeal in relation to experiments. This is of course supported by the fact that both absolute values of rate constants and correlations between kinetic and structural parameters in a number of systems have provided very good agreement with the theory (see below). Before we consider this we should, however, notice that the following features of the theory impose serious restrictions on the applicability to less 'conventional' systems such as strongly exothermic or low-temperature processes:

(1) Although the theory refers to adiabatic processes in the sense that the electron transfer probability is unity at the polarization $\vec{P}^*$, no explicit account of the adiabaticity effects on the activation energy is given. The calculation of this latter parameter (eqs.(1.24) and (1.25)) thus corresponds to the nonadiabatic limit. This question is discussed further in chapter 5.

(2) In practice only the activation energy is calculated, whereas quantum nuclear motion in intramolecular and medium modes is not incorporated. Such effects would be reflected in the pre-exponential factor and as a temperature dependent activation energy. In addition to shortcomings in the rationalization of high-temperature nuclear quantum effects (in particular kinetic isotope effects in proton transfer reactions) this essentially restricts the applicability of the theory to the 'normal' free energy range ($|\Delta G_0| < \lambda_0$) and to sufficiently high temperatures.

(3) The expression for the solvent reorganization energy (eq.(1.28)) refers to long-distance electron transfer, i.e. the solvent polarization from a given reactant is assumed not to be affected by the presence of the other reactant. In fact, the presence of this reactant ion excludes the corresponding volume from being repolarized and would thus give a smaller λ_0 than predicted by eq.(1.28) (see also chapter 7).

(4) The form of the Marcus theory commonly applied refers to a structureless dielectric medium characterized by a single nuclear mode and is an adaption of the dielectric continuum theory of Frohlich and Pekar. However, real media are characterized by a certain frequency dispersion, i.e. medium oscillators of different frequencies respond differently to the 'external' electric field of the ions. We shall see in later chapters that

incorporation of these effects provides a number of results (in particular free energy and temperature dependence) which are qualitatively different not only from the predictions of Marcus' theory but also from those of any other single-mode models.

The correlations most commonly studied experimentally with the aim of testing the theory of Marcus, and which are also valid in the high-temperature and adiabatic limits of the quantum formulation of rate theory, are the following:

(a) Free energy relationships (Bronsted relationships), i.e. relations between $\log k_r$ and ΔG_0 or between the logarithm of the current density and the overvoltage η for electrochemical processes for 'closely related' reactions.

The latter implies that λ_0 and all other parameters except ΔG_0 can be assumed constant throughout the series. Eq.(1.26) then predicts a quadratic relationship which becomes approximately linear with a slope of 0.5 (chemical or electrochemical transfer coefficient) when ΔG_0 or $e\eta$ is sufficiently small compared with λ_0.

Several examples of linear Bronsted relations for simple electron transfer reactions involving transition metal complexes have been reported. Thus, when one of the reactants possesses ligands in which substituents can be inserted, the only variable parameter in the series is ΔG_0, whereas the geometry of the reactants, and thus λ_0, w_r and w_p are approximately constant throughout the series. Common examples are the complexes $[Fe(phen)_3]^{3+/2+}$ and $[Ru(phen)_3]^{3+/2+}$ in which phen denotes 1,10-phenanthroline or various substituted phenanthrolines. Thus, reactions of $[Fe(phen)_3]^{2+}$ complexes with the strong oxidants Ce(IV) and Mn^{3+} (61), the reactions of $[Fe(phen)_3]^{3+}$ complexes with Fe^{2+} (62), and reactions of $[Ru(phen)_3]^{2+}$ with Ce(IV)(63) all display approximately linear relationships of the kind mentioned. Linear Bronsted and Tafel relations are also well-known for many proton transfer reactions involving carbon as a donor or acceptor atom(19) (see further chapter 6), for simple electrochemical processes(64) and for heterogeneous electron transfer between Ce(IV), $[Fe(CN)_6]^{3-}$ and $[Mo(CN)_8]^{4-}$ oxidants in aqueous solution and single srystals of various aromatic hydrocarbons (as measured by the rate of escape of the hole injected into the crystal)(65).

However, most of these reactions involve a quite drastic intramolecular reorganization in addition to the solvent reorganization. In the semiclassical formalism this would increase the effective value of λ (i.e. $\lambda_t = \lambda_o + \lambda_i$, where λ_i is the intramolecular reorganization energy) and thus the region of ΔG_0 over which the Bronsted or Tafel coefficient can be expected to be approximately constant. A closer analysis based on estimates of λ_i from spectroscopic data nevertheless shows that some curvature should be displayed (see chapter 4). When this is not the case one important reason is likely to be anharmonicity of the intramolecular modes which are reorganized, provided that this reorganization energy constitutes a considerable fraction of the total reorganization energy. (cf. chapter 4). For example, a representation of these modes by Morse potentials rather than by the harmonic potentials implicit in eq.(1.26) substantially increases the linearity. This is again associated with the fact that the Morse potential is itself approximately linear over quite wide intervals of free energy or overpotential.

An approximately quadratic dependence of the apparent activation energy on the free energy of reaction is well known for proton transfer reactions, in particular between oxygen- and nitrogen donors and acceptors(19,66,67). Due to the small proton transfer distance, λ_0 is also here small and the free energy interval over which a curvature in the Bronsted plot is manifested, is therefore correspondingly small. On the other hand, such a relationship for homogeneous and heterogeneous electron transfer reactions has only very recently been reported. Thus, homogeneous reactions involving transition metal complexes (followed by flow techniques)(68), aromatic hydrocarbons and corresponding anion radicals(69,70) and quinones, nitro compounds and their radicals(71) (followed by ESR and fluorescence quenching) have shown parabolic free energy relationships. The intramolecular reorganization energy of these reactants is presumably small, and so is λ_0 either because the reactants are large molecules (the transition metal complexes) or because the medium is apolar and relatively weakly coupled to the reactants. The curvature can therefore be observed even over relatively narrow free energy intervals. A low value of λ_0 is also the cause of the observation of a curved Tafel plot over a relatively narrow overpotential interval in the electrochemical reduction of several nitro compounds in apolar media(72). The curvature recently observed in the reduction of mercury ions at a mercury microelectrode(73) can be ascribed to the fact that the microcell construction applied made it possible to ignore the diffusion effects which would normally interfere with fast electrode processes, and to measure a curvature over several hundred millivolts.

(b) Relationships between the rate constants of a series of 'closely related' reactions with two different reagents of which one may be an electrode.

Provided that $\Delta G_0 \ll \lambda_0$, the ratio between the rate constants of the two reactants should be the same for all members of the series. Some verification of this effect has been reported(74), although the systems investigated (the reduction of cobalt(III) complexes at a mercury electrode and by homogeneous reductants) are known to deviate from the theory of Marcus in other respects. Moreover, comparisons of this kind are complicated by potential dependent double layer effects (dependence of w_r and w_p on η).

(c) Relation between the rate constant and optical and dielectric properties of the medium as expressed by the dependence of λ_0 on these parameters (eq.(1.28)).

One example, i.e. the reactions between several aromatic hydrocarbons and their radical anions in different alcohols have been found to exhibit the predicted relation between $\log k_r$ and the parameter c(75). Such comparisons are also difficult since solvent effects are reflected not only in the parameter c of λ_0 but also in both ΔG_0, w_r, and w_p.

(d) Application of a 'cross' relationship

$$k_{12} = (k_{11} k_{22} K_{12} f)^{\frac{1}{2}} \tag{1.29}$$

where

$$\ln f = (\ln K_{12})^2 / 4 \ln(k_{11} k_{22} / Z^2) \tag{1.30}$$

k_{12} is here the rate constant for reaction between an oxidant, Ox_1, and a reductant, Red_2, for two different redox couples. k_{11} and k_{22} are the rate constants for the corresponding homonuclear reactions, i.e. between Ox_1 and Red_1, and between Ox_2 and Red_2, respectively, and K_{12} is the equilibrium constant for the reaction.

This kind of relationship is frequently but occasionally incorrectly applied (in particular, the relation is invalid for strongly exothermic processes). Eqs.(1.29) and (1.30) are derived under the conditions that the work terms are small compared with the total activation energy and that the reorganization energy of both the solvent and intramolecular modes for the cross reaction is the average value of the corresponding quantities for the homonuclear reactions. The latter approximation is only valid for approximately thermoneutral processes ($\Delta G_0 < \lambda_0$). Several extensive tests of the cross relationship showing satisfactory agreement between theory and experiment have been reported. They include reactions of many transition metal complexes(76,77), reactions of substituted ferrocenes(78), reactions of ferrocenes with tri-p-anisylamine(79), and reactions between isolated biological redox components and small molecules(80,81) in aqueous solution.

The theory of Hush(82) is conceptually closely related to the early form of the theory of Marcus. A consistent application of the two theories would thus give essentially the same results, and for these reasons we shall refer to the literature for further discussion of this theory and some of the more subtle conceptual differences from the theory of Marcus(16,30). However, an alternative procedure was introduced by Levich and Dogonadze(25,50,51) who were the first to view the electron transfer as a quantum mechanical nonradiative electronic transition between manifolds of vibronic initial and final states. We shall provide a more detailed discussion of the formalism inherent in this work in subsequent chapters. At present we notice that like Marcus, Levich and Dogonadze in their first work viewed the reactant and product ions as structureless charge distributions embedded in a simple dielectric continuum. The reactants in the encounter complex were assumed to interact sufficiently weakly that the process could be described in terms of the time evolution of 'initially prepared' zero order states of noninteracting immobile ions by means of first order quantum mechanical perturbation theory. Thus, zero order Hamiltonian operators of the form

$$H_{a,b}(\vec{r},\vec{q}) = \frac{p_e^2}{2m_e} + V^e_{a,b}(\vec{r}) + V_{es}(\vec{r},\vec{q}) + \qquad (1.31)$$

$$+ H_s(\vec{q}) + V^s_{a,b}(\vec{q})$$

were defined. The first term is the kinetic energy of the migrating electron, and the second and third terms the energy of interaction of the electron with the ionic core of the donor or acceptor (corresponding to the subscripts a and b, respectively) and with the solvent, respectively. $H_s(\vec{q})$ is the solvent Hamiltonian in the absence of polarization (given by eq.(1.19)), and $V^{s}_{a,b}(\vec{q})$ the energy of interaction between the ions and the solvent. The stationary Schrodinger equation was furthermore solved within the framework of the Born-Oppenheimer approximation neglecting the dependence of the electronic wave functions on the nuclear kinetic energy operator. The perturbation inducing the electron transfer was thus assumed to be the electrostatic energy of interaction between the migrating electron and the acceptor ion, i.e. the term distingushing the Hamiltonian of eq.(1.31) from the total Hamiltonian of the system.

Postponing further discussion of the quantum mechanical approach to chapters 2 and 3 we notice here that in addition to the high-temperature result discussed above two other important results emerged from the early theory of Levich, Dogonadze, and their associates (i.e. before 1967). Firstly, they also considered the low-temperature limit of the rate expression. For $T \to 0$ all nuclear modes are 'frozen', and nuclear reorganization must therefore proceed by quantum mechanical nuclear tunnelling. Thus, at sufficiently low temperatures only exothermic processes may proceed by finite rates, and with vanishing activation energies. Secondly, the electron transfer formalism for homogeneous processes - for which only a single donor and acceptor electronic level are considered - was extended to electrochemical processes at metal and semiconductor electrodes(83,84) where the electronic structure of the 'substrate' electrode is of crucial importance for the phenomenology of the process (cf. the work of Gerischer(42). Thus, the extension of the theory of Marcus to electrochemical processes consists of a replacement of the collision number, reorganization energy, free energy of reaction, and work terms by their electrochemical analogues and inclusion of the effect of image forces on the solvent polarization. On the other hand, Dogonadze, Chizmadzhev, and Kuznetsov(83,84) did not discuss these features explicitly, but gave major attention to the electronic structure of the electrode. The electrochemical process was examined in the one-electron approximation, i.e.

viewing the overall process as a weighted average of independent 'microscopic' electron transfer steps to or from the individual electrode levels. The expressions for the current density, $i(\eta)$, therefore take the form

$$i_{cat}(\eta) = Ce\int_{-\infty}^{\infty} n(\varepsilon)\rho(\varepsilon)W_{cat}(\varepsilon,\eta)d\varepsilon \qquad (1.32)$$

$$i_{an}(\eta) = Ce\int_{-\infty}^{\infty} [1-n(\varepsilon)]\rho(\varepsilon)W_{an}(\varepsilon,\eta)d\varepsilon \qquad (1.33)$$

for the cathodic and anodic curent density, respectively. ε is the energy of an individual level in the metal or semiconductor, $\rho(\varepsilon)$ the level density, $n(\varepsilon)$ the Fermi distribution function, C the depolarizer concentration at the distance of ionic discharge, and $W_{an}(\varepsilon,\eta)$ and $W_{cat}(\varepsilon,\eta)$ the microscopic probability of electron transfer to or from the level ε. The integrand in eqs.(1.32) and (1.33) has a sharp maximum for certain values of $\varepsilon = \varepsilon^*$ and can thus be approximated by

$$i_{cat}(\eta) \approx Ce\rho(\varepsilon^*)n(\varepsilon^*)\Delta\varepsilon^* W_{cat}(\varepsilon^*,\eta) \qquad (1.34)$$

$$i_{an}(\eta) \approx Ce\rho(\varepsilon^*)[1-n(\varepsilon^*)]\Delta\varepsilon^* W_{an}(\varepsilon^*,\eta) \qquad (1.35)$$

where $\rho\Delta\varepsilon^*$ is the number of electronic levels effectively contributing to the integrals. For metals of small overvoltages ε^* coincides with the Fermi level, whereas for semiconductors ε^* coincides with the lower edge of the conduction band or the top edge of the valence band for electron and hole transfer, respectively. Just as the theory of Marcus, this whole formalism emphasizes the fundamentally similar nature of homogeneous and heterogeneous electron transfer processes. We shall provide further discussion of the implications of eqs.(1.34) and (1.35) in chapter 8.

Until about 1966 the theories of Marcus (Hush) and of Levich, Dogonadze and associates represented the two, alternative formulations of elementary condensed phase chemical processes within the concepts of dielectric continuum and multiphonon theory. A voluminous literature on further development - largely along the lines of the quantum theory of multiphonon processes initiated by Levich, Dogonadze and associates - has appeared since then. As a result, by the comprehensive theoretical framework and by the variability of the rate phenomena which can be rationalized, the quantum mechanical formulation of rate processes now constitutes a general theory of rate processes in condensed phases, in this way comparable to the transition state theory. We shall try to show this in what follows. However, at this stage we shall simply list some of the effects which have received a quantitative treatment in terms of the theory and to which we shall return in more detail later. Thus:

(A) Extension of the simple dielectric continuum model to include both frequency and space dispersion accounting for structural medium effects(85-87).

(B) Electronic - vibrational interaction in the limits of strong and weak coupling to the medium(88).

(C) General incorporation of both classical and quantum nuclear modes involving frequency and equilibrium coordinate shift as well as interconversion of modes(89-92).

(D) Within a semiclassical formalism incorporation of adiabaticity effects(93,94).

(E) Proton transfer reactions(95-97) and processes involving the transfer of heavier atomic groups, such as nucleophilic substitution (98) and inner sphere electron transfer reactions(99).

(F) Strongly exothermic processes(100-104).

(G) Higher order effects, i.e. electron transfer through intermediate states of real or virtual nature(104-107).

(H) Electron and atom group transfer processes in biological systems. This relates so far to an understanding of primary electron transfer reactions in bacterial photosynthesis(108,109), elementary steps in enzyme catalysis(110), recombination reactions of myoglobin and hemoglobin subunits with small molecules(111), and to certain aspects of membrane processes(112).

1.3 Chemical Reactions as a Class of Radiationless Processes

In subsequent sections we shall base our exposition of the quantum mechanical formalism essentially on the time evolution of zero order Born-Oppenheimer states corresponding to reactant and product molecules. The zero order states are not necessarily those corresponding to infinitely separated molecular species but may be modified by the potential energy of interaction between the reactants and products in their respective encounter complexes. In terms of scattering theory(113) these states represent ingoing and outgoing channels coupled by a transition operator, where in nearly all cases reduction to first order perturbation theory is necessary for further evaluation of the transition matrix elements.

In the nonadiabatic limit the rate expressions are subsequently represented in the form of statistically weighted products of a squared nondiagonal matrix element coupling the electronic wave functions of the donor and acceptor by the perturbation operator, and a Franck Condon factor for overlap of the nuclear wave functions. This is both formally and technically completely analogous to a 'general line shape function' as in optical and radiationless intramolecular electronic transitions. Conceptually this implies that the elementary chemical process, e.g. the electron transfer between solvated ions, can be viewed as the thermally averaged decay of an 'initially prepared' metastable state to a continuum of final states. In fact, the development of the quantum mechanical formalism for chemical rate processes has drawn heavily from the parallel development of the theory of molecular and solid state electronic transitions involving coupling to the nuclear modes of the reacting system.

We have already discussed the coupling between electrons and vibrational motion in the continuum approximation of Pekar and Frohlich. The same model, i.e. a single-mode approximation for the inertial polarization component was also applied by Huang and Rhys(114) in a consideration of both light absorption and nonradiative transitions in F-centres. An alternative approach to the study of the line shape of optical transitions was invoked by Lax(115) and O'Rourke(116) who considered the coupling of the electronic system to local modes characterized by configuration coordinates rather than the long-range coulomb interaction. In the single-mode approximation they obtained a line-shape expression similar to the one appearing in the chemical rate expressions.

Although on several occasions we shall refer to both molecular and solid state radiative and nonradiative processes, we shall not attempt to provide a systematic account of the development of these important areas. Reviews on both multiphonon effects in solid state processes such as the ones mentioned(117,118) and on molecular radiationless processes(2,3) are available. However, in order to give an impression of the variability of the phenomena which can be viewed under the formalism to be presented we shall briefly list those cases of molecular and condensed phase nonradiative relaxation processes which appear to have the closest conceptual similarity to what constitutes the topic of the present work, i.e. elementary chemical processes between sufficiently well defined molecular entities in condensed phases. Thus, molecular relaxation processes displaying this analogy include electronic and vibrational relaxation of excited states of large organic molecules(2,3). This may proceed either by the nuclear kinetic energy operator (in spin conserving processes) or by spin orbit coupling (involving spin states of different spin multiplicity). Furthermore, the following condensed phase processes should be listed in the present context:

(a) Thermal ionization and electron capture of semiconductor impurities. The former was considered also by Pekar(57) but a general formalism, also incorporated in electron transfer theory was developed by Kubo and Toyozawa(119), by Rickayzen, and by Krivoglaz(120).

(b) Electronic relaxation of impurity transition metal and rare earth metal ions and other localized 'impurity states' in crystals(5,117,121).

(c) Electronic energy transfer between localized states(122).

(d) Dynamics of electron localization in the limit of strong medium coupling. This was the subject of the theory of Pekar but it has received more recent study in the context of solvated electrons, reactions of trapped electrons in glasses, and polaron mobility(58) (see further chapter 4).

(e) Several solid state relaxation phenomena involving atom group transfer such as point defect diffusion, interstitial dipolar relaxation and ionic conductivity(5).

In the following chapters we shall aim at reviewing what we feel constitutes the major recent achievements in the quantum theory of elementary chemical processes giving the main emphasis to electron transfer. We shall thus aim at showing the character of the formalism as a unified rate theory and its application to a variety of apparently different classes of processes. We shall adopt the following sequence of topics:

(A) The dynamic role of the solvent and its endowing of the elementary rate processes with multiphonon character.

(B) The formal quantum mechanics and statistics of rate processes, and implications of the rate expressions.

(C) The effect of reorganization of intramolecular modes and relation to strongly exothermic processes, and the dynamics and reactions of trapped electrons.

(D) The semiclassical approximation in the adiabatic and nonadiabatic limits.

(E) Higher order processes and relation to Raman effect and antiferromagnetic coupling.

(F) Atom group transfer processes.

(G) Electrochemical processes.

(H) Application to biological systems.

Wherever possible we shall discuss recent experimental results in the light of the theory, in particular in so far as the experiments illustrate or verify the more subtle predictions of the theory which are not outcomes of semiclassical formulations.

2 MULTIPHONON REPRESENTATION OF CONTINUOUS MEDIA

2.1 Nature of Solvent Configuration Fluctuations

Ions and molecules embedded in a dielectric medium are subject to both short-range and long-range interactions with the medium molecules. The former refer to such effects as dispersion, exchange (or 'covalent'), surface tension (in the cavities occupied by the ions), and repulsive forces, whereas the latter are usually represented by coulomb interactions, but may also be pressure or elastic deformation fields. The solvation energies typically amount to several electron volts, and this obviously exerts a pronounced effect on the kinetics of the elementary reaction steps. In order to study these effects, ideally a given reaction should be followed both in the gas and solution phases. However, Such comparative experimental studies have only recently become possible, by the invention of new high-pressure mass spectrometric and flow techniques(123). The results are so far of a semiquantitative nature and refer only to proton transfer reactions. Less direct routes to the study of solvation effects requiring an elaborate theoretical framework must therefore be adopted.

The solvation effects are of two kinds. Firstly, solvation energies, free energies of reaction, and work terms are affected by equilibrium solvent interactions. Thus, in contrast to gas phase reactions, the solvated donor and acceptor electronic energy levels are usually different, even for exchange reactions with zero free energy of reaction. Secondly, as a radiationless process the electron transfer can only proceed when the donor and acceptor levels coincide. As we saw in chapter 1, this degeneracy is induced by thermal fluctuations in the local instantaneous solvent configuration leading to deviation from polarization equilibrium. In contrast to gas phase reactions, both of these effects lead to a high reaction probability also for nonresonant donor and acceptor levels.

A quantitative theoretical investigation of the dynamic role of the medium can follow two complementary approaches. In one approach the specific interaction between a given ion and a certain number, N, of discrete solvent molecules is calculated for different relative atomic configurations. The remaining macroscopic number may be incorporated within a continuum formalism. This is the approach taken for example by Muirhead-Gould and Laidler(124), and by Clementi and associates(125-127) in the calculation of free energies of hydration and optimal coordination number and geometry of simple ions in water. Ideally, N should be large but computational difficulties restrict N to small values, e.g. corresponding to the first solvation sphere. Adoption of these techniques is most convenient if the reacting molecules together with the first coordination sphere constitute well defined entities, i.e. if the solvent molecules in the first coordination sphere are located longer and possess properties markedly different from those of individual solvent molecules in the bulk solvent. We shall follow this 'semicontinuum' procedure in our subsequent analysis of the role of intramolecular modes in systems where these modes can be represented by simple model potentials.

In the present section we shall give attention to the alternative procedure which views the medium as a dielectric continuum interacting with the ionic charge distribution. Although this appears to violate the discreteness of structure of matter over molecular distances, the formalism is valid, both physically and technically, also over regions of molecular dimensions. Thus, the medium response to an electric field can be defined and characterized by a dielectric permittivity function, also over molecular dimensions, provided that the structural properties of the medium are incorporated in this function, i.e. by including its space and time dispersion. On the other hand, in order to estimate these effects, molecular interaction forces may have to be introduced, but this does not invalidate the continuum description.

We notice at first that an 'external' field $f(\vec{r},t)$ ($\vec{r}$ is the coordinate vector, and t the time) quite generally induces a reaction (a response) $\phi(\vec{r},t)$ in the medium. In the following we shall assume that the field and the response refer to the electric field $D(\bar{r},t)$ and the induced polarization, $P(\vec{r},t)$, i.e. the dipole moment per unit volume, in other words $D(\vec{r},t)$ has the form

$$\mathrm{div}\,\vec{D} = 4\pi\rho(\vec{r}) \qquad (2.1)$$

where $\rho(\vec{r})$ is the (permanent or induced) charge density at the point $\vec{r}$. However, as noted, the presence of molecules in the medium may be the source of other kinds of fields. For example, fields such as spherically symmetric pressure fields

$$f(\vec{r}) = P_0 a_0^2 / |\vec{r} - \vec{r}_0|^2 \qquad (2.2)$$

or the elastic deformation potential of Bardeen and Schockley(127)

$$f(\vec{r}) = C|\psi(\vec{r})|^2 \qquad (2.3)$$

would interact with the acoustical motion of the medium and cause a density response. P_0 is here a constant, a_0 the linear extension of the molecular pressure field source, and $\vec{r}_0$ its localization. Furthermore, C is an elastic coupling constant, and $\psi(\vec{r})$ the wave function of a localized 'impurity' electron. Finally, anticipating the discussion of chapter 9, $f(\vec{r})$ may be a field which causes a conformational change in a biological system of macromolecules.

In the dielectric continuum formalism the electric field, $\vec{D}(\vec{r},t)$ is usually taken to be the vacuum field corresponding to the same external charge distribution. Conditions for this assumption have been provided(130), and we shall assume that they are met in our further considerations. Thus, it differs from the electric field in the medium, $\vec{E}(\vec{r},t)$ by the induced polarization, i.e.

$$\vec{E}(\vec{r},t) = \vec{D}(\vec{r},t) - 4\pi \vec{P}(\vec{r},t) \tag{2.4}$$

This difference is expressed by the dielectric permittivity function. For a isotropic and uniform dielectric where the polarization at the point $\vec{r}$ and at time t depends only on the field at this same point and time (a 'local' dielectric), this relationship takes the form

$$\vec{D}(\vec{r},t) = \varepsilon(r)\vec{E}(\vec{r},t) \tag{2.5}$$

In general, however, the relation between $\vec{D}(\vec{r},t)$ and $\vec{P}(\vec{r},t)$ is more involved. Firstly, even in the absence of 'external' charges (from reactant and product ions) the local and instantaneous values of $\vec{P}(\vec{r},t)$ differ from zero due to small, thermally induced fluctuations from equilibrium, whereas the statistical average value of $\vec{P}$, i.e. $\langle\vec{P}(\vec{r},t)\rangle$, vanishes in the absence of external fields. As a consequence, the local dynamic polarization state of the medium is not completely characterized only by $\vec{P}(\vec{r},t)$, but requires also introduction of its time derivative, $\vec{P}(\vec{r},t)$.

Secondly, the relation between the polarization and the field is more complicated than implied by eq.(2.5) due to the interaction between the individual solvent molecules. If the field is sufficiently weak, the appropriate relationship is still approximately linear. However, due to the correlation between the medium molecules, the actual value of $\vec{P}(\vec{r},t)$ is determined by the field not only at the point $\vec{r}$ and at time t, but also at all other points, $\vec{r}'$, and times, t' within certain characteristic correlation ranges. This is expressed by writing eq.(2.5) in the form(85,130)

$$\vec{D}(\vec{r},t) = \int_{-\infty}^{t} dt' \int d\vec{r}'\, \varepsilon(\vec{r}-\vec{r}', t-t')\vec{E}(\vec{r}', t') \tag{2.6}$$

corresponding to

$$\langle\vec{P}(\vec{r},t)\rangle = \int_{-\infty}^{t} dt' \int d\vec{r}'\, G(\vec{r}-\vec{r}', t-t')\vec{D}(\vec{r}', t') \tag{2.7}$$

where $G(\vec{r}-\vec{r}', t-t')$ is a function which relates the polarization response to the electric field and which characterizes the medium space and time properties.

Eq(2.7) is an example of the general and very important problem of the response of a given physical system to an external time-dependent force(131). Furthermore, as written, eq.(2.7) (1) applies to an uniform medium, i.e. $G(\vec{r}-\vec{r}', t-t')$ depends specifically on $(\vec{r}-\vec{r}', t-t')$ rather than generally on $(\vec{r},\vec{r}', t-t')$. This assumption obviously requires reconsideration when ions are located in the medium and in particular for heterogeneous processes (electrochemical and membrane processes); (2) fulfills the causality principle, i.e. the occurrence of the perturbing force precedes that of the response which implies that $t' < t$; (3) incorporates implicitly damping effects in the linear response formalism via the correlation functions of the polarization vectors (see below).

In our further derivation of the medium Hamiltonian we transform the convolution integral of eq.(2.7) to Fourier space(85).

$$\langle P(\vec{k},\omega)\rangle = G(k,\omega)D(\vec{k},\omega) \tag{2.8}$$

$P(\vec{k},\omega)$, $G(\vec{k},\omega)$, and $D(\vec{k},\omega)$ are the Fourier transforms of $\vec{P}(\vec{r},t)$, $G(r,t)$, and $\vec{D}(\vec{r},t)$, respectively. Thus.

$$\langle\vec{P}(\vec{k},\omega)\rangle = V^{-\frac{1}{2}} \int d\vec{r} \int_{-\infty}^{\infty} dt \langle\vec{P}(\vec{r},t)\rangle e^{-i(\vec{k}\cdot\vec{r}\, -\, \omega t)} \tag{2.9a}$$

$$\vec{D}(k,\omega) = V^{-\frac{1}{2}} \int d\vec{r} \int_{-\infty}^{\infty} \vec{D}(\vec{r},t)\, e^{-i(\vec{k}\cdot\vec{r} - \omega t)} \qquad (2.9b)$$

and

$$G(k,\omega) = \int d(\vec{r}-\vec{r}\,') \int_{-\infty}^{\infty} d(t-t')\, G(\vec{r}-\vec{r}\,',t-t')\, e^{-ik(\vec{r}-\vec{r}\,')}$$

$$e^{i\omega(t-t')} \qquad (2.9c)$$

where the integration with respect to $\vec{r}$ is extended to the whole volume V of the medium. The dependence of $\vec{P}$, $\vec{D}$, and G on $\vec{r}$ and t is thus expressed equally well by their dependence on $\vec{k}$ and ω, and for this reason the dependence of these quantities on $\vec{k}$ and ω is referred to as their space and time dispersion, respectively. The response function is furthermore related to the dielectric permittivity function, $\varepsilon(k,\omega)$ by the equation

$$G(k,\omega) = \frac{1}{4} [1-\varepsilon(k,\omega)^{-1} \qquad (2.10)$$

At this point, before a formal derivation of the medium Hamiltonian by means of the linear response formalism it is appropriate to consider in qualitative terms the physical implications of the dependence of the permittivity on the time and space variables ω and $\vec{k}$. Our approach will furthermore both here and in the following be essentially phenomenological, i.e. we shall express the polarization properties in terms of an initially phenomenological and in fact macroscopic quantity, $\varepsilon(k,\omega)$. Information about $\varepsilon(k,\omega)$ can subsequently be gathered either from experimental data, or by the introduction of specific models for the microscopic properties of the medium in terms of which $\varepsilon(k,\omega)$ can be expressed.

We notice that the various kinds of polarization are associated with characteristic relaxation times. Thus if an alternating field in the form of a planar travelling wave(33,131)

$$\vec{D} = \vec{D}_o \exp[i(\vec{k}\cdot\vec{r} - \omega t)] \tag{2.11}$$

is imposed on the medium, the polarization will also vary periodically. However, generally $\vec{P}$ will lag behind $\vec{D}$, i.e. both $\langle\vec{P}\rangle$ and the internal (screened) electric field, $\langle\vec{E}\rangle$, are subject to a phase shift relative to $\vec{D}$ in addition to the screening. This is accounted for by means of the complex dielectric permittivity, $\varepsilon(k,\omega)$, viz.

$$\vec{D}_o \exp[i(\vec{k}\vec{r} - \omega t)] = \varepsilon(k,\omega)\ \langle\vec{E}\rangle = |\varepsilon(k,\omega)|e^{i\delta}\ \langle\vec{E}\rangle =$$

$$[\mathrm{Re}\,\varepsilon(k,\omega) + i\ \mathrm{Im}\,\varepsilon(k,\omega)]\ \langle\vec{E}\rangle \tag{2.12}$$

where Re and Im refer to the real and imaginary part, respectively, and the phase shift, δ , is related to $\varepsilon(k,\omega)$ by the equation $\mathrm{tg}\,\delta = \mathrm{Im}\,\varepsilon(k,\omega)/\mathrm{Re}\,\varepsilon(k,\omega)$.

The energy dissipated per unit time in the medium is proportional to $\mathrm{Im}\,\varepsilon(k,\omega)$, i.e. to $\sin\delta$. $\mathrm{Im}\,\varepsilon(k,\omega)$ is thus a measure of the amount of absorption of electromagnetic radiation or the oscillator strength at the particular frequency, whereas $\mathrm{Re}\,\varepsilon(k,\omega)$ is a measure of the amount of screening of the external field.

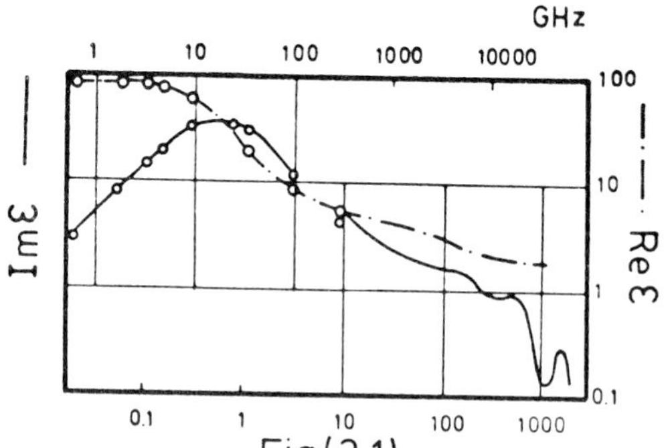

Fig(2.1)

Logarithmic plot of Im ε against frequency and wave number for H_2O at 293 °K. From microwave and submillimeter-wave data (132b).

The qualitative nature of the frequency dependence of both Re $\varepsilon(k,\omega)$ and Im $\varepsilon(k,\omega)$ (at fixed k -> 0) representative for a variety of liquid and disordered solids is pictured in fig (2.1). Im $\varepsilon(k,\omega)$ thus consists of several resonances of varying width separated by 'transparency' or 'nondissipative' zones. The resonances at the highest frequencies (in the UV region) correspond to transitions between different electronic levels. These bands are broad due to coupling to intramolecular vibrational modes. Several narrow resonances are located in the infrared region. These correspond to intramolecular stretching and bending modes (1500-3000 cm^{-1}) but also to motion of individual solvent molecules such as hindered translation and rotation (5-700 cm^{-1}), hydrogen bond bending and stretching (50-200 cm^{-1}) etc. Finally, a broad band is located at still lower frequencies where the absorption maximum is located at approximately 1 cm^{-1} (10^{11} s^{-1}, i.e. the microwave region) for liquid water and hydroxylic solvents.

The nature of the infrared and microwave bands is qualitatively different. The former have the character of narrow resonances. This is associated with a shift of the equilibrium position of the appropriate local mode and is usually represented by the form(56,133)

$$\mathrm{Im}\,\varepsilon(\omega) = [(\varepsilon_s - \varepsilon_o)/2]\left\{\frac{\omega}{[(\Omega_R-\omega)^2+\Gamma^2]} + \frac{\omega}{[(\Omega_R-\omega)^2+\Gamma^2]}\right\} \quad (2.13)$$

where Γ is a phenomenological damping coefficient. On the other hand, the latter is a broad band associated with the librational relaxation of water molecules or aggregates of water molecules through various temporary equilibrium positions. The presence of local barriers to the motion provides a different character for the frequency distribution, i.e. of the Debye form

$$\mathrm{Im}\,\varepsilon(\omega) = (\varepsilon_s - \varepsilon_o)\omega\tau_D/(1+\omega^2\tau_D^2) \qquad (2.14)$$

where τ_D is the Debye relaxation time (τ_D^{-1} is the maximum for the distribution(56). For this reason, in contrast to the infrared and UV bands, the maximum of this band is expected to be shifted drastically to much lower frequencies for the corresponding solids (being about 10^5 s^{-1} for ice (133)).

Fig. (2.1) can then be interpreted by noting that when the bands were completely separated the various transparency regions would correspond to screening of the electric field by polarization of different groups of modes. Thus, in the region between electronic and infrared absorption, the electronic but not the nuclear motion is able to follow the electric field. In the transparency zone between the infrared and Debye regions the atomic and electronic motion can follow the field, whereas the librational relaxation cannot. However, for water some overlap of the infrared and Debye bands occurs, and complete absence of absorption is in fact only observed in the visible region.

We can thus conclude that the continuum model allows for incorporation of medium frequency dispersion effects. Moreover, measurement of the dependence of the imaginary (dissipative) and real (transparent or nondissipative) components of the dielectric permittivity on the frequency provides experimental data for the frequency dispersion of the medium which can be associated with a particular kind of molecular dielectric response when the dissipative bands are sufficiently well separated.

Space correlation effects are expected when the external field has a scale of spacial variation comparable with the extension of the appropriate polarization mode. The physical origin of such effects is that individual medium molecules interact, by both electrostatic and exchange forces. As a consequence, a change of the electric field at the point $\vec{r}$ induces a polarization change not only at this point but also at other points within certain correlation lengths characteristic of the various interaction forces. These effects are incorporated in the continuum formalism by introducing correlation functions $S(\vec{r}-\vec{r}')$ which constitute a measure of the 'average excitation' of the medium in the appropriate correlation range and for the particular molecular motion(86,134).

A major qualitative effect of space correlation can immediately be noticed. Thus, consider the free energy of solvation of a simple spherical ion. For a structureless medium, i.e. in the absence of space correlation, this quantity, E^{B}_{sol}, is given by the Born formula(134).

$$E^{B}_{sol} = \frac{z_i^2 e^2}{2r_i}\left(1-\varepsilon_s^{-1}\right) \tag{2.15}$$

where z_i is the charge of the ion, and r_i its radius. This equation thus keeps all information about the medium in a single constant parameter, i.e. the bulk static dielectric constant ε_s. However, the Born equation gives hydration energies which are systematically higher than the experimental values(134). This is understandable if space correlation effects are important. Thus, the overall thermodynamic free energy of polarization E_{sol}, is the sum of the corresponding contributions from all different polarization types, E^{ν}_{sol}, i.e.

$$E_{sol} = \sum_{\nu} E^{\nu}_{sol} \tag{2.16}$$

If the correlation length of a given component $\lambda_\nu \ll r_i$, then the medium is effectively structureless with respect to this mode, i.e. $E^{\nu}_{sol} \propto z_i e / 2r_i$. If, on the other hand, $\lambda_\nu \gg r_i$, and the field otherwise not sufficiently strong to destroy the local structure, then $E^{\nu}_{sol} \propto z_i e / 2\lambda_\nu$, i.e. only the part of the medium which is more remote from the ion than the distance λ_ν effectively contributes to the polarization energy. If, for the sake of simplicity, correlation lengths around r_i are ignored, this means that strongly correlated polarization (for which $\lambda_\nu >> r_i$) contributes negligibly, and(134).

$$E_{sol} \approx \sum_{\nu \in \lambda_\nu < r_i} E^\nu_{sol} = \frac{z_i^2 e^2}{2r_i} (1-\varepsilon^{-1}_{eff}) \qquad (2.17)$$

Since E_{sol} is here smaller than E^B_{sol}, this implies that $\varepsilon_{eff} < \varepsilon_s$, in other words, incorporation of space correlation effects gives a smaller effective dielectric permittivity in regions within characteristic correlation distances from the ion (dielectric saturation).

In contrast to the frequency dependence of $\varepsilon(\omega)$, practically no data for the wave number dependence are available. This is associated with the fact that throughout the whole frequency range available for investigations of $\varepsilon(\omega)$, k (as determined by the relation ω = ck where c is the velocity of light) corresponds to wave lengths much larger than any conceivable correlation length for the polarization components. All experimental investigations therefore so far refer to the long-wave length limit ($k \to 0$). For practical purposes various model representations for $\varepsilon(k)$ have to be introduced. These functions are empirical but chosen in such a way as to ensure the correct limiting values at high and low k. Thus, for a polar solvent $\varepsilon(k) \to$ const = ε_s for $k \to 0$. All molecular aggregates are thus reoriented with the electric field and give a full contribution to the polarization in the limit of long wave lengths. On the other hand, when k^{-1} is smaller than some structural correlation length the corresponding field component changes sign within the correlated area. When the field is sufficiently weak that the local structure is maintained, this means that a smaller polarization is induced, and $\varepsilon(k)$ is therefore also smaller than $\varepsilon(0)$.

In the limit when $k \to \infty$, $\varepsilon(k) \to 1$. This would correspond to wave lengths smaller than all characteristic structural dimensions of the system and therefore to the absence of dielectric response. However, correlation effects in condensed phases are manifested over regions of molecular dimensions, and for this reason the limiting value of $\varepsilon(k) \to 1$ is not important. A simple and physically plausible (although still empirical) approximation is(135)

$$\varepsilon(k) = \varepsilon_o + \frac{\varepsilon_s - \varepsilon_o}{1 + a^2 k^2} \tag{2.18}$$

where $a = (\varepsilon_s/\varepsilon_o)\lambda$, the correlation length, and ε_o the shortrange dielectric constant, taken as 4.9 for water at room temperature and corresponding to both optical and infrared atomic components. This gives the following 'screening' function (cfr. eq.(2.6))

$$\varepsilon(\vec{r}-\vec{r}\,')^{-1} = \varepsilon_s^{-1}\left[1+\left(\frac{\varepsilon_s}{\varepsilon_o} - 1\right)e^{-|\vec{r}-\vec{r}\,'|/\lambda}\right] \tag{2.19}$$

The space polarization correlation functions contain all the information about the local structural properties of the medium. Consequently, for a highly structured medium such as water, the form of $S(\vec{r}-\vec{r}\,')$ is correspondingly involved. On the other hand, in view of the fact that the different kinds of nuclear motion are of widely different nature the associated correlation lengths can also be expected to differ widely and moreover to be mutually uncorrelated. This offers some simplification of the further analysis, i.e. by representing $S(\vec{r}-\vec{r}\,')$ in the form(134)

$$S(\vec{r}-\vec{r}\,') = \sum_\nu S_\nu(\vec{r}-\vec{r}\,') = \sum_\nu \langle P_\nu(\vec{r})P_\nu(\vec{r}\,')\rangle \tag{2.20}$$

where $S_\nu(\vec{r}-\vec{r}\,') \to 0$ for $|\vec{r}-\vec{r}\,'|/\lambda_\nu \to \infty$, and λ_ν is the correlation length associated with the particular polarization component ν. Thus, referring again to pure water, each of the three major groups of absorption bands is reasonably associated with a particular (approximately constant) correlation length. The librational motion between different temporary equilibrium positions thus extends over the region of hydrogen-bonded clusters of water molecules, i.e. approximately 10 A. on the other hand, the high-frequency infrared bands are associated with intramolecular atomic motion which is expected to have correlation lengths of molecular dimensions only, i.e. a few A, whereas the electronic correlation length is only some fraction of an A(134).

In view of the virtual absence of experimental data on the spacial dispersion of the dielectric permittivity function for systems appropriate to elementary chemical processes in condensed media, different empirical approaches must be applied. Commonly the correlators are approximated by some function decaying with $|\vec{r}-\vec{r}'|$ (e.g. exponential or Gaussian). These functions can in principle incorporate molecular properties of the medium molecules. The dielectric properties are subsequently expressed in terms of the parameters of the appropriate functions, and the parameter values in turn estimated by fitting to experimental data on for example solvation energies. Such an analysis has in fact provided the correlation lengths given above(134).

We shall now complete this section by giving a brief derivation of the Hamiltonian operator for the 'free' (i.e. not coupled to ionic charges) dielectric medium. In contrast to the Hamiltonian given in chapter 1 we shall formally incorporate frequency dispersion effects and show in a later section how the Hamiltonian parameters are related to the macroscopic medium dielectric properties.

The instantancous medium polarization, $\vec{P}$, can formally be viewed as being caused by an electostatic (so far fictitious) field $\vec{D}$. The potential energy of a uniform isotropic medium is then

$$U_p = -\int_0^{\vec{D}} \vec{P} d\vec{D} \tag{2.21}$$

Quite generally, any vector, $\vec{D}$, is completely defined by its div and rot by the equations

$$\vec{D} = \vec{D}^{\perp} + \vec{D}^{\parallel} \tag{2.22}$$

where

$$\text{div } \vec{D}^{\perp} = 0; \quad \text{rot } \vec{D}^{\perp} = \text{rot } \vec{D} = 0 \tag{2.23}$$

$$\text{div } \vec{D}^{\parallel} = \text{div } \vec{D} = 0; \quad \text{rot } \vec{D}^{\parallel} = 0 \tag{2.24}$$

and $\vec{D}^{\perp}$ and $\vec{D}^{\parallel}$ are called the transversal and longitudinal component, respectively (with respect to the direction of wave propagation, see belov). For an electrostatic field rot$\vec{E}$ = 0; if, moreover, the medium is uniform and isotropic, then rot$\vec{D}$ also vanishes due to the constancy of the permittivity of the medium, and for such a medium only the longitudinal component of $\vec{D}$ survives, i.e. $\vec{D} = \vec{D}^{\parallel}$, and $\vec{P} = \vec{P}^{\parallel}$.

We shall assume that these assumptions are valid in the following. Both $\vec{D}(\vec{r})$ and $\vec{P}(\vec{r})$ can then be expanded in Fourier series to which only the component in the direction of propagation contributes, i.e.

$$\vec{P}(\vec{r}) = \sum_{\vec{k}} \frac{\vec{k}}{|\vec{k}|} P_{\vec{k}} e^{i\vec{k}\cdot\vec{r}} \tag{2.25}$$

and

$$\vec{D}(\vec{r}) = \sum_{\vec{k}} \frac{\vec{k}}{|\vec{k}|} D_{\vec{k}} e^{i\vec{k}\vec{r}} \tag{2.26}$$

Where $P_{\vec{k}}$ and $D_{\vec{k}}$ are a set of coefficients characterizing the wave amplitudes. Since both $\vec{P}(\vec{r})$ and $\vec{D}(\vec{r})$ are real quantities, it also follows that $\vec{P}(\vec{r}) = \vec{P}(\vec{r})^*$ and $\vec{D}(\vec{r}) = \vec{D}(\vec{r})^*$, and consequently $P_{\vec{k}} = -P_{\vec{k}}^*$ and $D_{\vec{k}} = -D_{\vec{k}}^*$, where the superscript '*' indicates the complex conjugate.

In order to proceed further we notice that for crystalline media harmonic polarization waves are characterised by a few discrete stationary eigenfrequencies the number of which is determined by the number of atoms in the elementary cell, and which propagate through the crystal without damping. This is equivalent to a representation of P(r) in the form

$$\vec{P}(\vec{r}) = \sum_{\vec{k},\nu} \frac{\vec{k}}{|\vec{k}|} P_{\vec{k}\nu}\, e^{-i(\omega_{\vec{k}\nu} t - \vec{k}\cdot\vec{r})} \tag{2.27}$$

where ν characterizes the particular atomic motion and $\omega_{\vec{k}\nu}$ is the corresponding frequency. For a disordered medium, such as a liquid solution, the 'elementary cell' extends in principle over a macroscopic range and ν is now a continuous variable. In addition, harmonicity in the motion of individual solvent molecules can no longer be assumed, and for these reasons the approach represented by eq.(2.27) may therefore a priori seem less obvious. Nevertheless, provided that U_p depends quadratically on P (corresponding to a linear medium response to the external field, cfr. chapter 1) we can now show that this description remains valid. Thus decomposing the potential energy expression into wave components

$$U_p = \int_0^{\vec{D}} \vec{P} d\vec{D} d\vec{r} = \sum_{\vec{k}} \int_0^{D_{\vec{k}}} P_{\vec{k}}\, dD_{\vec{k}}^* = \sum_{\vec{k},\nu} \int_0^{\vec{D}_{\vec{k}}} P_{\vec{k}\nu}\, dD_{\vec{k}}^* \tag{2.28}$$

(where $\vec{P}$ and $\vec{D}$ refer to the longitudinal components only) we can incorporate the dispersion relations of the dielectric medium in the Hamiltonian function of the medium.

Moreover, the fictive external field is related to the induced polarization by the equation.

$$P_{\vec{k}\nu} = \frac{c_{k\nu}}{4\pi} D_{\vec{k}} \qquad (2.29)$$

where the, so far phenomenological, coupling constants $c_{k\nu}$ (cfr. eq. (1.16)) are a measure of the medium response (oscillator strength) at the paticular frequency $\omega_{\vec{k}\nu}$. The potential energy then takes the form

$$U_p = \sum_{\vec{k},\nu} \frac{4\pi}{c_{k\nu}} \int_0^{P_{\vec{k}\nu}} P_{\vec{k}\nu}\, dP^*_{\vec{k}\nu} \qquad (2.30)$$

Since U_p is a real quantity

$$U_p = \sum_{\vec{k},\nu} \frac{4\pi}{c_{k\nu}} \int_0^{P_{\vec{k}\nu}} \frac{1}{2}(P_{\vec{k}\nu}\, dP^*_{\vec{k}\nu} + P^*_{\vec{k}\nu}\, dP_{\vec{k}\nu}) = \sum_{\vec{k},\nu} \frac{2\pi}{c_{k\nu}} |P_{\vec{k}\nu}|^2 \qquad (2.31)$$

In the expression for the potential energy we see that the wave amplitudes, $P_{\vec{k}\nu}$, in fact now represent the dynamic (coordinate) variables. Similarly we introduce an additional set of variables, $\xi_{\vec{k}\nu}$, to characterize the kinetic energy of the medium polarization fluctuations. $\xi_{\vec{k}\nu}$ is related to the rate of change of the polarization amplitudes, and the total medium Hamiltonian function can then be written in the form

$$H_s = \sum_{\vec{k},\nu} \frac{2\pi}{c_{k\nu}} (|P_{\vec{k}\nu}|^2 + A_{\vec{k}\nu} |\xi_{\vec{k}\nu}|^2) \qquad (2.32)$$

Inserting $P_{\vec{k}\nu}$ and $\xi_{\vec{k}\nu}$ in Hamilton's equations of motion

$$\dot{P}_{\vec{k}\nu} = \partial H_s / \partial \xi_{\vec{k}\nu} \quad ; \quad \dot{P}^*_{\vec{k}} = \partial H_s / \partial \xi_{\vec{k}\nu} \qquad (2.33)$$

$$\dot{\mathcal{G}}_{\vec{k}\nu} = - \partial H_s / \partial P_{\vec{k}\nu} \quad ; \quad \mathcal{G}^*_{\vec{k}\nu} = - \partial H_s / \partial P^*_{\vec{k}\nu} \tag{2.34}$$

where the dot indicates differentiation with respect to time, we find that the equation of motion for $P_{\vec{k}\nu}$ is $\ddot{P}_{\vec{k}\nu} = -(4\pi^2/c_{k\nu}) A_{\vec{k}\nu} P_{\vec{k}\nu}$. $A_{\vec{k}\nu}$ is thus a direct measure of the frequency of the appropriate polarization component, i.e. $A_{\vec{k}\nu} = (c^2_{k\nu}/4\pi^2)\omega^2_{\vec{k}\nu}$, and the total Hamiltonian can be rewritten as

$$H_s = \sum_{\vec{k},\nu} \frac{2\pi}{c} \left(|P_{\vec{k}\nu}|^2 + \omega^{-2}_{\vec{k}\nu} |\mathcal{G}_{\vec{k}\nu}|^2 \right) \tag{2.35}$$

Finally, it is convenient to perform the following transformation by which the Hamiltonian is expressed by the real conjugate coordinate and momentum variables $q_{\vec{k}\nu}$ and $p_{\vec{k}\nu}$, i.e.(34)

$$P_{\vec{k}\nu} \rightarrow \left(\frac{\hbar\omega_{\vec{k}\nu} c_{k\nu}}{8\pi}\right)^{\frac{1}{2}} \begin{cases} q_{-\vec{k}\nu} + i q_{\vec{k}\nu} & k_\alpha > 0 \\ -q_{\vec{k}\nu} + i q_{-\vec{k}\nu} & k_\alpha \leq 0 \end{cases} \tag{2.36}$$

$$\mathcal{G}_{\vec{k}\nu} > i \left(\frac{\hbar\omega_{\vec{k}\nu} c_{k\nu}}{8\pi}\right)^{\frac{1}{2}} \omega_{\vec{k}\nu} \begin{cases} -p_{-\vec{k}\nu} + i p_{\vec{k}\nu} & k_\alpha > 0 \\ p_{\vec{k}\nu} + i p_{-\vec{k}\nu} & k_\alpha \leq 0 \end{cases} \tag{2.37}$$

where α (= x, y, z) refers to a component in $\vec{k}$-space, to give

$$H_s = \frac{1}{2} \sum_{\vec{k},\nu} \hbar\omega_{\vec{k}\nu} \left(q^2_{\vec{k}\nu} + p^2_{\vec{k}\nu} \right) \tag{2.38}$$

After transformation to quantum mechanical operators ($q_{\vec{k}\nu} \rightarrow q_{\vec{k}\nu}$, and $p_{\vec{k}\nu} \rightarrow -i\,\partial/\partial q_{\vec{k}\nu}$) this gives the medium Hamilton operator as

$$H_s = \frac{1}{2} \sum_{\vec{k},\nu} \hbar\omega_{\vec{k}\nu} (q_{\vec{k}\nu}^2 - \partial^2 / \partial q_{\vec{k}\nu}^2) \qquad (2.39)$$

Through eqs.(2.38) and (2.39) the medium is thus represented as a (practically infinite) set of independent harmonic oscillators along the collective normal coordinates $q_{\vec{k}\nu}$. This is a quite general result which is solely inherent in the assumption that the medium response depends in a linear fashion on the external forces. This implies that although the response functions of the polarization amplitudes to the field, $c_{k\nu}$, may well depend on both space and time coordinates, they are still independent of the field strength. The representation by eqs.(2.31) and (2.32) is thus very general and widely applicable, although for the same reasons not very diagnostic with respect to the nature of the field and the interaction forces.

We notice finally that polarization waves may interfere at given points in the medium. The assumption of linear interactions, i.e. sufficiently small amplitudes of the individual polarization components may therefore still give rise to large absolute polarization values.

2.2 Interaction with Ionic Charges

Up to this point we have discussed essentially bulk medium properties. Thus, eqs.(2.38) and (2.39) have not referred to particular ionic charges and are viewed as the Hamiltonian function and operator, respectively, of the pure medium in the absence of charges. Introduction of ionic charges into the medium, or the presence of a bulk metal or semiconductor phase such as in electrodic or membrane processes, disturbs the medium isotropy which we have assumed so far. Physically this is due to several different effects:

(a) In regions of the medium sufficiently remote from the ions (or the surface) the electrostatic field from the latter induces a finite average polarization. This is associated with the reorientation of the oligomeric clusters of solvent molecules in the medium and does not cause major changes in the local structure. The appropriate medium properties such as $\varepsilon(\vec{r}-\vec{r}', t-t')$ are therefore not disturbed significantly by this effect.

(b) On the other hand, close to the ions the structural effects depend strongly on their charge density. The electric field around large monocharged ions may still not be strong enough to modify the local solvent structure. The effect of the ionic field is then to induce an average polarization of basically the same kind as the one in more remote regions. In contrast, if the charge density is large, the local solvent structure is modified and a new structure established consisting of solvent molecules with properties (residence times, energies) different from those of the bulk. This must obviously be reflected in local changes of the physical parameters, $\varepsilon(\vec{r}-\vec{r}', t-t')$ in particular.

(c) Ions of dimensions which do not fit into the cavities of the local medium structure may exert a purely geometric 'structure-breaking' effect. This effect is observed for large ions such as Cs^+ and J^- and is reflected in both large ionic entropies, large diffusion coefficients of the local solvent molecules compared to bulk values etc. Other ions such as K^+, H_3O^+, NH_4^+ and OH^- are 'structure-making' species with corresponding effects on the appropriate physical quantities.

(d) The ions can finally establish proper chemical bonds with the nearest solvent molecules in such a way that the resulting complex constitutes a new well-defined chemical entity. The dynamics of the first coordination sphere is then most conveniently viewed as intramolecular motion and the continuum formalism applied to the higher coordination spheres.

The nonuniform character of media containing 'impurities' or phase boundaries is much more difficult to incorporate in the general continuum formalism than the space and time dispersion of uniform media. This is due to the fact that the response functions now take the more general form $G(\vec{r},\vec{r}'; t-t')$ and a simple relation to the dielectric permittivity such as eq.(2.10) no longer exists(136). We shall return to a few of these fundamental problems in chapter 8. We notice here that for practical purposes the nonuniform character of the medium is commonly handled in one of the following ways:

(A) The ions are assumed to be represented by charge distributions of sharp boundaries. The intramolecular modes of the ions are subsequently treated by appropriate model potentials, whereas uniform bulk medium properties are assumed to prevail up to the boundary. This is the simplest and most commonly applied approach which is also the basis of most of our following considerations. However, certain extensions can be added to this representation such as the incorporation of 'discrete' surface modes of the boundary, and coupling between intramolecular and medium modes.

(B) The ions in the medium may be viewed as a two-fluid system. Both the ions and the external solvent are thus viewed as continuous media and 'bulk' properties of each, subject to additional boundary conditions characteristic for the system, assumed(137). This 'hydrodynamic' approach is analogous to the liquid drop model for atomic nuclei, and expected to be more appropriate the larger the ions.

(C) The restriction of a sharp boundary may be relaxed by a 'soft-charge' description of the ion-medium interaction(138). Thus, the hydration energy may be calculated by explicit consideration of the interaction between the medium and both nuclear core charges and more diffuse electronic charge distributions represented by appropriate electronic wave functions. This is in fact similar to the view taken by Pekar and in later reports on the polaron and solvated electron problems(139,140). This approach provides a basis for the resolution of some fundamental problems concerning ionic dimensions of dissolved species and the relative orientation and distance between reactants in simple electron transfer reactions. However, generally the wave functions required are quite insufficiently accurately know except for the simplest cases.

We now return to a consideration of the modifications of the solvent Hamiltonian by the presence of the ions and at present only consider a single ion embedded in a continuous solvent. Our approach would also be valid for a solid-state impurity atom in a crystalline or amorphous medium provided that the electronic impurity level is located sufficiently far from the conduction and valence zones of the pure medium that interaction with the band levels can be ignored (a 'deep' impurity level(117,139)). The Hamiltonian function of the system, H_{si}, now has the form

$$H_{si} = H_i + H_s + H_{int} \tag{2.40}$$

where the first term is the (vacuum) energy of the ion, H_s is the Hamiltonian function of the pure solvent discussed above, and H_{int} is the interaction between the ion and the medium. We shall assume that short-range interactions are included either in the intramolecular structure of the ion or in some sharp ion-medium boundary, and although H_{int} typically corresponds to a strong polarization, we shall assume that it is still small compared with intramolecular electronic excitation energies of the ion. The medium is thus supposed not to induce intramolecular radiationless transitions, and in the following we shall consider the ground electronic state only. Anticipating the discussion of chapters 3 and 4 this means that the condition

$$\langle\alpha|H_{int}|\beta\rangle \ll \varepsilon^{o}_{\beta} - \varepsilon^{o}_{\alpha} \qquad (2.41)$$

where $|\alpha\rangle$ and $|\beta\rangle$ are the electronic wave functions of the ground and an excited state, respectively, and ε^{o}_{α} and ε^{o}_{β} the corresponding energies, must be valid. We shall finally assume that the linear approximation (eq.(2.21)) remains valid also in the presence of ions. H_{int} then takes the form (cfr. eq.(2.21))

$$H_{int} = - \int \vec{P}(\vec{r})\, \vec{D}\,(\vec{r};\, \vec{\xi}, \vec{Q})\, d\vec{r} \qquad (2.42)$$

where $\vec{D}$ refers to the field of the ions, $\vec{\xi}$ and $\vec{Q}$ are the coordinates of the electrons and nuclei of the ion, and $\vec{P}$ the polarization at the point $\vec{r}$.

In the general case the interaction energy thus depends on the molecular structure of the ions via the coordinate dependence of $\vec{\xi}$ and $\vec{Q}$. We can illustrate this coupling between the intramolecular and continuous solvent modes in the following way. Anticipating the results of the discussion of the Born-Oppenheimer approximation of the separation of electronic and nuclear motion, averaging of eq.(2.42) with respect to the electronic wave functions gives

$$H_{int} = - \int \vec{P}(\vec{r})\, \vec{D}\,(\vec{r};\vec{Q})\, d\vec{r} \qquad (2.43)$$

where

$$\vec{D}(\vec{r};\vec{Q}) = \langle\varphi^{o}(\vec{\xi},\vec{Q})|\vec{D}(\vec{r};\, \vec{\xi},\vec{Q})|\varphi^{o}(\vec{\xi},\vec{Q})\rangle \qquad (2.44)$$

$\varphi^{o}(\vec{\xi}, \vec{Q})$ is the solution of the stationary electronic Schrodinger equation

$$(H_i + H_{int})\varphi^{o}(\vec{\xi},\vec{Q}) = \varepsilon(\vec{Q})\varphi^{o}(\vec{\xi},\vec{Q}) \qquad (2.45)$$

and $\varepsilon(\vec{Q})$ the corresponding eigenvalue.

If the amplitudes of the intramolecular modes are sufficiently small, $\vec{D}(\vec{r},\vec{Q})$ can be expanded to first order in a series in $\vec{Q}$. This assumption is expected to be better the larger the distance of the point r from the ion. Thus,

$$\vec{D}(\vec{r};\vec{Q}) = \vec{D}(\vec{r};\vec{Q}_o) + \sum_s \left(\frac{\partial \vec{D}(\vec{r},\vec{Q})}{\partial Q_s}\right)_{Q_s = Q_{so}} (Q_s - Q_{so}) \qquad (2.46)$$

where s refers to the various components of $\vec{Q}$. This gives

$$H_{int} = H_{int}^{(1)} + H_{int}^{(2)} \qquad (2.47)$$

where $H_{int}^{(1)}$ is the interaction of the medium with the static 'impurity' field, i.e.

$$H_{int}^{(1)} = - \int \vec{P}(\vec{r})\vec{D}(\vec{r})d\vec{r} \; ; \; \vec{D}(\vec{r}) = \vec{D}(\vec{r};\vec{Q}_o) \qquad (2.48)$$

and $H_{int}^{(2)}$ the interaction with the impurity modes, i.e.

$$H_{int}^{(2)} = - \sum_s \int \vec{P}(\vec{r})V_s(\vec{r})Q_s d\vec{r} \; ; \; V_s = \left.\frac{\partial \vec{D}(\vec{r})}{\partial Q_s}\right|_{Q_s = Q_{so}} \qquad (2.49)$$

Most commonly, in applications of the theory to specific electron or atom group transfer systems attention is given only to $H_{int}^{(1)}$. This implies that all intramolecular modes are assumed to constitute a subset of normal system modes deconvoluted from all the solvent modes. The effect of $H_{int}^{(1)}$ can then be seen by expanding $\vec{P}(\vec{r})$ and $\vec{D}(\vec{r})$ exactly as in eqs.(2.25) and (2.26), and $H^{(1)}_{int}$ takes the form

$$H^{(1)}_{int} = - \sum_{\vec{k},\nu} u_{\vec{k}\nu} q_{\vec{k}\nu} \tag{2.50}$$

where

$$u_{\vec{k}\nu} = (\hbar\omega_{\vec{k}\nu} c_{k\nu} / 2\pi k^2)^{\frac{1}{2}} \left\{ \begin{array}{c} \mathrm{Im}(\vec{k}D_{\vec{k}}) \\ \\ \mathrm{Re}(\vec{k}D_{\vec{k}}) \end{array} \right\} \begin{array}{l} k_\alpha > 0 \\ \\ k_\alpha \leq 0 \end{array} \tag{2.51}$$

The different form of $u_{\vec{k}\nu}$ depending on the sign of k_α reflects the condition that the interaction energy must be a real quantity. Adding this term to eq.(2.39) gives finally for the total Hamiltonian of the ion and the solvent

$$H = H_i + \frac{1}{2}\sum_{\vec{k},\nu} \hbar\omega_{\vec{k}\nu} [(q_{\vec{k}\nu} - q_{\vec{k}\nu o})^2 - \partial^2/\partial q^2_{\vec{k}\nu}] - \frac{1}{2}\sum_{\vec{k},\nu} \hbar\omega_{\vec{k}\nu} q^2_{\vec{k}\nu o} \tag{2.52}$$

where

$$q_{\vec{k}\nu o} = - u_{\vec{k}\nu} / \hbar\omega_{\vec{k}\nu} \tag{2.53}$$

Provided that the polarization response to the electric field of the ions is linear (i.e. the interaction energy is linear in the polarization amplitude) the effect of the ion on the medium is then two-fold. Firstly, without frequency shifts the equilibrium polarization is shifted from the value of the pure solvent characterized by $q_{\vec{k}\nu} = 0$ to a value $q_{\vec{k}\nu} = q_{\vec{k}\nu o}$, and secondly, the free energy of solvation, $\frac{1}{2} - \sum_{\vec{k},\nu} \hbar\omega_{\vec{k}\nu} q_{\vec{k}\nu o}$ is added to the Hamiltonian.

We notice that appearence of terms quadratic in $q_{\vec{k}\nu}$, which is equivalent to a quadratic dependence of $\vec{P}$ on $\vec{D}$ would induce frequency shifts of the normal modes in addition to the equilibrium coordinate shifts. This would also be the effect of a nonvanishing $H^{(2)}_{int}$. This contribution would thus give additional terms of the form $a_{Qq}\ Q_s\ q_{\vec{k}\nu}$ where a_{Qq} are 'mixing' coefficients, in the Hamiltonian, in addition to the 'diagonal' terms (i.e. terms of the form $a_{QQ}\ Q_s^2$ and $a_{qq}\ q_{\vec{k}\nu}^2$). Diagonalization implies change of frequencies in both the ion and the medium, and the normal modes would represent a combined motion of atomic nuclei in both the two subsystems. Both quadratic effects and coupling between intramolecular and medium modes will therefore also induce change in the macroscopic parameters of the system.

2.3 Relation to Macroscopic Parameters

We have now provided a formal derivation of the ion-medium Hamiltonian under the assumption that the medium polarization response is a linear function of the electric field. However, in order to identify the nature of the associated oscillator parameters we must elucidate their relation to the physical parameters of the medium and to $\varepsilon(k,\omega)$ in particular. Such relations are denoted 'sum rules'(131).

We recall at first that for a structureless medium characterized by a single nuclear mode (cfr. chapter 1) the relation between $\vec{P}$ and $\vec{D}$ is given by eq.(1.16). The coupling constant $(4\pi)^{-1}(\varepsilon_0^{-1} - \varepsilon_s^{-1})$ is here independent on both $\vec{r}$ and t and the sum rules are simply $c_{k\nu} = c$, and $\omega_{\vec{k}\nu} = \omega$, where ω is the characteristic vibration frequency for the nuclear motion. In the general case, however, the polarization is represented by superposition of all different polarization branches (eq.(2.27)), i.e. frequency regions for which $\mathrm{Im}\,\varepsilon(k,\omega) \neq 0$, each having separate coupling constants. For crystalline materials the different polarization branches are characterized by well separated absorption peaks of $\mathrm{Im}\,\varepsilon(k,\omega)$, whereas amorphous materials such as liquids display broad, partially overlapping bands. As a consequence the sum rules are more involved than for the simple model system above.

The sum rules are generally derived by means of the fluctuation dissipation theorem which is a relation between the general response function (in 'spectral' or Fourier representation) and the space and time correlation function of the linear response (appendix 1). In the context of an electrostatic polarization field in a continuous dielectric, the resulting equation is

$$\frac{2\ \mathrm{Im}\ \varepsilon(k,\omega)}{\pi\,|\,\varepsilon(k,\omega)\,|^2} = \sum_{\nu} \omega\, c_{k\nu}\,[\,\delta(\omega - \omega_{\bar{k}\nu}) + \delta(\omega + \omega_{\bar{k}\nu})] \qquad (2.54)$$

or

$$\int_0^{\infty} \frac{2\ \mathrm{Im}\ \varepsilon(k,\omega)}{\pi\omega\,|\,\varepsilon(k,\omega)\,|^2}\, d\omega = \sum_{\nu} c_{k\nu} \qquad (2.55)$$

Recalling that $\mathrm{Im}\,\varepsilon(k,\omega)$ is a measure of the amount of absorption of electromagnetic radiation, we notice that eqs.(2.54) and (2.55) thus provide the interpretation of the 'force constants', $c_{k\nu}$, as oscillator strengths of the medium at the particular frequency $\omega_{\vec{k}\nu}$.

This view can be further illustrated by the derivation of a relation between $c_{k\nu}$ and the static dielectric permittivities in different nondissipative frequency regions analogous to eq.(1.16). These relations are valid provided that the absorption bands are well separated. If they overlap, a formally similar relation may still be established where, however, the values of the dielectric permittivities are now parameters to be fitted empirically to the actual absorption bands. Thus, for a group of absorption bands separated by nondissipative frequency regions (i.e. $\mathrm{Im}\,\varepsilon(\omega) = 0$) the following relation can be derived (appendix 1)

$$\varepsilon_{l+1}(k)^{-1} - \varepsilon_{l}(k)^{-1} = \sum_{\nu \in l} c_{k\nu}^{l} = \int_{\omega^{l}-\Delta\omega}^{\omega^{l}+\Delta\omega} \frac{2\,\mathrm{Im}\,\varepsilon(k,\omega)}{\pi\omega|\varepsilon(k,\omega)|^{2}}\,d\omega \qquad (2.56)$$

l refers to a particular absorption band, and to all modes within this band. The integration limits are located in the tranparency bands on each side of the absorption maximum (ω^{l}) and the corresponding values of the static dielectric permittivities in these two regions (for $\omega > \omega^{l}$ and $\omega < \omega^{l}$) are ε_{l+1} and ε_{l}, respectively. Complete summations and integrations over all frequency bands, i.e. from $0 < \omega < \infty$ gives

$$1 - \varepsilon(k,0)^{-1} = \sum_{\text{all}} c_{k\nu} = c(k) = \int_{0}^{\infty} \frac{2\,\mathrm{Im}\,\varepsilon(k,\omega)}{\pi\omega|\varepsilon(k,\omega)|^{2}}\,d\omega \qquad (2.57)$$

(cfr. eq.(2.10)). This provides furthermore a normalization of the oscillator strengths, i.e. $\sum_{\nu} f_{k\nu} = \sum_{\nu} c_{k\nu}/c(k) = 1$.

Eqs.(2.54)-(2.57) represent the formal relations between the Hamiltonian parameters of the medium and the macroscopic coupling properties (the sum rules). For example, eq.(2.57) together with eqs.(2.51)-(2.53) provides the relation between the equilbrium coordinate shift, $q_{\vec{k}\nu 0}$, and the solvation energy, i.e.

$$-E_{sol} = \frac{1}{2}\sum_{\vec{k},\nu} \hbar\omega_{\bar{k}\nu} q^2_{\vec{k}\nu o} = \frac{1}{8\pi}\sum_{\vec{k}} |D_{\vec{k}}|^2 \int_0^\infty \frac{2\ \mathrm{Im}\,\varepsilon(k,\omega)}{\pi\omega|\varepsilon(k,\omega)|^2} d\omega \qquad (2.58)$$

In particular, for a structureless medium with a single nuclear mode, eq.(2.56) and (2.58) give

$$-E_{sol} = \frac{1}{2}\hbar\omega q_o^2 = \frac{1}{8\pi}(1-\varepsilon_s^{-1})\sum_{\vec{k}} |D_{\vec{k}}|^2 = \qquad (2.59)$$

$$(1-\varepsilon_s^{-1})\frac{1}{8\pi}\int_0^\infty D^{-2} dV$$

i.e. the Born equation.

However, while the influence of the frequency dispersion of the medium is explicitly expressed by eqs.(2.53)-(2.57), the influence of the space dispersion is less obvious. To see the effect of the latter, explicit introduction of the space correlation functions of the polarization, i.e. as in eq.(2.22), or its Fourier transform, is required. With reference to appendix 1 the Fourier transform of the space correlation functions of the polarization fluctuations is related to $\varepsilon(k,\omega)$ by the general equation

$$F(\vec{k}) = \int S(\vec{r})e^{-i\vec{k}\cdot\vec{r}} d\vec{r} = \int \langle\vec{P}(\vec{r})\vec{P}(o)\rangle e^{-i\vec{k}\cdot\vec{r}} d\vec{r} =$$

$$= \frac{\hbar}{4\pi^2}\int_0^\infty \frac{\mathrm{Im}\,\varepsilon(k,\omega)}{|\varepsilon(k,\omega)|^2} \mathrm{cth}\frac{\hbar\omega}{2k_BT} d\omega \qquad (2.60)$$

For a particular polarization type, l, (cfr. eqs.(2.54) and (2.55)) this becomes

$$F_1(\vec{k}) = \frac{1}{8\pi}\sum_{\nu\in 1} c_{\vec{k}\nu}\hbar\omega_{\vec{k}\nu}\,\mathrm{cth}\,\frac{\hbar\omega_{\vec{k}\nu}}{2k_B T} \tag{2.61}$$

The physical effect of space correlation can then be noted by inserting eq.(2.61) in eq.(2.58) to give the following approximate expression for the solvation free energy, i.e. for sufficiently well separated absorption bands

$$-E_{sol} \approx \frac{1}{8\pi}\sum_{\vec{k}} |D_{\vec{k}}|^2 \sum_{\nu} c_{k\nu} = \frac{1}{8\pi}\sum_{\vec{k}} |D_{\vec{k}}|^2 \sum_{\nu} \frac{F_\nu(\vec{k})}{F_\nu(o)}\, c_{o\nu} = \tag{2.62}$$

$$\frac{1}{8\pi}\sum_{\vec{k}} |D_{\vec{k}}|^2 c(k)$$

where $c(k) = (1 - \varepsilon_s(k)^{-1})$.

For $k \to 0$ this expression becomes identical with the Born equation, eq.(2.15). For finite k, however, eq.(2.62) gives smaller values for $|E_{sol}|$. In view of the absence of experimental data on the space correlation this is illustrated most conveniently by choosing an empirical but plausible trial function for $S_\nu(\vec{r}-\vec{r}')$. The latter could for example be of the form

$$S_\nu(\vec{r}-\vec{r}') \approx e^{-|\vec{r}-\vec{r}_o|/\lambda_\nu}/|\vec{r}-\vec{r}_o| \;;\; F_\nu(\vec{k}) = F_\nu(o)/(1+k^2 r_o^2) \tag{2.63}$$

corresponding to an exponentially decaying and diffusely spreading effect, characterized by a given correlation length, λ_ν, for each polarization mode. Exactly this approach was the basis of a recent analysis of the hydration energies of alkali and halide ions(134). The Born equation gives too high values for these ions. On the other hand, application of eq.(2.62) with different kinds of trial correlation functions (step function, exponential function) for the three major polarization modes of water, i.e. electronic, infrared, and Debye, provided good agreement with the experimental data for a single set of correlation lengths. The actual values of the latter were 0.53 A, ~1 A, and 10 A in agreement with expectations from the physical nature of the three kinds of motion.

In summary, we have now presented a formalism for the dynamics of the polarization fluctuations in the external medium and their role in the coupling to the electronic energy levels of the separate ionic species which may participate in chemical reactions. Our approach has been essentially phenomenologic, i.e. we have derived a Hamiltonian operator which is characterized by some, initially unknown, parameters. The latter are subsequently identified by reference to the results of linear response theory and comparison with our knowledge regarding the physical nature of the medium. As a result, we can draw two important conclusions:

(1) thermal electron and atom group transfer processes (which are typically accompanied by strong medium reorganization) clearly possess multiphonon character, i.e. a multitude of relatively low frequency medium modes are excited during the process; (2) structural medium effects, as manifested by the polarization correlation functions, are generally expected to be strongly manifested for the more commonly studied dipolar media such as water.

Both of these effects are thus of crucial importance for the kinetics of elementary chemical processes in condensed phases. In particular, both effects are expected to be revealed when such processes are followed over large temperature intervals. Thus, frequency dispersion will affect the number of excitable phonon modes with changing temperature, and the solvent contribution to the activation energy is therefore a function of temperature. Furthermore, the amount of structuration of the medium is also dependent on temperature, and correlation lengths, in particular in the Debye region are longer the lower the temperature. We shall return to these important effects in later sections dealing with low-temperature elementary chemical processes.

3 QUANTUM MECHANICAL FORMULATION OF RATE THEORY

3.1 Elements of Scattering Theory

We shall now proceed to an outline of a formal quantum mechanical description of elementary chemical processes. Two approaches would be expected to provide the conceptual basis. Thus, chemical processes between separate species may be viewed as a scattering process, the time evolution of which is formulated in terms of the states of the infinitely separated reactants (33,113,141). On the other hand, relaxation processes in isolated molecular species or intramolecular processes in species trapped in solid media are usually viewed as the decay of a metastable excited 'zero-order' electronic state by the influence of an intramolecular or medium-induced perturbation (2-4). However, the essentials of both approaches consist in the choice of a suitable set of 'zero-order' functions of the system followed by a description of the time evolution of this set under the influence of some perturbation which arises from interactions which were not included in the zero order Hamiltonians. Considering at first the former approach we notice that the sequence of events in chemical reactions between mobile species can be divided in three stages (33,94). Firstly, in the 'infinitely remote' past the reactants are located sufficiently far from each other that all interreactant interactions can be ignored. In the intermediate stage a collision starts, i.e. the reactants approach eachother, and during the collision process the interaction induces the appropriate electronic and nuclear rearrangements. Finally, during the third stage the products are separated, and in the 'infinitely remote' future all interaction between the separated product molecules vanishes.

This division is closely analogous to the one commonly applied in the formalism of scattering theory in gas phase processes (113,141,142). With reference to the latter we shall denote the state of the system during the first and third stages of the process by the ingoing and outgoing reaction channel, respectively. The channel states are thus characterized by stationary wave functions corresponding to the bound states of the separated reactants and products. The channel wave functions moreover constitute two orthonormal sets. Since, however, the two channel Hamiltonian operators are different, the two sets of wave functions cannot be assumed to be mutually orthogonal. In this respect they differ from the zero order states of intramolecular processes.

In the formulation of the time evolution problem it is supposed that the stationary solutions to the channel Hamiltonians, H_c and $H_{c'}$

$$H_c = T_X + T_N + H_D + H_A + H_M + V_{DM} + V_{AM} + V_{AD} + V_{XA} + V_{XM} \tag{3.1}$$

$$H_{c'} = T_X + T_N + H_D + H_A + H_M + V_{DM} + V_{AM} + V_{AD} + V_{XD} + V_{XM} \tag{3.2}$$

are known. T_X and T_N are the kinetic energy of the transferring entity X(electron or atom group) and of all the other nuclei of the system, respectively, including both the donor and acceptor fragments and the external medium. H_D, H_A, and H_M are the Hamiltonian functions of the donor, the acceptor, and the medium, respectively, V_{DM} and V_{AM} the interactions which couple the donor and the acceptor to the medium, V_{XD}, V_{XA}, and V_{XM} the total interactions between X and the donor, the acceptor, and the medium, and V_{AD} the interaction between the donor and acceptor fragments which defines the geometry of the reaction complex. The total Hamiltonian of the system can then be written as

$$H = H_c + V_c = H_{c'} + V_{c'} \tag{3.3}$$

The total Hamiltonians thus include an additional term representing the interaction between the entity X and the acceptor ($V_c = V_{XA}$) or donor ($V_{c'} = V_{XD}$) fragment in the ingoing and outgoing channel, respectively. Since we shall discuss atom group transfer separately in chapter 6 we shall now explicitly refer to electron transfer, i.e. X now represents the transferring electron.

The problem of scattering theory is to find the probability of transition between the incoming and outgoing channels. This involves time evolution of nonstationary states and therefore requires the solution of the time dependent Schrodinger equation

$$i\hbar \frac{\partial \psi(t)}{\partial t} = H\psi(t) \tag{3.4}$$

with the initial limiting condition

$$\psi(t) \to \Phi_c \quad \text{for } t \to -\infty \tag{3.5}$$

Φ_c represents a set of stationary states in the ingoing channel c, i.e. solutions to the stationary Schrodinger equation of the Hamilton operator H_c. If the asymptotic solution to eq.(3.4) for $t \to +\infty$, and with the limiting condition of eq.(3.5) is subsequently expanded in a series of stationary outgoing channel wave functions each of which is normalized to unity, then the expansion coefficients are the probability amplitudes for transition to the corresponding particular state in the outgoing channel.

When calculating the transition probability, the 'scattering process' is viewed as a transformation which converts the 'infinitely remote' (in time and space) ingoing to the infinitely remote outgoing states. This is formally described by introducing a transformation operator (scattering matrix), S, (113,141)

$$\Psi(t_2) = S(t_2, t_1)\Psi(t_1) \tag{3.6}$$

which transforms the wave function at time t_1 to the wave function at time t_2. All information about the process is then contained in the initially 'black-box-like' scattering matrix, and the problem reformulated to a determination of the properties of this device.

We are moreover interested in the limits $t \to -\infty$ and $t \to +\infty$. Since $\Psi(-\infty)$ coincides with the given stationary state of the ingoing channel, Φ_c, then $\Psi(+\infty)$ is a superposition of wave functions of all outgoing channel states, and $\langle \Phi_{c'} | S(+\infty, -\infty | \Phi_c \rangle = S_{c'c}$ represents the probability amplitude for transition from a given state Φ_c in the ingoing channel c to a given state $\Phi_{c'}$ in the outgoing channel c'.

Since the chemical processes involve energy-conserving transitions between different ingoing and outgoing states, it is convenient to employ a different transition matrix, $\mathcal{T}$, the elements of which are related to those of the S-matrix by the equation

$$\langle c', b | S | c, a\rangle = \delta_{(c', b), (c, a)} - 2\pi i \delta(E_{c', b} - E_{c, a}) \tag{3.7}$$

$$\langle c', b | \mathcal{T} | c, a\rangle$$

where a and b refer to a particular energy level in the ingoing and outgoing channel, respectively. The delta function $\delta_{(c', b), (c, a)}$ vanishes whenever $c \neq c'$. On the other hand, the factor $\delta(E_{c', b} - E_{c, a})$ ensures that transitions between different states occurs only when their energies ($E_{c', b}$ and $E_{c, a}$) coincide. The probability per unit time of transition between the given set of states $|c, a\rangle$ and $|c', b\rangle$ in the ingoing and outgoing channels, W_{ba}, then takes the compact form(113)

$$W_{ba} = \frac{2\pi}{\hbar} |\langle c', b | \mathcal{T} | c, a\rangle|^2 \delta(E_{c', b} - E_{c, a}) \tag{3.8}$$

This equation represents a general formulation of the scattering problem for any chemical process. As a first step towards subsequent calculations we notice that eq.(3.8) is formally similar to the 'golden rule' of first order time dependent perturbation theory (141,142) which gives the probability per unit time of transition between two zero order states under the influence of a small perturbation, V. This expression thus differs from eq.(3.8) by a replacement of the total transition operator, $\mathcal{T}$, by the perturbation operator V, i.e. the first order expansion of $\mathcal{T}$ in this perturbation (see below)

$$W_{fi} = \frac{2\pi}{\hbar} |\langle f|V|i\rangle|^2 \delta(E_i - E_f) \qquad (3.9)$$

where $|i\rangle$ and $|f\rangle$ are the initial and final state wave functions and E_i and E_f the corresponding energies.

However, there is an important difference in the derivation of eqs.(3.8) and (3.9) apart from the fact that $\mathcal{T}$ includes interactions of all orders. Thus, eq.(3.9) rests on the assumption that the system performs transitions between the states $|i\rangle$ and $|f\rangle$ which belong to a single orthogonal set of wave functions. Eq.(3.9) derived from 'usual' perturbation theory therefore constitutes the ideological basis for the probability of electronic transitions between separate states inside either the ingoing or outgoing channel. On the other hand, applications to transitions from a particular ingoing to a particular outgoing channel state requires consideration of the mutual nonorthogonality between states from different channels. (see section 3.2).

We now invoke the exact expression for the elements of the $\mathcal{T}$-matrix for transition between any two channels c and c', i.e. (113).

$$\langle c',b|\mathcal{T}|c,a\rangle = \mathcal{T}_{c'c}(E_{c,a}) = V_c + V_{c'}(E_{c,a}+i\gamma-H)^{-1}V_c \qquad (3.10)$$

V_c and $V_{c'}$ represent the 'perturbations', i.e. the difference between the exact Hamiltonian H and the Hamiltonians, H_c and $H_{c'}$, of the ingoing and outgoing channel, respectively, i.e. $V_c = H - H_c$, and $V_{c'} = H - H_{c'}$. Addition of the small imaginary quantity $i\gamma$ implies a limiting process in which $\gamma \to 0+$ subsequent to the action of the operator. This compact form contains all information about the scattering process, but it is not very useful as it stands, since generally the 'true' eigenfunctions are unknown. For practical purposes the $\mathcal{T}$-matrix is usually calculated by an iterative procedure (the Born approximation), expanding the matrix in orders of the perturbations. Thus, we use the identity (113)

$$A^{-1} = B^{-1} + B^{-1}(B-A)A^{-1} \tag{3.11}$$

where A and B are two arbitrary operators. Putting $A = (E_{c,a} + i\gamma - H)$, $B = E_{c,a} + i\gamma - H_{c''}$, and $B-A = V_{c''}$, where the subscript, c", may refer not only to the ingoing and outgoing channels but also to any other channel through which the system may pass we obtain the following equation

$$(E_{c,a} + i\gamma - H)^{-1} = (E_{c,a} + i\gamma - H_{c''})^{-1} + (E_{c,a} + i\gamma - H_{c''})^{-1} V_{c''}(E_{c,a} + i\gamma - H)^{-1} \tag{3.12}$$

Inserting the left hand side of eq.(3.12) on the right hand side of the same equation gives the following infinite series expansion in V_c, $V_{c'}$, and $V_{c''}$

$$\mathcal{T} = V_c + V_{c'}(E_{c,a} + i\gamma - H_{c''})^{-1} V_c + V_{c'}(E_{c,a} + i\gamma - H_{c''})^{-1} V_{c''}(E_{c,a} + i\gamma - H_{c''})^{-1} V_c + \ldots \tag{3.13}$$

where in principle all possible reaction channels are included in the series. In particular, the first two matrix elements in the series are

$$\mathcal{T}^{(1)}_{c'c} = \langle c',b|V_c|c,a\rangle \tag{3.14}$$

$$\mathcal{T}^{(2)}_{c'c} = \langle c',b|V_{c'}(E_{c,a} + i\gamma - H_{c''})^{-1} V_c|c,a\rangle \tag{3.15}$$

The term $\mathcal{T}^{(1)}_{c'c}$ is usually interpreted as a direct transition from a given state in the ingoing to a given state in the outgoing channel induced by the perturbation V_c. The interpretation of the second order term can be illustrated by inserting a complete set of channel states $\sum_d |c'',d\rangle\langle c'',d| = 1$ to give

$$\mathcal{T}^{(2)}_{c'c} = \sum_d \langle c,b|V_{c'}|c\ ,d\rangle\langle c\ ,d|V_c|c,a\rangle / \tag{3.16}$$

$$(E_{c,a} + i\gamma - E_{c'',d})$$

This term is thus interpreted as a transition from $|c,a\rangle$ to $|c',b\rangle$ via a set of intermediate states $|c'',d\rangle$ (141-143). In particular, when the direct transition is forbidden, for example by a selection rule, the second term acquires a special importance being the first term of finite value in the series, provided that the corresponding matrix elements are nonvanishing.

The channel c″ may be identical with either the ingoing or the outgoing channel. This is the situation implicit in the majority of electron transfer theories which only incorporate these two channels. The appropriate intermediate states are then excited electronic states of these two channels, i.e. of the isolated donor and acceptor molecules. However, c″ may also represent a different reaction channel, and eq.(3.16) is then interpreted as electron transfer through a third chemical species present in the solution in addition to the donor and acceptor molecules.

This interpretation is particularly important for the application of the concepts of higher order processes to inner sphere and biological redox processes to which we shall return in chapters 7 and 9.

In our following discussion of elementary processes we shall mostly apply eq.(3.9) (eq.(3.14)), i.e. we shall assume that the perturbation interaction is sufficiently small that terms higher than first order can be ignored. This corresponds to the electronically nonadiabatic limit, and in many cases of condensed phase elementary chemical processes and intramolecular relaxation processes this is an adequate procedure. If the interaction is strong, corresponding to the limit of adiabatic processes, there is at present no practical way of performing the summation over all terms of eq.(3.13). We shall then adopt an alternative, procedure i.e. a semiclassical approach which will be further discussed in chapter 5.

We complete this section by providing the thermally averaged transition probability per unit time. Thus, eqs.(3.8), (3.9), and (3.14) represent the probability that the system is transferred between a particular pair of initial and final states. These expressions must be summed over all final states to give the probability that the system leaves the particular initial state of energy $E_{c,a}$. Moreover, although gas phase processes involving systems prepared in a given energy state have recently become experimentally tractable (144), in all condensed phase processes a rate constant or a radiative or nonradiative life time thermally averaged over all energy levels is measured. It is thus assumed that the processes are sufficiently slow compared with the vibrational relaxation times either in the medium or in the intramolecular modes that thermal equilibrium prevails cfr. eq.(1.6). This assumption is most likely to be valid for condensed phase processes and processes in large molecules, whereas electronic processes in small molecules at low pressures do not correspond to the 'statistical limit' and must be viewed within a single-level scheme (2-4). We shall thus calculate expressions such as eqs.(3.8) and (3.9) by means of a statistical distribution function. Since in principle our system is of quantum mechanical nature the latter is the Gibbs quantum statistical distribution function(145), i.e.

$$P_{a,v} = Z^{-1} \exp(-E^{v}_{c,a}/k_B T) \qquad (3.17)$$

$$Z = \sum_{a,v} \exp(-E^{v}_{c,a}/k_B T) = \exp(-G_c/k_B T) \qquad (3.18)$$

where Z is the statistical sum of the system in the ingoing channel, G_c the Gibbs free energy, and the summation is performed over all electronic (a) and vibrational (v) states. The thermally averaged first order transition probability per unit time is then

$$W_{c'c} = \frac{2\pi}{\hbar} Z^{-1} \sum_{a,v} \sum_{b,w} \exp(-E^{v}_{c,a}/k_B T) |\langle c',b|V_c|c,a\rangle|^2 \qquad (3.19)$$
$$\delta(E^{w}_{c',b} - E^{v}_{c,a})$$

where w refers to vibrational states in the outgoing channel. This expression constitutes the formal basis of our subsequent calculations.

3.2 Channel States and Nature of the Perturbation

We shall now take the 'golden rule' as the general starting point for a derivation of the rate expressions. In order to proceed further we shall have to specify the nature of both the zero order states, i.e. for chemical processes the electronic states of the individual reaction channels, and the nature of interactions between the different subsystems which induce the reaction. We shall perform this analysis in two steps. Firstly, we shall adopt a stationary scheme based on the Born-Oppenheimer separation of electronic and nuclear motion for the (stationary) ingoing and outgoing channels and in the absence of any interaction between the reacting molecules. Secondly, we shall describe the time evolution of the system by expanding the time dependent wave function in either of the zero order sets of ground and excited electronic states including the interaction left out of the channel Hamiltonians. During the latter procedure we shall incorporate the nonorthogonality of the two different basis sets explicitly. This generally provides a fairly complicated overall rate expression by introducing the probability of electron transfer not only from the donor to the acceptor ground levels but also between all excited levels as well as among levels inside each channel. Ultimately we shall therefore introduce the simplification of a two-level scheme involving only the donor and acceptor ground electronic levels. This is justified in view of the small value of the perturbation interactions compared with electronic excitation energies for molecular systems.

Turning at first to the stationary channel states we notice that in condensed phase processes all nuclear modes are essentially of vibrational nature. Thus, due to the trapping of the molecular reactants in temporary diffusional equilibrium positions, free translation and rotation are converted either to hindered, i.e. vibrational, motion, or to relaxational motion in which a reactant molecule passes energy barriers on its way towards an absolute (temporary) equilibrium orientation. The chemical processes can thus in a sense be viewed as intramolecular electronic transitions where the electrons are coupled to a very large number of both discrete molecular and nonlocal medium modes. In particular, as noted in chapter 2, the strong electronic coupling to the medium modes is included in the zero order states. This is in contrast to the medium effects in the life times of e.g. excited aromatic molecules (146) the decay of which is believed to be induced by meak electrostatic medium interactions.

We can then write the channel Hamiltonians in the form (cfr. eqs.(3.1) and (3.2))

$$H_{c,c'}(\vec{r},\vec{Q}) = H_N(\vec{Q}) + H_e(\vec{r}) + V_{eN}(\vec{r},\vec{Q}) \tag{3.20}$$

where $\bar{r}$ refers to the electronic coordinates, and $\vec{Q}$ to all intramolecular and medium nuclear coordinates. $H_N(\vec{Q})$ and $H_e(\vec{r})$ are the nuclear and electronic Hamiltonian, respectively, and $V_{eN}(\vec{r},\vec{Q})$ the interaction between the electrons and the nuclei.

The appearance of the term $V_{eN}(\vec{r},\vec{Q})$ in eq.(3.20) means that the variables of the electronic and nuclear subsystems cannot be separated. In order to proceed we shall exploit the large difference in the masses and velocities of the electrons and the nuclei. Thus, the motion of the nuclei corresponds to excitation in the infrared region of electromagnetic radiation, whereas electronic motion corresponds to excitation in the ultraviolet region. Due to the much lower characteristic nuclear velocity compared with the velocity of the electrons, the latter move in an essentially constant external field determined by the instantaneous values of the nuclear coordinates and depending parametrically on the latter. This is the physical contents of the Born-Oppenheimer adiadatic approximation (147) which we invoked in chapter 1 to illustrate the donor-acceptor level degeneracy and the role of the solvent fluctuations in electron transfer reactions.

Following the procedure of the Born-Oppenheimer approximation we first solve the stationary 'electronic Schrodinger equation'

$$(H_e + V_{eN})\psi(\vec{r},\vec{Q}) = \mathcal{E}(\vec{Q})\psi(\vec{r},\vec{Q}) \tag{3.21}$$

where $\mathcal{E}(\vec{Q})$ is the electronic energy, and the parametric dependence on $\vec{Q}$ is explicitly shown. If l refers to a particular electronic state, and $\psi_l(\vec{r},\vec{Q})$ represents the corresponding electronic wave function, then all the $\psi_l(\vec{r},\vec{Q})$'s constitute a complete orthonormal set of functions, and the total channel wave function, Ψ, of the system is

$$\Psi(\vec{r},\vec{Q}) = \sum_l \chi_l(\vec{Q})\,\psi_l(\vec{r},\vec{Q}) \tag{3.22}$$

where the coefficients, $\chi_l(\vec{Q})$ are the nuclear wave functions to be determined subsequently by solution of the complete wave equation. Eq.(3.22) thus implies that the electrons follow the nuclear motion and adjust themselves to the instantaneous nuclear coordinate values.

By inserting eq.(3.22) in the complete stationary (channel) Schrodinger equation

$$H_{c,c'}\,\Psi = E\,\Psi \tag{3.23}$$

the latter can be rewritten in the (so far accurate) form

$$\begin{aligned} &\sum_l (\mathcal{E}_l + H_N - E)\chi_l \psi_l = \\ &\sum_l \psi_l [(\mathcal{E}_l + H_N - E)\chi_l - \psi_l H_N \chi_l + H_N \chi_l] = 0 \end{aligned} \tag{3.24}$$

Multiplying from the left by $\psi_n(\vec{r},\vec{Q})^*$ (the complex conjugate of $\psi_n(\vec{r},\vec{Q})$) and integrating with respect to $\vec{r}$ subsequently gives the following set of coupled equations

$$(H_N + \mathcal{E}_n - E)\chi_n + \sum_l L_{nl}\chi_l = 0 \tag{3.25}$$

where

$$L_{nl} = \int \psi_n^* [H_N, \psi_l] d\vec{r} \qquad (3.26)$$

and [,] denotes the commutator. L_{nl} represents the matrix elements of the nonadiabaticity operator, L, and their explicit form is given below. Within the Born-Oppenheimer approximation the terms containing these matrix elements are regarded as small to the same extent as the rate of change of the nuclear coordinates. Neglecting these terms altogether gives the set of nuclear wave functions, χ_{nv}, of the 'pure' Born-Oppenheimer states by the equations

$$[H_N(\vec{Q}) + \mathcal{E}_n(\vec{Q}) - E_{nv}]\chi_{nv}(\vec{Q}) = 0 \qquad (3.27)$$

where v refers to the nuclear quantum numbers. In this stage the 'effective' potential energy thus arises not only from the nuclear interaction energy, but also from the electronic energy, $\varepsilon_n(\vec{Q})$. This is related to the fact that the nuclei effectively interact with a delocalized electronic charge distribution of density $|\psi_n(\vec{r},\vec{Q})|^2$, determined by the total electronic energy. Together with the nuclear potential energy, $\varepsilon_n(\vec{Q})$ therefore determines a potential energy surface, i.e. an energy function of the coordinates $\vec{Q}$ in a manydimensional space for each electronic state n. Potential energy surfaces generally play a crucial role in the description of both chemical and radiative and nonradiative intramolecular processes. They determine equilibrium configurations of the system, the barriers over or through which it must pass etc. Estimates of the form of potential energy surfaces are therefore decisive for the calculation of kinetic parameters of reacting systems.

However, the 'pure' Born-Oppenheimer states are not exact solutions. Inclusion of the nonadiabaticity operator, L, implies that the electronic and nuclear motion can no longer be separated, and in terms of the zero order wave functions this means that there is now a finite probability of radiationless transitions between the zero order states induced by the operator L. The explicit form of the matrix elements of the latter is

$$L_{nl} = \frac{1}{2}\int \psi_n^* \left[2\frac{\partial\psi_1}{\partial Q}\frac{\partial}{\partial Q} + \frac{\partial^2\psi_1}{\partial Q^2}\right]d\vec{r} = L_{nl}^{(1)} + L_{nl}^{(2)} \tag{3.28}$$

where

$$L_{nl}^{(1)} = \int\left(\psi_n^* \frac{\partial\psi_1}{\partial Q} d\vec{r}\right)\frac{\partial}{\partial Q} = J_{nl}\frac{\partial}{\partial Q} \tag{3.29}$$

and

$$L_{nl}^{(2)} = \frac{1}{2}\int \psi_n^* \frac{\partial^2\psi_1}{\partial Q^2} d\vec{r} \tag{3.30}$$

which is obtained from eq.(3.26) by noting that the potential energy contribution from the commutator vanishes. This expression can be generalized by summation over several nuclear modes for which the matrix elements of L assume finite values.

L_{nl} is thus a measure of the dependence of the electronic wave functions on the nuclear coordinates. When this dependence is weak, $L_{nl}^{(2)}$ can be neglected compared with $L_{nl}^{(1)}$, which can be transformed to the following form commonly met (2-4). By differentiating eq.(3.21) with respect to Q we obtain for the particular state l

$$H_e\frac{\partial\psi_1}{\partial Q} + \frac{\partial V_{eN}}{\partial Q}\psi_1 + V_{eN}\frac{\partial\psi_1}{\partial Q} = \frac{\partial\varepsilon_1}{\partial Q}\psi_1 + \varepsilon_1\frac{\partial\psi_1}{\partial Q} \tag{3.31}$$

Multiplication from the left by ψ_n^*, integration with respect to r, and rearrangement subsequently gives

$$(\varepsilon_n - \varepsilon_l)\int \psi_n^* \frac{\partial \psi_l}{\partial Q}\, d\vec{r} + \int \psi_n^* \frac{\partial V_{eN}}{\partial Q} \psi_l \, d\vec{r} = \frac{\partial \varepsilon_l}{\partial Q}\delta_{nl} \quad (3.32)$$

which provides the following relations:

(a) for n = l

$$\frac{\partial \varepsilon_l}{\partial Q} = \left\langle \frac{\partial V_{lN}}{\partial Q} \right\rangle_l \quad (3.33)$$

where $\langle\ \rangle_l$ denotes averaging with respect to the electronic state l.

(b) for n ≠ l

$$L_{nl}^{(1)} = \left[\int \psi_n^* \frac{\partial V_{eN}}{\partial Q} \psi_l \, d\vec{r} / (\varepsilon_n - \varepsilon_l)\right] \frac{\partial}{\partial Q} \quad (3.34)$$

We notice that $L_{nl}^{(1)}$ is large in regions of the potential energy surfaces where $\varepsilon_l \approx \varepsilon_n$, i.e. near the intersection of the zero order surfaces. This is commonly expressed as the break-down of the Born-Oppenheimer approximation.

This kind of perturbation was firstly suggested by Robinson and Frosch (148) and is now generally believed to be responsible for intramolecular transitions between electronic states having the same spin multiplicities (internal conversion) in both large isolated and solute molecules (2-4,149). If the wave functions ψ_l and ψ_n belong to different spin multiplicities the matrix element vanishes but in such cases the transition may still proceed as a second order process by spin orbit coupling via other electronic states (148) (intersystem crossing, cf. the second order processes represented by eq.(3.16), and chapter 7). However, in the context of chemical processes these perturbations all correspond to transitions within each reaction channel, whereas perturbations of a different nature are responsible for reaction between two reaction centres involving more than a single channel. Quite generally this involves all potential interaction energy contributions left out of the zero order channel Hamiltonians, e.g. electrostatic electron and core interactions between the centres (electron and atom group transfer, polaron mobility, diffusion), electrostatic dipole interactions (long-range electronic excitation energy transfer) etc.

We shall see this in more detail by considering explicity the time dependence of the system in a way analogous to that of Holstein for the polaron mobility problem (58,91). Considering for the sake of simplicicty only a single ingoing and outgoing channel which we shall denote by i and f, respectively, we can write the total time dependent wave function of the system as

$$\Psi(\vec{r},\vec{Q},t) = \sum_{l} \chi_{il}(\vec{Q},t)\,\psi_{il}(\vec{r},\vec{Q}) + \sum_{\lambda} \chi_{f\lambda}(\vec{Q},t)\,\psi_{f\lambda}(\vec{r},\vec{Q}) \qquad (3.35)$$

where l and λ now refer to electronic states in the ingoing and outgoing channels, respectively. Inserting eq.(3.35) in the time dependent Schrodinger equation (eq.(3.4)) and ignoring the nonadiabaticity operator gives

$$i\hbar\Big(\sum_{l} \dot{\chi}_{il}\psi_{il} + \sum_{\lambda} \dot{\chi}_{f\lambda}\psi_{f\lambda}\Big) = \sum_{l}(\psi_{il}H_{iN}\chi_{il} + \varepsilon_{il}\chi_{il}\psi_{il} + V_i\chi_{il}\psi_{il}) + \sum_{\lambda}(\psi_{f\lambda}H_{fN}\chi_{f\lambda} + \varepsilon_{f\lambda}\chi_{f\lambda}\psi_{f\lambda} + V_f\chi_{f\lambda}\psi_{f\lambda}) \qquad (3.36)$$

where V_i and V_f are the appropriate perturbations in the ingoing and outgoing channel, respectively (V_{eA} and V_{eD}, cf. eqs.(3.1) and (3.2)), and the dot indicates differentiation with respect to time. Multiplying from the left by ψ_{im}^* and $\psi_{f\mu}^*$, where m and μ are particular electronic states in the ingoing and outgoing channel, respectively, and integrating with respect to r, gives

$$i\hbar\,\dot{\chi}_{im} + i\hbar\sum \dot{\chi}_{f\lambda} S^{m\lambda}_{if} = (H_{iN} + \varepsilon_{im})\chi_{im} + \sum_{l} V^{ml}_{i}\chi_{il} + \sum [S^{m\lambda}_{if}(H_{fN} + \varepsilon_{f\lambda})\chi_{f\lambda} + V^{m\lambda}_{fi}\chi_{f\lambda}] \tag{3.37}$$

and

$$i\hbar\sum_{l} \dot{\chi}_{il} S^{\mu l}_{fi} + i\hbar\,\dot{\chi}_{f\mu} = (H_{fN} + \varepsilon_{f\mu})\chi_{f\mu} + \sum V^{\mu\lambda}_{f}\chi_{f\lambda} + \sum_{l} [S^{\mu l}_{fi}(H_{iN} + \varepsilon_{il})\chi_{il} + V^{\mu l}_{i}\chi_{il}] \tag{3.38}$$

where we have introduced the overlap integrals, S, for states belonging to different channels and the perturbation matrix elements with the subscripts indicating the channels involved, and the superscripts the electronic states.

Multiplication of eq.(3.38) by $S^{m\mu}_{if}$, summation with respect to μ, and insertion in eq.(3.37) gives (150)

$$i\hbar\sum_{l}(\delta_{ml} - \sum_{\mu} S^{m\mu}_{if}S^{\mu l}_{if})\dot{\chi}_{il} = \sum_{l}[(\delta_{ml} - \sum_{\mu} S^{m\mu}_{if}S^{\mu l}_{fi})(H_{iN} + \varepsilon_{il}) + V^{ml}_{i} - \sum_{\mu} S^{m\mu}_{if}V^{\mu l}_{i}]\chi_{il} + \sum_{\lambda}(V^{m}_{f} - \sum_{\mu} S^{m\mu}_{if}V^{\mu}_{f})\chi_{f\lambda} \tag{3.39}$$

or in matrix form

$$i\hbar\,(1 - S^{+}S)\dot{\chi}_{i} = [(1 - S^{+}S)(H_{iN} + \varepsilon_{i}) + V^{ii}_{i} - S^{+}V^{fi}_{i}]\chi_{i} + (V^{if}_{f} - S^{+}V^{ff}_{f})\chi_{f} \tag{3.40}$$

where $S = S^{fi}$ is the overlap matrix for the two channels, and $S^{+} = S^{if}$. An analogous procedure gives for χ_f

$$i\hbar\,(1 - SS^{+})\dot{\chi}_f = [(1 - SS^{+})(H_{fN} + \varepsilon_f) + V_f^{ff} - S\,V_f^{if}]\chi_f + \quad (3.41)$$

$$+ (V_i^{fi} - S\,V_i^{ii})\chi_i$$

In the simplified case of a single donor and acceptor level only, which is the model applied in nearly all treatments of homogeneous electron transfer processes eqs.(3.39)-(3.41) take the form

$$i\hbar\,\dot{\chi}_i = [(H_{iN} + \varepsilon_i) + (V_i^{ii} - S_{if}\,V_i^{fi})(1 - S^2)^{-1}]\chi_i + \quad (3.42)$$

$$+ [(V_f^{if} - S_{if}\,V_f^{ff})(1 - S^2)^{-1}]\chi_f$$

and a similar equation in which the indices 'i' and 'f' are inverted. We have here omitted the indices of the electronic levels. These equations now have the following implications

(a) The general expression for the time evolution of the system takes a rather complicated form. Thus, the perturbations induce coupling of a given initial state not only to all states of the outgoing channel but also to all other states of the ingoing reaction channel.

(b) The perturbation interactions are seen to play a dual role. Firstly, the ingoing channel states are modified (distorted) by the perturbations. Thus, the dominating term inside the first bracket (summation) on the right hand side of the equations are the 'diagonal' matrix elements of the perturbation which represent a measure of the distortion of each electronic level. However, these terms do not involve coupling to the outgoing channel states, i.e. the chemical process, and they are most conveniently included in a modified channel Hamiltonian which thus includes all coulomb and exchange terms in the first brackets. This also implies that the work terms discussed previously may be included in the modified channel Hamiltonians (151), the corresponding stationary wave equations of which in the two-level scheme are

$$[(H_{iN} + \varepsilon_i) + (V_i^{ii} - S_{if} V_i^{fi})(1 - S^2)^{-1}]\chi_{iv} = E^o_{iv} X^o_{iv} \qquad (3.43)$$

and a similar equation for which the indices 'i' and 'f' are inverted, and the vibrational quantum number 'v' replaced by that of the final state, 'w'.

Secondly, the terms which induce the transitions between different channel states are the second terms (sums) of the right hand side of eqs.(3.39)-(3.42). By expanding the nuclear wave functions in the form (58,91)

$$\chi_i(\vec{Q},t) = \sum_v C_{iv}(t)\chi^o_{iv}(\vec{Q}) \exp(-iE^o_{iv} t/\hbar) \qquad (3.44)$$

$$\chi_f(\vec{Q},t) = \sum_w C_{fw}(t)\chi^o_{fw}(\vec{Q}) \exp(-iE^o_{fw} t/\hbar) \qquad (3.45)$$

inserting eqs.(3.44) and (3.45) in eq.(3.42), exploiting eq.(3.43), and following the approach of 'conventional' time dependent perturbation theory (141) with respect to the nuclear wave functions we finally obtain a scheme identical to the latter provided that the 'effective' perturbation operator, V_{eff}, is

$$V_{eff} = (V_f^{if} - S^{if} V_f^{ff})(1 - S^2)^{-1} \qquad (3.46)$$

For the two-level scheme the golden rule expression thus takes the form

$$W_{fi} = \frac{2\pi}{\hbar} \sum_{v} \sum_{w} \exp(-\beta E^{o}_{iv}) |\langle \chi^{o}_{fw} | V_{eff} | \chi^{o}_{iv} \rangle|^{2} \delta(E^{o}_{fw} - E^{o}_{iv}) \quad (3.47)$$

which will be the basis of most of our further discussion. The electronic wave functions are here included in V_{eff}, and we notice that the latter can be interpreted as the matrix element with respect to the initial and final electronic states of the total perturbation, corrected for the perturbation averaged with respect to either of the zero order initial or final electronic states, i.e. $(1-S^2)V_{eff} = \langle i|V_f - \langle f|V_f|f\rangle|f\rangle = \langle f|V_i - \langle i|V_i|i\rangle|i\rangle$. The latter contribution is thus a measure of the 'distortion' of the donor and acceptor orbitals caused by the interaction between the reactants, and this composite character of V_{eff} implies that even when $V_{i,f}$ is large, the approach based on perturbation theory may still be adequate.

In subsequent calculations it is now common to invoke the Condon approximation according to which V_{eff} can be factored out to give the following rate expression(33,34,91,115-117)

$$W_{fi} = \frac{2\pi}{\hbar} |V_{eff}|^{2} \sum_{v} \sum_{w} \exp(-\beta E^{o}_{iv}) |\langle \chi^{o}_{fw} | \chi^{o}_{iv} \rangle|^{2} \delta(E^{o}_{fw} - E^{o}_{iv}) \quad (3.48)$$

where $|V_{eff}|^2$ is taken at the nuclear coordinates where its value is maximum. The justification for this step is that while the nuclear wave functions are highly localized in the nuclear coordinate space, both the electronic wave functions and the perturbation depend relatively weakly on the nuclear coordinates (25). However, this assumption is still subject to some reservation depending on both the particular nuclear coordinates and the nature of the perturbation. Thus:

(A) Considering at first electron and atom group transfer between two centres and related processes (electronic energy transfer, diffusion, polaron mobility) we notice that while the matrix elements of the electronic wave functions may depend weakly on intramolecular and medium modes this is not so for the nuclear coordinates which determine the relative orientation and distance between the centres. We have taken this effect into account by our discussion in chapter 1. However, as noticed there and as shown in general terms by Dogonadze and Kuznetsov(26), the Condon approximation can be relaxed when the dependence of the electronic coupling matrix element on the appropriate nuclear coordinates is available with sufficient accuracy. For the processes we are considering at present this is generally not the case, however, although we shall discuss certain attempts towards a calculation of this non-Condon effect below. The effect has been studied in simple terms (152) by using a one-dimensional displaced oscillator model for the appropriate mode. Thus, for an exponential dependence of the matrix element on the nuclear coordinate (152a) low-frequency nuclear modes, corresponding to interreactant motion, the non-Condon effects result in a small positive correction to both the activation energy and the transfer coefficient, whereas the pre-exponential factor is determined by the electronic coupling matrix element at its maximum value in the absence of non-Condon effects. If the mode has a high frequency, the non-Condon effect gives an additional factor less than unity in the overall pre-exponential factor.

(B) While the non-Condon effects in radiative processes and in electron transfer may thus be of minor importance (apart from those arising from interreactant motion), they may seriously affect intramolecular processes induced by the nonadiabaticity operator (eq.(3.29)). This is associated with the fact that in contrast to radiative processes, for which the electronic transitions occur for nuclear coordinate values close to the initial state equilibrium, the nonradiative matrix elements assume their largest values near the crossing of the potential energy surfaces of the initial and final states. In these cases the nuclear coordinate dependence of the electronic wave functions is itself directly responsible for the process, and inclusion of non-Condon effects may give results which are orders of magnitude larger than those obtained within the Condon scheme (153).

Incorporation of the non-Condon effect on internal conversion provides rather cumbersome results for the corrections to the Condon scheme, and disentanglement from the formal scheme can only be performed in the form of numerical estimates and for weak electronic-vibrational coupling (153a) (i.e. a small relative displacement of the initial and final state potential energy surfaces). Thus, the results for the decay of a single harmonic 'vibrationless' level (the low-temperature limit) obtained by Nitzan and Jortner(153a) could be written in the form

$$W_{fi} = W_{fi}^{cond}\, \xi \tag{3.49}$$

where W_{fi}^{Cond} is the result obtained by the Condon approximation, and ξ a correction factor which contains contributions from both the energy denominator and the electronic wave functions of the matrix element (cfr. eqs.(3.34)). Numerical estimates of ξ in a single displaced mode system revealed both that W_{fi} can be expected to be higher than W_{fi}^{Cond} by two or three orders of magnitude, and that ξ varies approximately as the square of the energy gap between the initial and final states. Absolute values of nonradiative decay rates are therefore not well reproduced by the Condon approximation, and the energy gap law is expected to be modified relative to the predictions of the Condon approximation (see below). On the other hand, both the dependence on the potential surface displacement and the deuterium isotope effect are little affected by the non-Condon factors. ξ also varies only little for thermally excited vibrational levels in the initial state which means that the temperature dependence of the decay rate would not be significantly affected.

In our further discussion we shall occasionally refer to both intramolecular radiationless and radiative processes. However, since we shall deal primarily with chemical processes, we shall take eq.(3.48) and the Condon approximation as the basis of our approach. We shall then complete this section by noting the formal similarity between the rate expressions for chemical and radiationless processes considered so far and the line shape functions for optical electronic transitions. Thus, the radiative transition probability for an absorbing centre in a medium of refractive index n (and for a dipole allowed transition) at the photon energy $h\nu$, handled within the Condon approximation, is (115)

$$\Gamma(h\nu) = \frac{2\pi}{\hbar}\left(\frac{8\pi^3 \nu C}{3hnc}\right) |\langle\psi_n|\mu|\psi_1\rangle|^2 Z^{-1} \sum_v \sum_w \exp(-\beta E^\circ_{iv}) \tag{3.50}$$

$$|\langle\chi_{fw}|\chi_{iv}\rangle|^2 \, \delta(E^\circ_{iv} - E^\circ_{fw} \pm h\nu)$$

where μ is the electronic dipole moment between the initial and final states, C the concentration of absorption centres, and the minus and plus signs in the delta function refer to photon emission and absorption, respectively. Comparison between eqs.(3.48) and (3.50) shows firstly that the nuclear part (line shape) of a nonradiative process is completely analogous to the absorption or emission process at zero photon energy for optical electronic transitions between bound states coupled to the nuclear modes. This is again valid for both 'deep' impurity electrons in the band gap of insulators or semiconductors, where the electrons are coupled to both local discrete modes and delocalized lattice modes, and for electronic transitions in molecular systems. Secondly, the functional dependence of $\Gamma(h\nu)$ on $h\nu$ is identical to the free energy relationship (or energy gap law) for a chemical process, the free energy relationship for the latter thus representing a kind of 'chemical spectroscopy'(101). Conclusions about line shape functions, the role of low-frequency medium modes and high-frequency intramolecular modes, vibrational resonance structure in the free energy relationship etc. to be discussed in the following are therefore valid for both radiative and nonradiative processes.

3.3 Evaluation of Transition Matrix Elements

We have now briefly suveyed the formal theoretical framework of scattering theory and perturbation theory within which we shall view the elementary chemical processes. We have also defined the zero order channel states and the nature of the coupling between the states which leads to the chemical process. Finally, we have noted formal analogies and differences between chemical processes of the electron and atom group transfer type and intramolecular radiative and nonradiative processes. We shall now transform the general expression for the transition probability, eq.(3.48), to a physically transparent expression for the rate constant, directly applicable to a concrete system provided that the appropriate spectroscopic and thermodynamic data are available. We shall do this with particular reference to the electron and atom group transfer systems, i.e. in view of the continuous and disordered nature of the surrounding medium, we shall view the chemical process as an electronic transition accompanied by the excitation of a multitude of nuclear modes (a multiphonon process). For the same reason we shall calculate the rate of a thermally averaged process, i.e. in contrast to decay rates of spectroscopically prepared 'single levels'. In addition to these general features we shall invoke the following plausible assumptions:

(A) We shall consider the irreversible decay of thermally averaged independent levels. The irreversibility is justified in view of the diffusion of the mobile product species apart, after the dissipation of the activation energy, or by the time limit of the particular experiment. Moreover, the independence of the level decays corresponds to the limit of nonadiabatic processes. We shall subsequently discuss nonadiabaticity criteria with reference to both experimental data and to a semiclassical formulation of the rate theory.

(B) The coupling between the reactants and the medium provides the thermal averaging. However, the polarization dynamics implies that the medium is not only an inert heat bath but part of the reacting system, constituting together with the reacting species a decaying 'supermolecule'.

(C) In the present chapter we shall largely give attention to the medium modes, whereas discrete intramolecular modes will be explicitly discussed in chapter 4 (an exception is the intramolecular decay processes discussed in section 3.5.2). In most common chemical processes the medium modes are moreover strongly coupled to the electronic system, corresponding to a large horizontal displacement of the appropriate potential energy surfaces.

(D) Although our results in principle will be valid for all kinds of interaction fields, we shall basically have electrostatic fields in mind (cf. chapter 2). Moreover, we shall consider linear effects only, i.e. we shall represent the medium as a set of independent harmonic oscillators along the collective coordinates $q_{\vec{k}\nu}$ which are related to the polarization components. This means that the electronic energy (eqs.(3.21) and (3.27)) takes the form

$$\mathcal{E}_1(q_{\vec{k}\nu}) = \mathcal{E}_{1o} + \sum_{\vec{k},\nu} u_{\vec{k}\nu}\, q_{\vec{k}\nu} \tag{3.51}$$

where $\mathcal{E}_{lo}$ is the 'pure' electronic energy, i.e. the eigenvalue of the operator $H_e(\vec{r})$. More specifically, this energy now refers to the electronic state of the ions, whereas the energy of the medium electrons is included in the energy of the continuous medium modes.

(E) Insertion of eq.(3.51) together with eqs.(2.39) and (3.27) gives the stationary Schrodinger equation for the nuclear subsystem, i.e. corresponding to a set of independent harmonic oscillators. The nuclear wave functions therefore take the form (141,142)

$$X_{1v} = \exp[-(q^{1}_{\vec{k}\nu} - q^{1}_{\vec{k}\nu o})^2/2](\pi^{\frac{1}{2}} 2^v v!)^{-\frac{1}{2}} H_v(q^{1}_{\vec{k}\nu} - q^{1}_{\vec{k}\nu o}) \quad (3.52)$$

where H_v is the Hermite polynomial of degree v, and the superscript '1' refers to the appropriate electronic state. We shall thus (i.e. within the two-level scheme) view the chemical process as electronic transitions between two many-dimensional paraboloidal potential energy surfaces

$$U_i(q_{\vec{k}\nu}) = U_{io} + \frac{1}{2}\sum_{\vec{k},\nu} \hbar\omega_{\vec{k}\nu}(q_{\vec{k}\nu} - q^{i}_{\vec{k}\nu o})^2 \quad (3.53)$$

$$U_f(q_{\vec{k}\nu}) = U_{fo} + \frac{1}{2}\sum_{\vec{k},\nu} \hbar\omega_{\vec{k}\nu}(q_{\vec{k}\nu} - q^{f}_{\vec{k}\nu o})^2 \quad (3.54)$$

where the subscripts 'i' and 'f' refer to the initial and final electronic states, respectively, U_o $(= \varepsilon_o - \frac{1}{2}\sum_{\vec{k},\nu} \hbar\omega_{\vec{k}\nu} q^2_{\vec{k}\nu o})$ to the electronic energies including solvation energies and all energies of interaction between the reactant species not leading to the chemical process. However, although the appropriate coordinates are essentially of a 'collective' nature, this representation to a certain extent conceals the continuous medium response. The continuous nature of the medium spectrum and the frequency dispersion have some very important qualitative implications for the two fundamental phenomenological rate laws, i.e. the Arrhenius and the Bronsted relationships, in particular at low temperatures. For this reason, subsequent to our harmonic oscillator representation, we shall briefly outline a different formalism which avoids the explicit use of harmonic oscillators and expresses the rate constant directly in terms of the medium response function $\varepsilon(k,\omega)$.

3.3.1 Harmonic Oscillator Representation

We now consider the general expression for the rate constant, eq.(3.48). Expressions of this type were calculated early by various authors in the context of both optical absorption of colour centres (trapped electrons) in polar crystals(114-117) and thermal processes involving electron capture by 'deep' impurities (i.e. far from the valence and conduction bands) in semiconductors(120). The first quantum mechanical formulation of the rate of simple electron transfer processes by Levich and Dogonadze(51) also closely followed the same approach. While these early theories largely focused on electronic-vibrational coupling which involved equilibrium normal coordinate displacement only (even though frequency dispersion was also considered,(120)), a general theory which is valid also for both frequency shift and mode mixing and which in principle does not depend on the harmonic approximation was introduced by Kubo and Toyozawa(119). This general approach can be used to calculate both optical line shapes (temperature broadening and maximum shifts) and nonradiative decay rates. In the context of chemical processes this method was firstly introduced by Dogonadze, Kuznetsov and Vorotyntsev to incorporate all the effects of intramolecular and solvent restructuration(90).

The rate expressions have a rather involved appearance in the general case, but for several special cases they can be recast in simple and physically transparent forms. This relates for example to the high-temperature limit where all the nuclear modes behave classically, to the low-temperature limit where the modes in either the initial or the final electronic state are 'frozen' at their ground vibrational levels, generally for displaced modes with no frequency shifts etc.

In order to calculate the rate expression of eq.(3.48) we shall represent the delta function in terms of a Fourier integral, i.e.(154)

$$\delta(E_{fw} - E_{iv} + \Delta E) = \frac{\beta}{2\pi i}\int_{-i\infty}^{i\infty} \exp[-(E_{fw} - E_{iv} + \Delta E)\beta\alpha]\,d\alpha \quad (3.55)$$

where E_{iv} and E_{fw} now refer to the vibrational energies, and $\Delta E = U_{fo} - U_{io}$ is the energy gap, i.e. the difference between the minima of the initial and final state potential energy surfaces, including the difference between the zero point vibrational energies. This transforms eq.(3.48) to the form

$$W_{fi} = \frac{\beta |V|^2}{i\hbar} Z^{-1} \sum_{v_\varkappa} \sum_{w_\varkappa} \int_{-i\infty}^{i\infty} d\alpha \int_{-\infty}^{\infty} \prod_\varkappa dq^i_\varkappa \int_{-\infty}^{\infty} \prod_\varkappa ds^i_\varkappa \exp[-\beta\alpha\Delta E +$$

$$h v_\varkappa (1 - \alpha) + h w_\varkappa \alpha] \prod_{w_\varkappa} \chi_{fw_\varkappa}(q^f_\varkappa - q^f_{\varkappa o}) \prod_{v_\varkappa} \chi_{iv_\varkappa}(q^i_\varkappa - q^i_{\varkappa o})$$

$$\prod_{w_\varkappa} \chi_{fw_\varkappa}(s^f_\varkappa - q^f_{\varkappa o}) \prod_{v_\varkappa} \chi_{iv_\varkappa}(s^i_\varkappa - q^i_{\varkappa o}) \quad (3.56)$$

where we have introduced the variables s to distinguish the separate integrations of the two factors in the Franck Condon nuclear overlap function and the notation $V = V_{eff}$. We notice that the vibrational motion in both the initial and final states is represented as independent oscillators characterized by the index $\varkappa$, i.e. so far only separability of the modes, but not harmonicity is assumed. Eq.(3.56) also incorporates both mode mixing (i.e. a change of normal modes from $q^i_\varkappa$ to $q^f_\varkappa$ when going from the initial to the final state), frequency shift, and equilibrium coordinate shift. This is indicated by the different superscripts 'i' and 'f' for these two states.

The main problem is now to calculate the double sum over $v_\varkappa$ and $w_\varkappa$ in eq.(3.56). This is possible by noting that the order of the summation with respect to $v_\varkappa$ and $w_\varkappa$ and the integration with respect to α can be inverted, and that for harmonic oscillator functions closed expressions are available for sums of the form(117a,155)

$$\sum_{r=0}^{\infty} t^r \chi_v(x)\chi_v(y) = \pi^{-\frac{1}{2}}(1 - t^2)^{-\frac{1}{2}}\exp\left\{\frac{2xyt - (x^2+y^2)t^2}{1 - t^2}\right\} \quad (3.57)$$

Thus, the following equations now refer to harmonic oscillator modes only. Inserting eq.(3.57) in eq.(3.48), putting $Z = \prod_\kappa \exp(-\beta v_\kappa \hbar\omega_\kappa^i) = \prod_\kappa [1-\exp(-\beta\hbar\omega_\kappa^i)]^{-1}$ and taking $q_{\kappa o}^i = 0$ as the equilibrium value of q_κ^i the following rate expression is obtained

$$W_{fi} = \frac{\beta|V|^2}{i\hbar}\left\{\prod_\kappa [1 - \exp(-\beta\hbar\omega_\kappa^i)]\right\} \int_{-i\infty}^{i\infty} d\alpha \int_{-\infty} \prod_\kappa dq_\kappa^i \int \prod_\kappa ds_\kappa^i$$

$$\pi^{-\frac{1}{2}} [1 - \exp(-2\hbar\omega_\kappa^f\alpha)]^{-\frac{1}{2}} \{1-\exp[-2\hbar\omega_\kappa^i(1-\alpha)]\}^{-\frac{1}{2}} \exp(-\beta\alpha\Delta E)$$

$$\exp\left\{ - (\xi_\kappa^f - q_{\kappa o}^f)^2 \text{th}\frac{\beta\hbar\omega_\kappa^f}{2}\alpha - \frac{1}{4}\eta_\kappa^{f\,2} \text{cth}\frac{\beta\hbar\omega_\kappa^f}{2}\alpha - \right. \quad (3.58)$$

$$\left. -\xi_\kappa^{i\,2} \text{th}\frac{\beta\hbar\omega_\kappa^i}{2}(1 - \alpha) - \frac{1}{4}\eta_\kappa^{i\,2} \text{cth}\frac{\beta\hbar\omega_\kappa^i}{2}(1 - \alpha)\right\}; \qquad \xi_\kappa = \tfrac{1}{2}(q_\kappa + s_\kappa); \quad \eta_\kappa = q_\kappa - s_\kappa$$

We now invoke the simplifying and for the medium very plausible assumption that all the normal modes and frequencies are identical in the initial and final states, i.e. $q_\kappa^i = q_\kappa^f = q_\kappa$ and $\omega_\kappa^i = \omega_\kappa^f = \omega$. We do this for the sake of convenience, since the effects of frequency shift and mode mixing can be adequately incorporated in the theory(90). We notice that the assumption of no mixing of intramolecular modes is likely to be well justified for a number of outer sphere electron transfer processes and processes involving the transfer of small atom groups such as protons.

However, such processes are often subject to intramolecular frequency shifts. With this assumption eq.(3.58) can be rewritten as

$$W_{fi} = \frac{\beta |V|^2}{i\hbar} \left\{ \prod_{\varkappa} [1 - \exp -\beta\hbar\omega_{\varkappa})] \right\} \int_{-i\infty}^{i\infty} d\alpha \quad I \tag{3.59}$$

where

$$I = \int_{-\infty}^{\infty} \prod_{\varkappa} dq_{\varkappa} \int_{-\infty}^{\infty} \prod_{\varkappa} ds_{\varkappa} \pi^{-1} [1 - \exp(-2\beta\hbar\omega_{\varkappa}\alpha)]^{-\frac{1}{2}} \{1 - \tag{3.60}$$

$$\exp[-2\beta\hbar\omega_{\varkappa}(1 - \alpha)]\}^{-\frac{1}{2}} \exp\left\{ -\beta\alpha\Delta E - (\xi_{\varkappa} - q_{\varkappa o})^2 \quad th\frac{\beta\hbar\omega_{\varkappa}\alpha}{2}\right.$$

$$\left. - \frac{1}{4}\eta_{\varkappa}^2 cth\frac{\beta\hbar\omega_{\varkappa}\alpha}{2} - \xi_{\varkappa}^2 \quad th\frac{\beta\hbar\omega_{\varkappa}}{2}(1 - \alpha) - \frac{1}{4}\eta_{\varkappa}^2 cth\frac{\beta\hbar\omega_{\varkappa}}{2}(1 - \alpha)\right\}$$

The integrations over $\xi_{\varkappa}$ and $\eta_{\varkappa}$ can now be performed accurately, since the exponent in eq.(3.58) is a quadratic form with respect to these variables. The result is seen to be

$$W_{fi} = \frac{\beta |V|^2}{i\hbar} \int_{-i\infty}^{i\infty} d\alpha \exp\left\{ -\beta\alpha\Delta E - \sum_{\varkappa} \frac{sh\frac{\beta\hbar\omega_{\varkappa}\alpha}{2} sh\frac{\beta\hbar\omega_{\varkappa}(1 - \alpha)}{2}}{sh\frac{\beta\hbar\omega}{2}} q_{\varkappa o}^2 \right\} \tag{3.61}$$

The integration with respect to α can generally be performed by the saddle point method (see below). However, at first we shall consider the limiting case of low frequencies (high temperatures). Thus, provided that the inequalities $\beta\hbar\omega_{\varkappa}\alpha \ll 1$, and $\beta\hbar\omega_{\varkappa}(1-\alpha) \ll 1$ are both valid, we can replace the hyperbolic sine and tangent functions by their arguments. Eq.(3.61) then takes the form

$$W_{fi} = \frac{\beta|V|^2}{i\hbar}\int_{-i\infty}^{i\infty}\exp[-\beta\alpha\Delta E - \sum_{\kappa}\frac{1}{2}\beta\hbar\omega_{\kappa}\alpha(1-\alpha)q_{\kappa o}^2] \qquad (3.62)$$

which can be integrated accurately to give

$$W_{fi} = (\pi\beta/\hbar^2 E_r)^{\frac{1}{2}}|V|^2\exp[-\beta(E_r + \Delta E)^2/4E_r] \qquad (3.63)$$

where we have introduced the notation $E_r = \frac{1}{2}\sum_{\kappa}\hbar\omega_{\kappa}q_{\kappa o}^2$ (see below). In the high-temperature limit a 'conventional' rate equation is thus obtained which contains an activation energy, $(E_r + \Delta E)^2/4E_r$, and a pre-exponential factor.

More generally, integrals of the form

$$I = \int_{-\infty}^{\infty}\exp[-A(\alpha)]d\alpha \qquad (3.64)$$

where the function $A(\alpha)$ has a sharp minimum, can be approximated by the saddle point method (156). The saddle point, α^*, is found from the equation

$$\partial A(\alpha)/\partial\alpha\,|_{\alpha=\alpha^*} = 0 \qquad (3.65)$$

subsequent Taylor series expansion

$$A(\alpha) \approx A(\alpha^*) + \frac{1}{2}A''(\alpha^*)(\alpha - \alpha^*)^2 \qquad (3.66)$$

gives

$$I \approx (2\pi/|A''(\alpha^*)|)^{\frac{1}{2}}\exp[-A(\alpha^*)] \qquad (3.67)$$

The validity condition for eq.(3.67) is that $|A'''(\alpha^*)| \ll |A''(\alpha^*)|^{\frac{3}{2}}$, or

$$\left|\sum_{\varkappa}(\beta\hbar\omega_{\varkappa})^{3}\,\frac{q^{2}_{\varkappa o}}{\mathrm{sh}\,\frac{\beta\hbar\omega_{\varkappa}}{2}}\,\mathrm{sh}\frac{\beta\hbar\omega_{\varkappa}}{2}(1-2\alpha)\right| \ll \left|\sum_{\varkappa}(\beta\hbar\omega_{\varkappa}^{2})^{2}\,\frac{q^{2}_{\varkappa o}}{\mathrm{sh}\frac{\beta\hbar\omega_{\varkappa}}{2}}\,\mathrm{ch}\frac{\beta\hbar\omega_{\varkappa}}{2}(1-2\alpha)\right|^{\frac{3}{2}} \tag{3.68}$$

This condition is generally valid in the high-temperature limit. For example, for thermoneutral processes $\alpha^{*} = 0.5$ (_E = 0, see below), and eq.(3.68) is valid whenever $(\beta\hbar\omega_{\varkappa})^{4} \ll \beta E_{r}$, a condition which is comfortably met for all chemical processes of interest. However, in the low-temperature limit, for which $\beta\hbar\omega_{\varkappa} \gg 1$, this procedure is only convenient for certain ranges of the parameters. Consideration of the following variants is thus appropriate:

(a) $\beta\hbar\omega_{\varkappa}(1-2\alpha) \gg 1$, or $\alpha < 0.5$.

The latter inequality implies that $\Delta E < 0$ (see below), and the validity of eq.(3.68) is equivalent to the condition $|\Delta E| \gg \hbar\omega_{\varkappa}$. Following the procedure outlined we find α^{*} from the equation

$$\sum_{\varkappa}\frac{1}{2}\hbar\omega_{\varkappa}q^{2}_{\varkappa o}\;\exp(-\beta\hbar\omega_{\varkappa}\alpha^{*}) = |\Delta E| \tag{3.69}$$

and W_{fi} takes the form

$$W_{fi} = (2\pi/|A''(\alpha^{*})|)^{\frac{1}{2}}|V|^{2}/\hbar\;\exp[-\beta\alpha^{*}\Delta E\; - \tag{3.70}$$

$$\frac{\mathrm{sh}\frac{\beta\hbar\omega_\varkappa}{2}\alpha^* \ \mathrm{sh}\frac{\beta\hbar\omega_\varkappa}{2}(1-\alpha^*)}{\mathrm{sh}\frac{\beta\hbar\omega_\varkappa}{2}} q_{\varkappa o}^2]$$

For a single mode $\alpha^* = -\ln(|\Delta E|/E_r)/\beta\hbar\omega_\varkappa$. W_{fi} then takes the more transparent forms

(1) For $\alpha < 0$, i.e. $|\Delta E| > E_r$ or strongly exothermic processes

$$W_{fi} = (2\pi/\hbar\omega_\varkappa|\Delta E|)^{\frac{1}{2}} |V|^2/\hbar \exp(-q_{\varkappa o}^2/2) \tag{3.71}$$

$$\exp(-\gamma|\Delta E|/\hbar\omega_\varkappa)$$

where $\gamma = \ln(|\Delta E|/E\ )-1$. This is the 'energy gap law' which is known in the theory of intramolecular radiationless decay of excited electronic levels and which shows an approximately exponential dependence of the decay rate on the energy gap $|\Delta E|$ (2-5).

(2) For $0 < \alpha < 0.5$, i.e. for weakly exothermic processes

$$W_{fi} = (2\pi/\hbar\omega_\varkappa|\Delta E|)^{\frac{1}{2}}|V|^2/\hbar \exp(-\tfrac{1}{2} q_{\varkappa o}^2) \tag{3.72}$$

$$\exp[-(\gamma + 1)|\Delta E|/\hbar\omega_\varkappa]$$

(b) $-\beta\hbar\omega_\varkappa(1-2\alpha) \gg 1$, or $\alpha > 0.5$.

This corresponds to endothermic processes, and eq.(3.68) again implies that $\Delta E \gg \hbar\omega_\varkappa$. Following the same procedure as before the rate expression is formally identical to eq.(3.70) but the equation from which α^* is determined differs from eq.(3.69) by a replacement of α^* with $1-\alpha^*$. For a single mode

(1) For $\alpha > 1$, i.e. for strongly endothermic processes

$$W_{fi} = (2\pi/\hbar\omega_\varkappa|\Delta E|)^{\frac{1}{2}}\,|V|^2/\hbar\,\exp(-\beta\Delta E)\,\exp(-q_{\varkappa o}^2/2) \tag{3.73}$$

$$\exp(-\gamma\Delta E/\hbar\omega_\varkappa)$$

(2) for $0.5 < \alpha < 1$, i.e. for weakly endothermic processes

$$W_{fi} = (2\pi/\hbar\omega_\varkappa|\Delta E|)^{\frac{1}{2}}\,|V|^2/\hbar\,\exp(-\frac{1}{2}q_{\varkappa o}^2)\,\exp(-\beta\Delta E) \tag{3.74}$$

$$\exp[-(\gamma + 1)\Delta E/\hbar\omega_\varkappa]$$

For a chemical reaction involving strong excitation in a single high-frequency mode the activation energy is thus identical to the energy gap. This is understandable in view of the fact that excitation to a level $m \approx \Delta E/\hbar\omega_\varkappa$ is required for the reaction to proceed. On the other hand, due to the large frequency, tunnelling from this level is favoured relative to further thermal excitation. Moreover, we notice that eqs.(3.71)-(3.74) fulfill the principle of microscopic reversibility, i.e. the ratio between the two rate constants is $\exp(-\beta\Delta E)$.

(c) If, finally, $\beta\hbar\omega_\varkappa(1-2\alpha) \ll 1$, and $\beta\hbar\omega_\varkappa \gg 1$, or $\alpha \approx 0.5$ and $\Delta E \approx 0$, then integration of eq.(3.61) gives

$$W_{fi} = \frac{2\pi}{\hbar}|V|^2\,\exp(-\frac{1}{2}\sum_\varkappa q_{\varkappa o}^2) \tag{3.75}$$

In the general case, when reorganization of high-frequency modes is important, justification for the application of the saddle point method requires a more comprehensive analysis (157). We shall prefer, however, to give a different derivation in line with earlier calculations on optical transitions (115,116) and thermal electron transfer (51,91,95,96). Thus, we can firstly rewrite the exponent of eq.(3.61) in the form

$$- \sum_{\varkappa} \frac{q_{\varkappa o}^{2}}{2} \,\mathrm{cth}\frac{\beta\hbar\omega_\varkappa}{2} - \beta\alpha\Delta E + \sum_{\varkappa} \frac{q_{\varkappa o}^{2}}{4\,\mathrm{sh}\frac{\beta\hbar\omega_\varkappa}{2}} \left\{ \exp\left[\frac{\beta\hbar\omega_\varkappa}{2}(2\alpha - 1)\right] + \exp\left[\frac{\beta\hbar\omega_\varkappa}{2}(1 - 2\alpha)\right] \right\} \qquad (3.76)$$

Expanding the exponential function in eq.(3.61) gives

$$\exp\left\{ -\beta\alpha\Delta E - \sum_{\varkappa} \frac{\mathrm{sh}\frac{\beta\hbar\omega_\varkappa}{2}\alpha\,\mathrm{sh}\frac{\beta\hbar\omega_\varkappa}{2}(1-\alpha)}{\mathrm{sh}\frac{\beta\hbar\omega_\varkappa}{2}} \right\} = \exp\left\{ -\beta\alpha\Delta E - \right. \qquad (3.77)$$

$$\left. \sum_{\varkappa} \frac{q_{\varkappa o}^{2}}{2}\,\mathrm{cth}\frac{\beta\hbar\omega_\varkappa}{2} \right\} \prod_{\varkappa} \left\{ \sum_{k=0}^{\infty} \sum_{l=0}^{\infty} \frac{1}{k!}\frac{1}{l!} \left(\frac{q_{\varkappa o}^{2}}{4\,\mathrm{sh}\frac{\beta\hbar\omega_\varkappa}{2}} e^{-\frac{\beta\hbar\omega_\varkappa}{2}} \right)^{k+l} e^{l\beta\hbar\omega_\varkappa} \right.$$

$$e^{\beta\hbar\omega_\varkappa (l - k)\alpha}$$

Insertion of this in eq.(3.61) and using the integral representation of the delta function (eq.(3.55)) gives subsequently

$$W_{fi} = \frac{2\pi|V|^{2}}{\hbar} \sum_{k=o} \sum_{l=o} \frac{1}{\omega_\varkappa} \exp\left[- \sum_{\varkappa} \frac{q_{\varkappa o}^{2}}{2}\,\mathrm{cth}\frac{\beta\hbar\omega_\varkappa}{2} \right] \prod_{\varkappa} \frac{1}{k!}\frac{1}{l!} \qquad (3.78)$$

$$\left(\frac{q_{\varkappa o}^{2}}{4\,\mathrm{sh}\frac{\beta\hbar\omega_\varkappa}{2}} e^{-\frac{\beta\hbar\omega_\varkappa}{2}} \right)^{k+l} e^{l\beta\hbar\omega_\varkappa}\, \delta(k - l + \Delta E/\hbar\omega_\varkappa)$$

This result can be transformed to an expression originally derived by Lax(115) and by Levich and Dogonadze(51). Thus, introducing the modified Bessel function, $I_m(z)$, where $z = \frac{q_{\varkappa o}^2}{2}\text{csch}\frac{\beta\hbar\omega_\varkappa}{2}$, and (154)

$$I_m(z) = \left(\frac{z}{2}\right)^m \sum_{k=o} \frac{(z/2)^k}{k!(m+k)!} \tag{3.79}$$

eq.(3.78) can be rewritten in the form

$$W_{fi} = \frac{2\pi}{\hbar^2}|V|^2 \prod_\varkappa \sum_{m=-\infty}^{\infty} \frac{1}{\omega_\varkappa} \exp[-z\text{ch}\frac{\beta\hbar\omega_\varkappa}{2} + m\beta\hbar\omega_\varkappa] \tag{3.80}$$

$$I_m(z)\delta(m - \Delta E/\hbar\omega_\varkappa)$$

m thus acquires the physical meaning of being the total number of vibrational quanta (phonons) involved in the process. This equation has several limiting and more transparent forms:

(a) For $z \ll 1$ (the weak-coupling limit) I_m can be approximated by the first term in the infinite sum eq.(3.79), i.e. $(\frac{z}{2})^m/m!$. Since moreover $I_m(z) = I_{|m|}(z)$, eq.(3.78) takes the form

$$W_{fi} = \frac{2\pi}{\hbar^2}|V|^2 \tilde{\prod_\varkappa} \sum_{m=-\infty}^{\infty} \frac{1}{\bar{\omega}_\varkappa} \exp[-(m+|m|)\beta\hbar\omega_\varkappa/2]\frac{1}{m!}\left(\frac{q_{\varkappa o}^2}{2}\right)^{|m|} \tag{3.81}$$

$$\exp\left(-\frac{q_{\varkappa o}^2}{2}\right)\delta(m - \Delta E/\hbar\omega_\varkappa)$$

This is the low-temperature weak-coupling limit of the rate expression for discrete vibration frequencies. We notice that the 'activation energy' is zero and ΔE, respectively, for exothermic and endothermic processes, respectively (cf.eqs.(3.71)-(3.74)). The rate expression furthermore contains Franck Condon nuclear overlap functions $\frac{1}{m!}\left(\frac{q_{\varkappa 0}^2}{2}\right)^{|m|} e^{-q_{\varkappa 0}^2/2}$. Finally, the free energy relationship displays a resonance character which, however, is not manifested unless the appropriate mode is of local nature (i.e. not for a continuous manifold of high-frequency medium modes).

(b) The maximum contribution to the sum is given by terms for which $m\hbar\omega_\varkappa \approx \Delta E$. If $|m| >> 1$, we can express $m!$ by Stirling's formula(154). For a single mode of frequency $\omega_\varkappa$ this gives

$$W_{fi} \approx (2\pi/\hbar\omega_\varkappa|\Delta E|)^{\frac{1}{2}} \frac{|V|^2}{\hbar} \exp(-q_{\varkappa 0}^2/2) \exp(-\gamma|\Delta E|/\hbar\omega_\varkappa) \quad (3.82)$$

i.e. the energy gap law already obtained earlier.

The rate expressions derived above all refer to the simplified cases where both normal modes and frequencies are identical in the initial and final states. As long as these modes refer to the external medium this representation is adequate. However, intramolecular modes commonly undergo frequency shifts in addition to equilibrium coordinate shifts, even to an extent where a particular mode which behave classically in the initial state displays quantum behaviour in the final state, or vice versa. Furthermore, in many of the most thoroughly investigated chemical processes such as nucleophilic substitution and inner sphere electron transfer, changes of normal modes in the final state relative to the initial state occurs. These effects can be incorporated in the formalism outlined above (90), but since they refer to intramolecular modes, we shall postpone a discussion to chapters 4 and 5. At present we shall discuss some implications of the formalism outlined so far. Thus:

(A) The role of low- and high-frequency modes, corresponding to $\beta\hbar\omega_\varkappa(1-\alpha^*)$, $\beta\hbar\omega_\varkappa\alpha^* << 1$, and the inverse inequality, respectively, is fundamentally different. The low-frequency modes define the activation energy of the process, the quantity $(E_r + \Delta E)^2/4E_r$

being the saddle point of the manydimensional intersection surface between the initial and final state surfaces spanned by the low-frequency coordinates. On the other hand, the high-frequency modes proceed from their initial to their final state equilibrium value by a subbarrier (tunnelling) motion. Thus, the quantity $(m!)^{-1}(q_{\kappa 0}^2/2)^{|m|}\exp(-q_{\kappa 0}^2/2)$, which appears in eq.(3.81), is the square of the overlap integral of the ground and m'th level harmonic oscillator wave functions in the 'classically forbidden' region. This quantity is shown in chapter 6 to be identical to the Gamov tunnelling factor for a barrier consisting of two intersecting parabolae and an effective particle mass coinciding with the reduced mass of the appropriate mode, thus stressing the nonclassical motion of these modes.

The sequence of events during the process when both high and low-frequency modes are present is thus that firstly the system arrives to the saddle point region of the classical low-frequency modes. This motion requires an activation energy determined by the relative position of the potential energy surfaces of the classical modes. In this region the high-frequency nuclear modes are reorganized from their initial to their final state equilibrium configuration with a probability given by the appropriate Franck Condon nuclear overlap factor, and the electron subsystem is reorganized with a probability given by the electronic coupling matrix element. As long as $\Delta E < E_r^{cl}$ the high-frequency modes in their lowest vibrational state give the dominating contributing to the overall rate. This is the situation prevailing in the majority of 'common' chemical processes. On the other hand, when $\Delta E > E_r^{cl}$ the total rate expression assumes larger values when Franck Condon factors corresponding to excited levels of the quantum modes are inserted.

(B) We have considered the nonadiabatic limit, i.e. we have assumed that the electronic coupling matrix element, $|V|$, is small. This would commonly be expected to be revealed by small values of the pre-exponential factor in a 'phenomenological'

Arrhenius relationship. However, we have also seen that this factor may contain contributions from high-frequency modes as well, a small pre-exponential factor thus being indicative of reorganization of the 'total' quantum subsystem only. The important question of the degree of adiabaticity of real chemical processes is in fact by no means settled, but we shall discuss a few systems below where rather unambiguous conclusions about the nonadiabaticity of the process can be drawn.

(C) The equations of the previous section contain important quantitative information relating to the two most important phenomenological kinetic laws, i.e. the Bronsted free energy relationship (energy gap law) and the Arrhenius temperature relationship. Considering at first the former, we see that in the presence of classical modes only, a parabolic relationship between the activation energy (or $\ln W_{fi}$) and ΔE is predicted, identical with the predictions of the theory of Marcus. In this case the free energy relationship is characterized solely by the parameter $E_r^{cl} = \frac{1}{2}\sum \hbar\omega_\kappa q_{\kappa 0}^2$, where the summation runs over all classical nuclear modes. The latter may consist of both intramolecular and medium modes. The physical meaning of the contribution of the former is seen by noting the relationship between the dimensionless equilibrium coordinate shift, $q_{\kappa 0}$, and the real coordinate shifts, $r_{\kappa 0}$, i.e. $q_{\kappa 0} = (\mu\omega_\kappa/\hbar)^{\frac{1}{2}} r_{\kappa 0}$, where μ is the reduced mass associated with the normal coordinate r_κ. This contribution to E_r^{cl} is thus the total energy required for the reorganization of the system from its initial to its final state equilibrium value. The contribution of the medium has an analogous meaning, i.e. being the total (free) energy required to reorganize the medium modes from their initial to their final state equilibrium value. However, the relation between $q_{\kappa 0}$ and the medium characteristics is here less transparent being expressed by the Fourier components of the medium polarization (chapter 2 and section 3.4).

In view of the definition of the medium modes via the medium polarization, the harmonic approximation is likely to be adequate for these modes. On the other hand, a strong reorganization of low-frequency intramolecular modes would often require that anharmonicity is incorporated. This will modify the parabolic free energy relationships predicted for the simple displaced oscillator model. Thus, as noted in chapter 1, if the chemical process involves bond stretching represented by Morse potentials, the free energy plots become 'flatter' since these potentials are themselves almost linear over quite large intervals. If modes of 'moderately' large frequencies are also reorganized, excited states of these modes may contribute to the total rate expression which will then consist of a sum of expressions such as eq.(3.63) weighted by the Franck Condon factors of the high-frequency modes. This will also decrease the curvature of the overall free energy relationship. We can therefore conclude that free energy relationships for elementary chemical processes in the 'normal' free energy range, for which $|\Delta E| < E_r^{cl}$, are not diagnostic with respect to deductions about the nature of the nuclear modes.

As $\Delta E \to E_r^{cl}$, or $\Delta E \to -E_r^{cl}$, then $E_A^{cl} \to E_r^{cl}$, and $E_A^{cl} \to 0$, respectively. These two situations correspond to the minimum of the final state potential energy surface being located on the initial state surface and vice versa, and the corresponding processes are commonly named barrierless and activationless, respectively(29,90). In the two cases the 'activated' state configuration thus coincides with the equilibrium configuration of the final and initial state, respectively. As the numerical value of ΔE increases still further, excited vibrational states of the high-frequency modes begin to participate. This means that the free energy plot falls off less rapidly than predicted by the parabolic relationship. In this region the free energy relationship is therefore more informative with respect to the nature of the appropriate modes (see further chapter 4). Thus, in the limit of strongly exothermic processes we can see (eq.(3.82)) that the higher the frequency, the more it contributes to the sum over the final vibrational states (or the better the mode can accept the energy dissipated).

The free energy relationship is commonly characterized by the phenomenological Bronsted coefficient $\alpha^* = -k_B T d\ln W_{fi} / d(\Delta E)$. For purely classical modes we see that $\alpha^* = \frac{1}{2} + \Delta E/2E_r$. Both in this case and for strongly exothermic and endothermic processes (eqs.(3.71)-(3.74)) this parameter is seen to coincide with the saddle point of the integration variable α. Thus $\alpha^* \to 0$, and $\alpha^* \to 1$ in the activationless and barrierless region, respectively.

(D) On the basis of the theory outlined above three characteristic temperature ranges in the Arrhenius relationship are expected. At very low temperatures all modes are 'frozen', i.e. they remain in their ground vibrational levels. Only exothermic processes can then occur, with zero activation energy and large negative apparent activation entropies due to the quantum motion of the nuclear modes. If different modes in the system are characterized by sufficiently widely separated vibration frequencies, there will be an intermediate temperature region in which the low-frequency modes are thermally excited and contribute to a measurable activation energy, whereas the high-frequency modes remain in their ground vibrational states, and the corresponding tunnelling factors would contribute to the apparent activation entropy. This situation is expected to be of importance in particular for proton transfer reactions in which the high-frequency modes associated with the motion of the proton remain 'frozen' even at room temperature, whereas reorganization of low-frequency medium modes represents the origin of the activation energy. Finally, at sufficiently high temperatures all modes will be excited and contribute 'classically' to the activation energy.

The Arrhenius temperature dependence should thus ideally reveal an appreciable structure and consequently offer some criteria as to the nature of the appropriate nuclear modes. Moreover, investigation of chemical processes over a considerable temperature range, in which a given set of modes changes from being largely of quantum to being largely of classical nature should offer a posibility of separating the electronic and the (quantum) nuclear contributions in the pre-exponential factor. In practice such analysis is troubled by difficulties associated with the fact that the spectrum of medium modes contains a continuous manifold of components. As the temperature is increased, the number of medium modes which behave classically also increases practically continuously over wide temperature intervals and the resulting phenomenological activation energy is itself a monotonously increasing function of the temperature. A quantitative separation of the medium modes from local modes thus requires additional assumptions about the frequency dispersion of the former (cf. the next section).

3.4 The Role of a Continuous Vibration Spectrum

The formalism developed so far is applicable to electron transfer systems in which the electron is coupled to an arbitrary number of nuclear harmonic modes which undergo an equilibrium coordinate shift. In cases where the appropriate nuclear modes refer to intramolecular motion they can be directly related to the molecular geometry, vibration frequencies and other quantities which are in principle experimentally available. However, if the modes refer to the disordered or crystalline medium, this identification is less direct in view of the collective nature of the medium modes.

It is often useful to introduce the collective medium response and the continuous frequency distribution directly into the expressions for the rate constants. For common chemical processes in the high-temperature and normal free energy regions this does not have any major effects. On the other hand, at low temperatures and for strongly exothermic processes it is essential to consider explicitly the effect of the medium frequency distribution. This is because the reorganization of quantum modes are important in these regions, and the presence of a continuous manifold of such modes from the medium will strongly affect the kinetic relationships.

Frequency dispersion was firstly incorporated in the thermal electron transfer formalism by Ovchinnikov and Ovchinnikova (158) and investigated in the context of several of the phenomena mentioned above by Dogonadze and his associates (85,87,88,104,109). If we restrict ourselves to electrostatic interaction between the electronic charges and a continuous polar medium, and recall the relationship between the normal coordinates and the dielectric permittivity from eq.(2.59), it is plausible that the general rate equation equivalent to eq.(3.61) can be written in the form (87,88)

$$W_{fi} = \frac{\beta |V|^2}{i\hbar} \int_{-i\infty}^{i\infty} d\alpha \exp[-\beta\alpha\Delta E - \Phi_m(\alpha)] \tag{3.83}$$

where

$$\Phi_m(\alpha) = \frac{1}{\hbar} \int_{-\infty}^{\infty} \frac{1}{\omega} \mathcal{E}_r(\omega) \frac{\mathrm{sh}\frac{\beta\hbar\omega}{2}\alpha \, \mathrm{sh}\frac{\beta\hbar\omega}{2}(1-\alpha)}{\mathrm{sh}\frac{\beta\hbar\omega}{2}} d\omega \tag{3.84}$$

We have introduced the reorganization energy at the frequency ω, $\mathcal{E}_r(\omega)$ by the equations

$$\varepsilon_r(\omega) = \frac{1}{4\pi^2} \sum_{\vec{k}} |D^f_{\vec{k}} - D^i_{\vec{k}}|^2 \frac{\mathrm{Im}\,\varepsilon(k,\omega)}{|\varepsilon(k,\omega)|^2} \qquad (3.85)$$

$$E_r^m = \int_0 \varepsilon_r(\omega)\, d\omega \qquad (3.86)$$

where E_r^m is the total medium reorganization energy. We notice that a rigorous derivation of eqs.(3.83) and (3.84) has been given (88). In view of this, we shall here take eq.(3.61) and the relation

$$\frac{1}{2} \sum_{\vec{k},\nu} \hbar\omega_{\vec{k}\nu}\, q^2_{\vec{k}\nu o} = \frac{1}{4\pi^2} \sum_{\vec{k}} |D^f_{\vec{k}} - D^i_{\vec{k}}|^2 \int \frac{d\omega}{\omega} \frac{\mathrm{Im}\,\varepsilon(k,\omega)}{|\varepsilon(k,\omega)|^2} \qquad (3.87)$$

which is obtained by combining eqs.(2.56), (2.58), and (2.59), as sufficient justification for the validity of eqs.(3.83) and (3.84).

While eqs.(3.83) and (3.84) represent the general theory of frequency dispersion of continuous media in the electron transfer system, the subsequent analysis depends on the properties of the system as expressed by the dependence of $\varepsilon(k,\omega)$ on ω. For example, inserting the particular form given in eq.(2.54) and assuming that $\hbar\omega_{\vec{k}\nu} \ll k_B T$, transform eq.(3.83) to eq.(3.61) and subsequently to the high-temperature rate expression derived previously. In the general case, however, a concrete form of $\varepsilon(k,\omega)$, either known from experimental data, or a form which is plausible in view of the particular nature of the medium must be inserted and the integrations over ω and α subsequently performed numerically. We shall illustrate this below by choosing a Debye frequency distribution which was discussed in chapter 2 and which is known to reproduce the dielectric properties of water and other hydroxylic solvents (56,158,159).

However, it is possible to derive certain limiting formulae of somewhat greater simplicity. We shall thus briefly, without giving the actual derivations (88) consider the following features of particular importance for the frequency dispersion in electron transfer systems:

(a) The vibration spectrum ($\mathrm{Im}\,\varepsilon(k,\omega)/|\varepsilon(k,\omega)|^2$) of water can be (empirically) reproduced by a sum of three Debye frequency distributions with characteristic maximum frequencies and molar absorbances(158). This functional dependence can be inserted in eq.(3.84) and the integrations over ω performed. It turns out that the result practically coincides with the result which is obtained by dividing the integration interval in two regions, i.e. from zero to $\omega_{cl} = k_B T/\hbar\alpha(1-\alpha)$, and from $k_B T/\hbar\alpha(1-\alpha)$ to infinity, and replacing the hyberbolic functions by their low- and high-temperature forms, respectively in the two regions(87). This gives a rate constant of the form

$$W_{fi} = (\pi\beta/\hbar^2 E_r^{cl})^{\frac{1}{2}} \exp(-\sigma) \exp[-(E_r^{cl} + \Delta E)^2/4E_r^{cl} k_B T] \qquad (3.88)$$

where

$$E_r^{cl} = \frac{1}{4\pi^2} \sum_{\vec{k}} |D_{\vec{k}}^f - D_{\vec{k}}^i|^2 \int_0^{\omega_{cl}} \mathrm{Im}\,\varepsilon(k,\omega)/|\varepsilon(k,\omega)|^2 \, d\omega \qquad (3.89)$$

and

$$\sigma = \frac{1}{4\pi^2} \sum_{\vec{k}} |D_{\vec{k}}^f - D_{\vec{k}}^i|^2 \int_{\omega_{cl}}^{\infty} \mathrm{Im}\,\varepsilon(k,\omega)/|\varepsilon(k,\omega)|^2 \, d\omega \qquad (3.90)$$

Several important conclusions can be drawn from this result. Firstly, $\omega_{cl} = k_B T/h\alpha(1-\alpha)$ represents the 'effective' limit between quantum and classical (or low- and high-temperature) behaviour of the nuclear medium modes. In particular, for thermoneutral processes for which $\alpha = 0.5$, this limit is $4k_BT$. Secondly, the modes which contribute to the classical medium reorganization energy are those for which the frequencies are located in the interval from zero to ω_{cl} (cf.eq. (3.89)). For water the absorption of these modes constitute about 82 % of the total absorption of all modes(87,158), i.e. the integral of $\mathrm{Im}\,\varepsilon(k,\omega)/|\varepsilon(k,\omega)|^2$ over the interval from zero to ω_{cl} amounts to about 82 % of the same integral taken from zero to the edge of the onset of electronic absorption. Since the latter integral equals $\varepsilon_0^{-1} - \varepsilon_s^{-1}$ (cf. chapter 2), we see that if this factor is used to calculate E_r^{cl}, (eq.(1.28)) such as in the theory of Marcus, it should be corrected by the factor 0.82. Thirdly, the remaining part of the medium modes contribute to the rate expression via the tunnelling factor, $\exp(-\sigma)$. Since the absorption of these modes constitute a relatively minor fraction of the total number of modes which are reorganized, they may be ignored in many practical applications of the theory to processes which proceed in the 'normal' temperature and free energy regions. However, at low temperatures and for strongly exothermic processes, for which $\alpha \rightarrow 0$, and therefore $\omega_{cl} \rightarrow 0$, the region of classical frequencies becomes small, and it is then essential to analyze explicitly the role of the 'high-frequency' medium modes.

(b) The continuous nature of the medium modes implies that thermal electron transfer reactions display several interesting features at very low temperatures (88,109,111). This is due to the fact that in the solid state in which these processes necessarily occur, comparable fractions of the medium oscillators ($\hbar\omega_M \approx$ 10-100 cm^{-1}) behave as quantum and as classical oscillators. Furthermore, since the number of quantum oscillators increases with decreasing temperature, whereas the number of classical oscillators decreases, the Bronsted and the Arrhenius relationships are qualitatively different at low and at high temperatures.

Eqs.(3.83) and (3.84) were analyzed in detail by Dogonadze and his associates (87,88,104,157). Rather than to reproduce their calculations here we shall provide some of the particularly useful limiting formulae, a discussion of their physical implications, and a consideration of the expected general temperature dependence of the rate constant. We shall choose the particular example of a Debye frequency distribution of the form given in eq.(2.13), and we shall not complicate the discussion by incorporating discrete high-frequency modes or exothermicity effects in addition to the effects of the continuous nature of the medium vibration spectrum.

We notice at first that at high temperatures in e.g. aqueous solution, the limit between quantum and classical oscillators is located at a value where most of the medium modes are classical, and where the maximum for absorption is located far to the low-frequency side of $k_B T/\hbar\alpha(1-\alpha)$ (87). The number of classical oscillators, and therefore E_r^{cl}, is thus only weakly dependent on the temperature in this region. This situation is quite different at low temperatures, where $k_B T/\hbar\alpha(1-\alpha)$ is now located at the 'ascending' part of the (Debye) absorption band (cf. fig.2.1). In this temperature region the number of classical oscillators increases with increasing temperature giving a similar increase of E_r^{cl}. At sufficiently low temperatures this dependence is linear and will be reflected in an almost temperature independent rate constant. Although the activation energy has a finite value, the apparent temperature dependence thus vanishes, because the temperature dependent activation energy, when inserted in an Arrhenius equation, gives a temperature independent rate constant.

At low temperatures we thus expect three classes of processes(88). The reorganization energy of the classical modes may be large compared with the 'energy gap', $|\Delta E|$. In the limit when $T \to 0$ this region vanishes altogether, and at low temperatures it is expected to be small. The rate expression (for a Debye distribution) is

$$W_{fi} = \frac{\beta|V|^2}{\hbar} (\hbar\Omega_D/E_r^m)^{\frac{1}{2}}(\pi/\gamma\beta\hbar\Omega_D)^{2E_r^m/\pi\hbar\Omega_D} \exp(\tfrac{1}{2}\beta|\Delta E|) \tag{3.91}$$

where Ω_D is the Debye frequency, and $\ln\gamma = c \approx 0.56$ is Euler's constant (155). Secondly, when strong medium coupling prevails (which is expected in the majority of chemical processes), the energy gap is typically much larger than the classical reorganization energy, but lower than the total medium reorganization energy. This gives the rate expression

$$W_{fi} = \frac{2|V|^2}{\hbar|\Delta E|}\left(\frac{E_r^m}{\hbar\Omega_D}\right)\left[\frac{\pi e|\Delta E|}{2\gamma E_r^m}\right]^{2E_r^m/\hbar\Omega_D} \tag{3.92}$$

i.e. a temperature-independent rate constant. Finally, as $|\_E| \to E_r^m$, the rate expression takes the form

$$W_{fi} = \frac{(2\pi)^{1/2}|V|^2}{\hbar}[\hbar\Omega_D E_r^m \ln[2E_r^m/\pi\hbar\Omega_D]]^{-\frac{1}{2}} \tag{3.93}$$

$$\exp\left\{-(E_r^m - |\Delta E|)^2/E_r^m\hbar\Omega_D \ln[4E_r^m/\pi(E_r^m - |\Delta E|]\right\}$$

i.e. also an activationless rate constant.

These equations reveal several interesting features of low-temperature processes. Firstly, ΔE appears in these equations as $|\Delta E|$ which reflects the fact that because of the freezing of all modes in their initial vibrational ground state, only exothermic processes can proceed. Secondly, a temperature dependence of the rate constant is only displayed for very small values of the energy gap (eq.(3.91)). However, the dependence differs appreciably from a normal Arrhenius temperature behaviour by the strong power dependence on T ($2E_r^m/\pi\hbar\Omega_D$ is a large quantity).

We have already noticed that this reflects two effects, i.e. that the activation energy increases with increasing T making βE_A^{cl} approximately temperature independent, and the fact that the number of quantum modes decreases with increasing temperature. This again implies that the tunnelling factor $\exp(-\sigma)$ increases. At very low temperatures the interpretation of the temperature dependence is thus quite different from the high-temperature interpretation.

The energy gap dependence is also different from the high-temperature behaviour of this relationship. For small values of $|\Delta E|$, i.e. $|\Delta E| < E_r^{cl}$, we notice that even though a 'conventional' Arrhenius temperature dependence is not expected, the Bronsted relationship between the rate constant and the energy gap is still valid in a formally conventional form (eq.(3.91)). When $|\Delta E| > E_r^{cl}$ but still lower than the total medium reorganization energy, the apparent activation energy vanishes, and the energy difference, $|\Delta E| - E_r^{cl}$, is now dissipated among 'high-frequency' oscillators. As a result, the tunnelling factor, $\exp(-\sigma)$, increases with increasing $|\Delta E|$. This region is represented by eq.(3.92) and is seen to display a high power law energy gap dependence. It is of interest to notice that the apparent Bronsted coefficient,

$$\alpha = dE_A/d|\Delta E| + k_B T d\sigma/|\Delta E| \tag{3.94}$$

now reflects the variation of the tunnelling factor with the energy gap since the apparent activation energy vanishes. This is in contrast to the high-temperature interpretation of the Bronsted relationship according to which the energy gap largely affects the activation energy.

While the interpretation of the low-temperature results does not present any fundamental difficulties as long as $|\Delta E|$ remains finite, it is less obvious for 'symmetric' reactions for which $\Delta E = 0$. According to eq.(3.91), $W_{fi} \to 0$ as a power law in T when $|\Delta E| \to 0$. This result is understandable if the frequency dispersion includes modes of very low frequency ($\omega \to 0$) subject to reorganization. Thus, for a given mode the barrier height for harmonic potentials and in a symmetric reaction is $\frac{1}{8}\hbar\omega_{\varkappa} q^2_{\varkappa 0}$, and the tunnelling factor $\exp(-\sigma) = \exp(-q^2_{\varkappa 0}/2)$ (cf. the discussion above). However, while the barrier height has a constant value, the 'effective' barrier width, as represented by $q^2_{\varkappa 0}/2$, depends on the frequency. Recalling the definition of $q_{\varkappa 0}$ from chapter 2(eq.(2.36)) we see that $q_{\varkappa 0}$ is inversely proportional to the square root of $\omega_{\varkappa}$. Hence, $q_{\varkappa 0} \to 0$ when $\omega_{\varkappa} \to 0$, and therefore $\exp(-\sigma) \to 0$. This conclusion rests on the presence of medium oscillators of sufficiently low frequencies, such as in the Debye distribution. If the frequency dispersion is cut off by a lower finite value, the rate constant remains finite for T, $|\Delta E| \to 0$ due to the finite value of $\exp(-\sigma)$.

We whall complete the present section by showing some numerical results relating to the frequency dispersion problem. Thus, while the interpretation of the rate expressions in the limits of low and high temperatures is clear in qualitative terms, most experimental data on low-temperature chemical processes do not correspond to these limits but rather to some intermediate temperature region (e.g. 77° K), for which the general equations (3.83) and (3.84) must be applied.

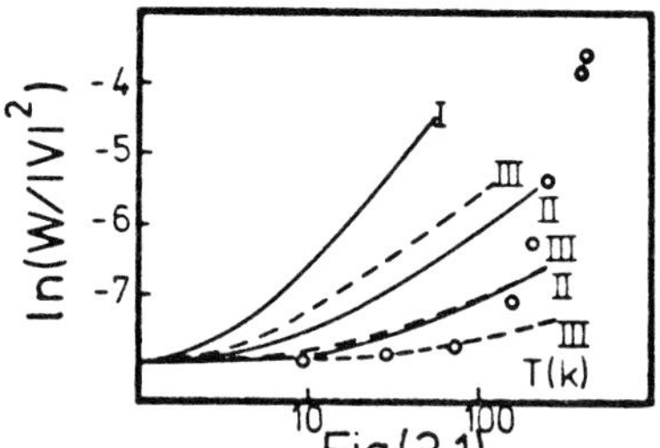

Fig(3.1)

$\ln(W_{fi}/|V|^2)$ plotted against lnT. W_{fi} given by eq.(3.83). ( ) E_r^m = 0.35 ev, (----) E_r^m = 0.30 ev. I: Ω_D = 100 cm^{-1} ; II: Ω_D = 200 cm^{-1} ; o experimental points of DeVault and Chance (chapter 9).

Fig.(3.1) shows representative plots of $\ln W_{fi}/|V|^2$ against lnT normalized to the same value for T->0 (this value is the low-temperature rate constant of the cytochrome c-bacteriochlorophyll electron transfer, cf. chapter 9). These calculations refer to a Debye frequency distribution, and the parameters E_r^m and Ω_D correspond to a fairly strong medium coupling and representative solid-state phonon frequencies. We see that the qualitative conclusions reached above are borne out, i.e. primarily the one concerning a wide activationless low-temperature region followed by an 'activated' high-temperature region and a transition region around the temperature $\hbar\Omega_D/2k_B$.

3.5 Relation to Experimental Data

Although our discussion so far is valid for any category of multiphonon electronic relaxation process, we have focused on chemical electron transfer and given explicit attention to the medium modes only.

We shall see in the following chapter that some of the most convincing experimental tests of the fundamental conclusions of the theory can be extracted from investigations on systems where electronic coupling to both intramolecular and medium modes is important. In order to interprete these data we shall thus at first have to specify quantitatively the role of each of these subsystems in the process. We shall perform such an analysis in chapter 4. At present it is appropriate to search for experimental data which can illustrate some general features of the theory which do not in the same way require specification of the nature of the nuclear modes or which can support our previous statements about the formalism as a kind of 'unified' rate theory applicable to a variety of different processes.

In this context we recall that although the Bronsted and Arrhenius relationships emerge naturally from our theoretical framework, they are most commonly insufficiently diagnostic with respect to the molecular mechanism. Linear or curved free energy relationships at room temperature can thus be interpreted in terms of any reasonable pair of intersecting potential energy surfaces. If these relationships should be of major value for the interpretation of the detailed mechanism they must therefore be investigated under 'unusual' conditions, e.g. for strongly exothermic processes or at low temperatures.

With this in mind we shall now discuss a few experimentally investigated systems which point to the following important general features of the theory:

(1) The electronic factor in the rate expressions; (2) the energy gap law, temperature dependence and deuterium isotope effect in 'intramolecular' electronic processes; (3) the relationship between the chemical rate expressions and the line shape of optical transitions in selected condensed phase systems.

3.5.1 The Electronic Factor

We have considered electron transfer processes in the nonadiabatic limit, i.e. we have assumed that the electronic factor, V_{eff}, in the rate expression is small and must be included explicitly (cf. chapter 5). The important question of the possible nonadiabaticity of simple electron transfer processes between mobile reactants has in fact been extensively discussed without, however, so far leading to definite conclusions. Thus, it has been suggested (16) that the critical value for the perturbation matrix element is so small (≈0.01 ev) that all common chemical processes must be adiabatic. On the other hand, this estimate is based on the Debye librational motion of the solvent being the only nuclear subsystem. If high-frequency intramolecular modes are also reorganized, the critical value is higher amounting to some tenths of an electron volt (chapter 5). Also, we have seen that the overall interaction between the reactants leads both to deformation of the donor and acceptor electronic levels and to a chemical reaction, and a separation of the two effects requires a more elaborate analysis.

Accurate calculation of the coupling matrix element is probably still beyond the capability of quantum chemistry. An attempt towards this aim was made by Dolin, Dogonadze and German (160) who calculated the functional dependence of both the various contributions to V_{eff} and the overall rate constant on the interreactant distance for the electron exchange between MnO_4^- and MnO_4^{2-}. This system is conveniently simple, and the geometries of both oxidation states are known from crystallographic data. They used a fixed relative orientation corresponding to maximum overlap of the donor and acceptor orbitals, a 13-electron set of electronic wave functions of the Slater type, and included all electrostatic interactions between the (13) valence electrons and the nuclei. The absolute values of V_{eff} are insufficiently accurate for the nonadiabaticity criterion but a valuable result of the calculations was the observations of a narrow (≈ 0.5 A) effective width of the interreactant distance R (cf. eq.(1.10)) and that the electronic factor decreases more rapidly with increasing R than the electrostatic repulsion increases, giving maximum weight to the contact distance.

Information about the nonadiabatic character of particular chemical processes is generally reflected in small values of the pre-exponential factor of the rate expressions. However, as we have seen, this factor may contain the Franck Condon nuclear overlap integrals of high-frequency modes as well, and even though these factors could be calculated if the necessary spectroscopic data were available, this is rarely possible in practice. Moreover, reactions between mobile species may be subject to steric requirements which are reflected in small values of the reaction volume (eq.(1.12)). Conclusions about electronic nonadiabaticity in simple electron transfer reactions must therefore be based on comparison among data from a considerable variety of processes to see if any particular process may show 'unusually' low values of the pre-exponential factor.

Taube(161) and Chou, Creutz and Sutin(162) have recently discussed several such criteria. They seem to be grouped in the following categories:

(a) Comparison of specific rates of self-exchange in the systems $[Mn(CNR)_6]^{3+/2+}$ and $[Fe(phen)_3]^{3+/2+}$ where R is the ethyl or the tert-butyl group and phen 1,10-phenanthroline or substituted derivatives. In the former systems the rate constants ($dm^3 mol^{-1} s^{-1}$) are $64\cdot10^4$ and $4.0\cdot10^4$ for the two substituents, i.e. the more bulky the substituent the lower the rate. For the second group of systems the rate constants are $6\cdot10^6$, $17\cdot10^6$, $8\cdot10^6$ and $0.4\cdot10^6$ for the unsubstituted, the 3,4,5,8-methyl, the 4,7-phenyl, and the 4,7-cyclohexyl derivatives, respectively. As the ligands become more voluminous the radii of the reactants increase, whereas the solvent reorganization energy and consequently the activation energies are expected to decrease. For adiabatic processes the rates would thus be expected to increase with increasing ligand size. On the other hand, the larger the ligand the longer is the electron transfer distance and therefore the lower the rate if the electronic factor is important. This may be the effect seen in the former group of systems. The results from the second group imply that the smallest complexes react adiabatically, whereas nonadiabatic effects are displayed for the more bulky groups.

(b) Comparisons based on Marcus 'cross-relation' (eq.(1.29). This relation has been extensively tested experimentally (chapter 1) and quite good agreement between theoretical and experimental values of the rate constants is often found. However, there is a tendency that rate constants for the cross reactions calculated on the basis of experimental values of the self-exchange rate constants are higher than the ones observed experimentally, and the discrepancy increases with increasing exothermicity of the process.

Nonadiabaticity has been suggested as a possible cause for the discrepancies observed. This would for example make the low value of the self-exchange rate constant of the $Eu^{3+/2+}$ couple (k_{obs} = $2 \cdot 10^4$ $dm^3 mol^{-1} s^{-1}$) as compared with the $Fe^{3+/2+}$ reaction (k_{obs} = 4 $dm^3 mol^{-1} s^{-1}$) understandable. Even though the former ions are larger and should consequently show a smaller activation energy, the rate is much smaller. Nonadiabaticity in the $Eu^{3+/2+}$ self-exchange reaction is also plausible since this process involves well shielded f donor and acceptor orbitals which would give a small electronic overlap compared with the d orbitals involved in the second couple. However, the effects in other systems are less easy to interprete unambiguously (161,162), and in addition to nonadiabaticity they probably involve one or more of the following factors: (1) the cross relationship is valid for intramolecular harmonic potentials only. However, anharmonicity effects are important for some of the systems in which large coordinate shifts occur, and more so the more exothermic the process (chapter 4). (2) The validity of the cross relationship requires that the work terms are constant and that the mechanism is identical for both self-exchange and mixed reactions. However, when one reactant is an aquo ion and the other one e.g. a tris-phen complex, inspection of molecular models indicates that the large ligand molecules may be 'stacked' in a parallel fashion in the collision complex. Since electron density is also delocalized onto the ligands, the nature of the donor and acceptor orbitals may be different in the two sets of reactants, being largely of the d type for aquo ions and of π type for the complexes which contain the organic ligands. (3) The discrepancies are frequently located in the activation enthalpy indicating that other effects than nonadiabaticity may be responsible for the observed discrepancies.

(c) Nonadiabaticity effects are expected to be much more unambiguously manifested in solid state electron transfer between fixed donor and acceptor centres, even though the electron transfer distance may vary according to some distribution law. Provided that the distribution law is known (e.g. a random distribution) the decay pattern of the reactants can furthermore give information about the electronic wave functions in the actual medium (163).

It was recognized early (24,161) that several interpretational ambiguities might be removed by measuring the rates of intramolecular electron transfer processes in which the donor and acceptor centres are located at positions known from the molecular structure. Reports on such measurements showing indications of nonadiabatic effects in the intramolecular electron transfer have recently appeared (161,164). The most comprehensive data refer to the electron transfer from Ru(II) to Co(III) in molecules of the type

$$[(NH_3)_5Co^{III}L\cdots\cdots LRu^{II}(NH_3)_4SO_4]^{4+} \tag{3.95}$$

where the complex is decomposed to $[Co(H_2O)_6]^{2+}$ subsequent to the electron transfer. L....L represents 4,4'-bipyridyl (DBP), its dimethyl derivative (DMBP) and several related bidentate ligands of variable length and rigidity (-S- (DPS), -CH=CH- (DPEy), $-CH_2-$ (DPMa), and $-CH_2-CH_2-$ (DPEa) inserted between the rings). The results of these investigations can be summarized as follows:

(1) The intramolecular rate constants vary from $44\cdot10^{-3}$ s^{-1} to $1.0\cdot10^{-3}$ s^{-1} in the order of complexes given, i.e. remarkably

little. Moreover, the activation enthalpy is nearly constant varying from 20.3 kcal to 18.6 kcal. Since the redox potentials for the Ru(II)/Ru(III) 'end' of the molecule are nearly constant, this result is understandable in view of the approximately constant intramolecular reorganization and the relatively weak dependence of the solvent reorganization energy on the electron transfer distance.

(2) The activation entropies are approximately zero except for the last two complexes in the series. This is in marked contrast to the large negative values commonly observed for bimolecular electron transfer processes between 2+ and 3+ charged mobile ions. The constancy is also remarkable in view of the different coupling between the two ends of the molecules (for example, while the two pyridine rings in 4,4'-bipy are coplanar, they are almost perpendicular to eachother in the dimethyl derivative due to the interference of the methyl substituents) and suggests that the intramolecular electron transfer in these complexes proceeds adiabatically.

(3) The activation entropies for the DPMa and DPEa complexes are somewhat lower ($- S_A \approx 9$ cal K^{-1}) than for the previous complexes. For the DPEa complex this effect is likely to be associated with the flexibility of the $-CH_2-CH_2-$ bond system providing a direct electron transfer route between the two centres, i.e. bypassing the $-CH_2-CH_2-$ entity. This is not possible for the DPMa complex which thus shows indication of nonadiabatic electron transfer.

(4) The rate differences parallel with the intensities of the intervalence optical transitions of the corresponding Ru(II)-Ru(III) complexes (cf. the discussion following eq.(3.50) and section 3.5.c). These intensities are a measure of the coupling between the donor and acceptor centres.

3.5.2 Intramolecular and Medium-induced Electronic Relaxation.

Electronic relaxation processes in single molecular centres of large molecules or solid-state 'impurities' have been the subject of much recent experimental and theoretical investigation (2-5,165). Many data for such processes are illustrative with respect to the conclusions of the general theory outlined above, and several tests of the theoretical relationships derived (e.g. the exponential energy gap law) have so far only been obtained from studies of such processes. We shall now give a brief account of some of these data. Since, however, our main object is chemical processes in a more 'conventional' sense, our account of this important field must necessarily be rather schematic, and we shall have to refer to the literature for details (2-5).

The electronic relaxation processes, internal conversion and intersystem crossing, were discovered by studies of the luminescence processes of large molecules in condensed media. Such studies provide data on (thermally averaged) radiative (emission) life times, $\tau_{rad}(T)$, and quantum yields, $\varphi(T)$. Separation of radiative and nonradiative decay mechanisms could subsequently be obtained from the relationships

$$\varphi(T) = k_{rad}/(k_{rad} + k_{nr}) \tag{3.96}$$

$$\tau_{rad}(T) = (k_{rad} + k_{nr})^{-1} \tag{3.97}$$

where the subscripts 'rad' and 'nr' refer to the radiative and nonradiative mechanism, respectively and the k's to the overall rate constants. When commonly $\varphi(T) < 1$, this implies a nonradiative component in the observed life time.

The nonradiative rate constant, k_{nr}, is expressed by the golden rule equation

$$k_{nr} = \frac{2\pi}{\hbar}\sum_{v}\sum_{w} p_{iv} \left|\langle fw|H'|iv\rangle\right|^2 \delta(E_{iv} - E_{fw}) \qquad (3.98)$$

This equation implies that the system decays to a practically continuous manifold of final vibrational states, a condition which is valid when the decaying molecule is sufficiently large or coupled to a medium. p_{iv} is the probability that a given initial vibrational-electronic state is populated, and H' the perturbation which induces the transition between the zero order states. This perturbation depends on the nature of these states. It is taken as the spin-orbit coupling for intersystem crossing, or as the nuclear momentum operator (eqs.(3.26)-(3.30)) for internal conversion in large molecule (148,149). In medium-induced processes it may be the dipole interaction with the nearest medium molecules (145) or the coupling operator between intramolecular and medium nuclear modes (cf. chapter 2). Whatever the nature of H' it is assumed to be negligible compared with the Hamiltonian of the radiation field which prepares the excited zero order states, i.e. it is only important when this field is switched off.

Provided that the interaction between the zero order states is small (a small splitting in the intersection region of the potential energy sufaces corresponding to the zero order states) the nonradiative decay mode is appreciable compared with the radiative mode (in the opposite case the decay is purely radiative). If the Born-Oppenheimer and Condon approximations can furthermore be invoked, (eq.(3.98)) can subsequently be evaluated by procedures closely related to the one outlined above.

We can thus again represent the rate expression as a product which contains a matrix element coupling the electronic wave functions and a Franck Condon nuclear overlap factor of all modes which undergo displacement or frequency shift. However, the different nature of the perturbations may now provide some modifications from the results derived above. We can see this in the case of internal conversion, where the modifications are due to the presence of additional factors of the form $|\sum_{k=1}^{p} \langle \chi_{fw} | \partial/\partial Q_k | \chi_{iv} \rangle|^2$ in the rate expression (166-168)

$$W_{fi} = \frac{2\pi}{\hbar} \sum_v \sum_w p_{iv} \left| \prod_{k=1}^{p} \langle \chi_{iv} | \partial/\partial Q_k | \chi_{fw} \rangle \right|^2$$

$$\left| \prod_{j \neq k} \langle \chi_{iw_j} | \chi_{fw_j} \rangle \right|^2 \delta(E_{iv} - E_{fw} - \Delta E) \qquad (3.99)$$

$$v = \sum_j v_j \quad ; \quad w = \sum_j w_j$$

The number of 'promoting' modes, p, i.e. the ones for which the matrix elements of the nuclear momentum operator is appreciable, is usually believed to be small compared with the number of 'accepting modes', i.e. those which are subject to a finite equilibrium coordinate or frequency shift (3,167). There are not many data, however, which can illustrate the nature of these modes. Generally the nature of the promoting modes must be determined by the symmetries of both the electronic and nuclear wave functions which are coupled by the operator $\partial/\partial Q_k$(149,166,167). This implies that the appropriate initial and final state vibrational wave functions cannot both be totally symmetric (166,167). Analysis based on spectroscopic data for $\pi \rightarrow \pi^*$ transitions in benzene, naphthalene, anthracene and other aromatic compounds has shown that equilibrium nuclear coordinate shifts in the excited state relative to the ground electronic state are only likely to occur in the symmetric C-H and C-C stretching modes, and that the 'reduced' displacements, $q_{ko}^2/2$, furthermore have small values (of the order of unity)(167,169). These modes are therefore likely to constitute the majority of the accepting modes. The promoting modes must then be found among the non-totally symmetric C-H bending and C-C skeletal deformation modes, i.e. for the promoting modes in these systems the equilibrium coordinate shift vanishes. This has the important implication (in the harmonic approximation) that the vibrational quantum numbers of a particular promoting mode in the initial and final state must differ by one and that the 'effective energy gap is modified by this amount (165,167). However, in the general case, and in particular for electronic processes in solids where the promoting modes are those of the lattice phonons, these modes may also be subject to equilibrium coordinate shifts (5,170,171).

The radiationless processes presently discussed are thus essentially intramolecular processes, and the coupling between the electronic states induced by intramolecular modes. The analogous transitions of the same molecular entities embedded in an 'inert' medium are not necessarily strongly affected by the latter, at least not for large molecules for which 'internal' thermal equilibrium is likely to prevail. Thus, there is evidence for the absence of drastic medium effects on the absolute rate values for intersystem crossing in benzene and naphtalene when these molecules are dissolved in a variety of different hydrocarbon and hydroxylic solvents (172), i.e. solvents which contain no heavy atoms. These processes are therefore commonly viewed as electronic relaxation in 'isolated' molecules coupled to a heat bath. This is in contrast to electron transfer and solid-state relaxation processes induced by the lattice modes (170,171), where the role of the medium is not only to provide a heat bath, but where fluctuations in the medium modes themselves induce the process. In these processes the medium modes therefore rather constitute 'intramolecular' modes of a 'supermolecule' and provide a multitude of accepting and possibly promoting modes.

We now summarize experimental data from multiphonon intramolecular and solid-state relaxation processes of particular relation to elementary chemical processes:

(a) The exponential energy gap law, eq.(3.71), is well documented experimentally for the low-temperature (77° K) decay of the lowest triplet state to the ground state of several families of aromatic hydrocarbons $C_x H_y$ dissolved in inert media (165,167,173). This temperature corresponds to the ground vibrational state of the initial electronic state. From the experimental data, separate values of γ in the range 0.5-1.3 and of the pre-exponential factor in the range 10-10^3 s^{-1} could be obtained (167). Using an average value of $\Delta E \approx 10^4$ cm^{-1} and values of the equilibrium coordinate shifts from spectroscopic data ($\sum_\kappa q^2_{\kappa 0}/2$ = 0.13-0.42), and assuming that the dominating accepting mode is the C-H stretching (cf. above) values of the spin orbit coupling matrix element in the order 10^{-2} cm^{-1} is found (167).

A similar energy gap law has been observed for the internal conversion from the second to the first excited singlet state of a series of substituted azulenes dissolved in cyclohexane and giving values of the internal conversion coupling of approximately $10^3 cm^{-1}$ (174). An exponential energy gap law has also been observed for the low-temperature (4.2 ° K) electronic relaxation of various excited states of Dy and Nd rare earth ionic dopants in different crystalline lattices (5,170,171,175). The energy gaps are here about an order of magnitude smaller than for the aromatic hydrocarbons (10^3 cm^{-1} vs. 10^4 cm^{-1}) and correspond to transitions between f orbitals of the localized 'impurities'. These orbitals are well shielded from the surroundings, and only small values of the coordinate shifts are therefore expected. Analysis of the experimental energy gap plots gave values of about 1 cm^{-1} , 200-300 cm^{-1} , and 0.1-0.2 for the electronic coupling factor, the frequency of the dominating accepting mode, and the effective nuclear coordinate shift, respectively (5,170,171). These values are compatible with low frequency values of promoting and accepting phonon modes.

(b) The rate of the electronic relaxation of excited states of crystal impurity rare earth ions corresponding to 4f-4f or 5d-4f transitions (5,170,171) or transition metal ions (121) (Co^{2+} and Ni^{2+} in KMgF crystals) generally display an activationless behaviour for temperatures lower than 50-100 °K and increases with increasing temperature at higher temperatures.

This behaviour is understandable in view of the fact that these processes involve many promoting and accepting modes of relatively low frequencies (100-250 cm^{-1}). Although these modes are vibrationally excited at temperatures in the 'activationless' temperature region, the frequency dispersion of the medium is therefore expected to invoke a behaviour qualitatively similar to the one represented in fig.(3.1) (cfr. section 3.4).

The triplet state life time of benzene and n-dodecylbenzene shows a similar pattern (176). However, the somewhat sharper transition between the activationless and activated regions for these systems is likely to be associated with the thermal excitation of discrete intramolecular modes of somewhat higher frequencies (cfr. chapters 4 and 9).

(c) The energy gap law for strongly exothermic processes provides an interpretation of the deuterium isotope effect for electronic-vibrational relaxation of large organic molecules. We have thus seen that the dominating accepting modes are the C-H stretching which should therefore give rise to an isotope effect when D is substituted for H. The ratio of the decay rates of the perhydrated and perduterated compounds is (165,167)

$$W_H/W_D \approx \exp\left[\frac{\Delta E}{\hbar}\left(\frac{\gamma_D}{\omega_D} - \frac{\gamma_H}{\omega_H}\right)\right] \tag{3.100}$$

where the index 'H' and 'D' refers to the appropriate isotope. In the limit of the applicability of eq.(3.71) this equation thus shows that an 'appreciable 'normal' isotope effect ($W_H/W_D >$ 1) is expected which increases with increasing energy gap. This is borne out by experimental data for the decay rates of the lowest excited triplet state of several aromatic compounds (crysene, pyrene, benzanthracene) and their perdeuterated anal-oogues(165b,167,177). Moreover, studies of the isotope effect has provided information about the nature of the promoting modes(178). It was thus observed that the decay rates of the lowest triplet of naphthalene is approximately inversely proportional to the number of deuterium substituents but does not depend on the position of substitution. Since the electronic factor, $\langle f|\partial/\partial Q_k|i\rangle$, is expected to depend on both the isotope and the position of substitution if the C-H modes are the promoting modes, the absence of any such dependence implies that these modes cannot be promoting.

3.6 Lineshape of Optical Transitions.

We noticed in chapter 1 that the quantum theory of chemical processes was initiated by the close formal relationship between multiphonon optical processes, such as in colour centres in alkali halide crystals, and thermal electron transfer processes. Our previous calculations therefore also predict several properties of optical line shape functions for electronic transitions between local levels which are coupled to either discrete or lattice phonon modes. We shall summarize briefly some of the important features of optical line shape relating to this analogy and discuss a few systems where it appears to be particularly illustrative. For this discussion a single-mode description is adequate but our predictions can be extrapolated to systems where several modes are coupled to the electronic subsystem. Thus:

(1) The calculation scheme outlined above gives an optical line shape of a Gaussian form sufficiently close to the maximum and 'strongly' coupled modes (157). The line shape expression is

$$W_{fi} = W_{fi}^{max} \exp[-E_r + \Delta E \pm h\nu)^2/\Delta_m] \qquad (3.101)$$

where '+' and '-' refer to emission and absorption, respectively. The halfwidth, Δ_m, is given by

$$\Delta_m = 2\hbar\omega_\kappa E_r \coth\frac{\beta\hbar\omega_\kappa}{2} \qquad (3.102)$$

and the maximum value, W_{fi}^{max}, is

$$W_{fi}^{max} = 2\pi^{\frac{1}{2}}|d|^2/\hbar\Delta_m \qquad (3.103)$$

where d is the transition dipole matrix element, and the remaining quantities have been defined before.

These equations imply that 'vertical' transitions from nuclear configurations close to equilibrium in the initial state to a vibrationally excited final state, dominate. We notice also that in the high-temperature limit the expression coincides with our previously derived high-temperature rate expressions for thermal electron transfer except for the different electronic coupling factor. These equations are thus equivalent to free energy relationships for the thermal processes, in the strongly exothermic region which corresponds to the region around the absorption maximum for the optical process. Finally, the equations predict the expected broadening of the line width with increasing temperature but no shift of the maximum. The latter effect is expected, however, if the transition is accompanied by frequency shift in the nuclear modes, if several modes of different frequencies are present, or if anharmonicity effects are important.

(2) In the high-temperature region the line width increases with increasing temperature by the relationship $\Delta_m = (2k_BTE_r)^{\frac{1}{2}}$. At

room temperature such a dependence is thus expected when coupling to low-frequency modes dominate. High-frequency (or low-temperature) effects are manifested by asymmetry in the absorption bands, i.e. a longer 'tail' towards the high-energy region, in a way analogous to what is predicted for the free energy relationships of the corresponding thermal processes. Asymmetry effects are thus not only expected when the nuclear modes are subject to frequency shifts or anharmonicity. This can also be seen by inspection of the low-temperature form of the rate expression for the thermal electron transfer (eq.(3.81)).

In chapter 4 we shall discuss the influence of low- and high-frequency modes on the 'line shape' of both optical and thermal electronic transitions on the basis of numerical calculations. At present we shall give a brief discussion of certain optical processes which appear to be particularly closely related to thermal electron transfer between separated donor and acceptor centres. This class of optical processes is commonly named intervalence transitions and refers to compounds which contain different oxidation states of the same element, and where the electronic transition is associated with a characteristic interaction between the two separated centres.

Mixed valence compounds occur in a variety of different areas of which the better known are mixed oxide compounds (bronzes of tungsten, molybdenum etc, oxide minerals, e.g. of iron), polynuclear and cluster metal complexes, and biological compounds (for example copper proteins)(179). Depending on the amount of interaction between the two centres, these materials often reveal interesting spectral and conduction properties which in turn can be exploited to elucidate this interaction. For example, if there is no interaction, the properties of the mixed compounds is expected to be just the combination of the independent contributions of the components (class I compounds). If there is a small interaction, individual metal centres may still be distinguished by different metal-ligand bond lengths, vibration frequencies etc. but some cooperative effects may be revealed by for example electronic conduction by thermally activated 'hopping', or by strong absorption bands in the visible, near ultraviolet, or near infrared regions which correspond to electron transfer from one centre to another (class II). Finally, the two sites may be identical, indicating 'complete' electronic delocalization. Such compounds would also show new absorption bands (class III)(179).

The relationship between thermal and optical electron transfer in the limit of weak interaction (class II) was studied firstly by Hush(180,181), and more generally by Dogonadze and his associates (157). Hush pointed out the simple relationship expected between the band width and position of the absorption maximum for a given intervalence transition, and the activation energy and energy gap for the analogous thermal process, when all nuclear modes behave classically. Thus, provided that the splitting of the zero order surfaces can be ignored, the following relationship is obtained (181)

$$E_A^{th} = (h\nu_{max})^2/4(h\nu_{max} - \Delta E) \qquad (3.104)$$

For $\Delta E = 0$ this simplifies to $E_A^{th} = h\nu_{max}/4$. Furthermore, if the band shape is Gaussian the following relations between the (decadic) molar extinction coefficient at the maximum frequency, ε_{max}, the oscillator strength, f, and the band width, Δ (cm^{-1}), could be obtained (181)

$$\varepsilon_{max} = 10^9 f/4.6\Delta_m \qquad (3.105)$$

$$\nu_{max} = h(\Delta_m)^2/16\ln 2 k_B T \qquad (3.106)$$

No data for both optical and thermal electron transfer referring to the same chemical process appear to have been reported. However, several recent studies on the intervalence absorption of mixed Ru(II)/Ru(III) complexes offer a rationalization, although incomplete, in terms of this analogy. These compounds are binuclear entities of the following general types

$$(NH_3)_5Ru^{II}\ pyr\ Ru^{III}(NH_3)_5$$

I

$$(bipy)_2ClRu^{II}\ pyr\ Ru^{III}Cl(bipy)_2 \qquad (3.107)$$

II

$$(NH_3)_5Ru^{II}\ L.....L\ Ru^{III}Cl(bipy)_2$$

III

where pyr is pyrazine coordinated to both metal centres via its nitrogen atoms, and L.....L refers to the bidentate ligands also discussed in section 3.5.1. We consider at first the groups II and III(182):

(1) The complexes undergo reversible one-electron transfer reactions to form well-defined entities with the ruthenium oxidation numbers in the combinations [2,2], [2,3], and [3,3]. The electrochemical potentials are very similar to those of the corresponding monomers.

(2) Several absorption peaks in the visible and ultraviolet regions, corresponding largely to metal-ligand transitions, appear in the binuclear complexes, essentially as the sum of the spectra of the analogous monomers, although shifts of the band maxima may occur.

(3) Except for a band at 1599 cm^{-1} in the [2,3] compound, which is believed to arise as a result of the break-down of the D_{2h} symmetry of the coordinated pyrazine, the infrared spectra of the symmetric complexes show the bands characteristic of the analogous monomeric Ru(II) and Ru(III) complexes. The asymmetric complexes are less conclusive in this respect.

(4) The most interesting feature is the appearance of a broad absorption band in the visible or near infrared region for the [2,3] complexes. Neither the [2,2] nor the [3,3] compounds show this effect. For the symmetric compound II the maximum is located at 1300 cm^{-1} ($\mathcal{E}$ = 455 M^{-1} cm^{-1}), whereas the location is in the region 680-960 cm^{-1} (where $\mathcal{E}$ varies from 300 to 530 M^{-1} cm^{-1}) for the unsymmetrical complexes. These lower frequencies are compatible with an intervalence transition, since both the energy gap and the 'Franck Condon effect' contribute to the absorption energy for the unsymmetrical complexes. Within the experimental accuracy the band shape is Gaussian and the relation between the band width and the band maximum frequency is that predicted from the high-temperature form of eq.(3.101).

The evidence listed suggests that the near infrared or visible absorption band for the [2,3] compound corresponds to optical electron transfer between two weakly interacting redox centres. According to the theory of Hush this would offer two additional tests of the internal consistency of the interpretation. If coupling to the medium is important, then the maximum of the optical transition is directly related to the medium reorganization energy and should display a strong dependence on parameters which affect the latter. In particular, using the simple model outlined in chapter 1, the energy of the optical transition should depend linearly on ($\varepsilon_0^{-1} - \varepsilon_s^{-1}$), which can be varied by choosing different solvents, and an inverse dependence on the internuclear distance which varies according to the nature of the bridge ligand.

Both of these predictions are borne out by experimental data for the two groups of complexes (182d-f) thus giving additional support to the suggestion of intervalence optical electron transfer. However, the relations have one inconsistency which has not been satisfactorily disentangled. A quantitative agreement between the theory and the experimental data requires an additional solvent-independent classical intramolecular reorganization energy of 10-16 kcal(182f). This conclusion would not be strongly modified if the data were analyzed in terms of the theory outlined in the present chapter. However, this mode has not been identified in a way which is compatible with the known crystallographic and spectroscopic properties of Ru complexes (ammine complexes), according to which very small ($\approx$ 0.04 A) bond length changes accompany the change in oxidation state (183) and would give much smaller values of the intramolecular reorganization energy.

While the bipy complexes thus display features characteristic of distinct centres, the properties of certain other [2,3] Ru-complexes show evidence of indistinguishable metal centres, even though the picture for these ions is not totally consistent(184-186). This refers to the compounds I in which the bridge ligand is either cyanogen (-NC-CN-)(185) or pyrazine(186), and in particular the latter compound has been studied in considerable detail as a prototype of the group of mixed-valence binuclear Ru complexes (187-191). Thus, both the metal-ligand distances (189), a NH_3 'rocking' frequency at 800 cm^{-1}, and a metal-ligand stretching frequency at 449 cm^{-1} (188) are intermediate between those of Ru(II) and Ru(III). This is most simply explained by a delocalized ground state. The intervalence band is located at 1570 cm^{-1}, but it is both narrower, and much more intense (ε = 5000 M^{-1} cm^{-1}) than those for the bipy complexes. In addition, it displays a pronounced asymmetry towards the high-frequency region, the position of the band maximum depends very weakly on the solvent, and the band shape position is reported to be essentially independent of the temperature in the region 77-298 °K (186). These features have the following implications: (1) the coupling to the solvent is weak. However, the absence of solvent dependence on the band position does not necessarily imply electron delocalization. Thus, the pyrazine ligand is small and the two metal centres 'entangled', so the simple Hush model cannot be expected to be valid. According to this model the field contributions from each centre is spherically symmetric and unaffected by the presence of the second centre. The model can be corrected by invoking a more realistic charge distribution which takes into account the smaller polarizability of the space occupied by the metallic centres and the screening from the outer solvent (60); (2) the geometry and the vibration frequencies of the [2,3] compound suggests that electron delocalization in the ground state occurs; (3) the asymmetry of the absorption band and the independence of the temperature suggests electronic coupling (nuclear reorganization) to an intramolecular mode of quite high frequency ($\approx$ 1000 cm^{-1}), the nature of which, however, is so far not easy to envision.

However, other evidence, based on the temperature dependent paramagnetism and EPR and NMR line broadening data for the [2,3] compound (191) seems to be compatible with localized electronic states in this complex. The overall picture of the electronic structure of [2,3] compounds of type I is therefore by no means settled.

4 THE EFFECT OF INTRAMOLECULAR MODES

4.1 Special Features of Electron Transfer Processes

In chapter 3 we derived general rate expressions for elementary condensed phase (multiphonon) chemical processes. Our point of departure was first order perturbation theory, corresponding to the electronically nonadiabatic limit, and we obtained explicity rate equations essentially for a two-level system (initial and final electronic state) and the harmonic approximation. Moreover, we analyzed the general features of low- and high-temperature behaviour of the nuclear modes, and we finally discussed several different kinds of experimental data to the extent that they illustrate the fundamental conclusions drawn from the theory.

In the present chapter we shall more explicitly incorporate the molecular structural aspects of the reactants. We shall thus consider the electron transfer between two solvated ions or molecules and incorporate the coupling to both of the following two sets of nuclear modes:

(a) Discrete high-frequency modes characterized by the nuclear coordinates Q_c and the vibrational frequencies ω_c. In common electron transfer reactions these modes do not necessarily have the same equilibrium configurations and frequencies in the two electronic states, nor are they necessarily adequately represented by the harmonic approximation. The high-frequency modes correspond for example to metal-ligand vibrations in the first coordination sphere of metal complexes ($\omega_c \approx$ 300-500 cm^{-1}) or to the C-C ($\omega_c \approx$ 1000-1500 cm^{-1}) and C-H ($\omega_c \approx$ 3000 cm^{-1}) modes of organic molecules. Thus, at room temperature $\hbar\omega_c > k_B T$ and the modes should a priori be treated quantum mechanically.

(b) Low-frequency modes of the outer medium, characterized by the normal coordinates q_κ and frequencies ω_κ. Provided that the medium responds in a linear fashion to the field of the reactants these modes were seen (chapter 2) to be adequately represented by the harmonic approximation. We have also seen that for many common electron transfer reactions the frequency dispersion can be neglected and the medium modes approximated by an average frequency $\langle\omega_\kappa\rangle \approx 1\ \mathrm{cm}^{-1}$ corresponding to the Debye region. These modes are therefore adequately treated within the framework of the classical approximation. However, for strongly exothermic processes, and in particular for low-temperature processes, the medium frequency dispersion and the quantum behaviour of some medium modes must be properly incorporated.

When the equilibrium nuclear configuration and the frequencies are identical in the initial and final states the molecular structural aspects can be disregarded, and the electron transfer probability is determined by the classical modes only. If we also disregard the medium frequency dispersion, this gives the Gaussian 'line shape' and quadratic energy gap law (eq.(3.63) firstly derived by Marcus using a classical approach and which is well obeyed by a number of electron transfer systems in the 'normal' energy gap region (chapter 1).

Configurational changes in the intramolecular nuclear structure are expected to be manifested in the following ways:

(1) The experimental activation energy will be higher than that obtained from eq.(3.63) with E_r calculated from Marcus' relation (eqs.(1.27) and ((1.28)) including corrections for frequency dispersion effects and for reduction of polarization in the space occupied by the reactants. Marcus has provided an extension of the classical treatment to account for this effect (49c), and semiclassical or quantum treatments of the configurational changes of the intramolecular modes have also been introduced by several people for both inner and outer sphere electron transfer and ligand substitution processes (93,94,98,99).

(2) The activation energy should exhibit a marked temperature dependence. Thus, at sufficiently low temperatures all high-frequency modes are 'frozen' and do not contribute to the activation energy, which is then solely determined by the medium modes. In contrast, at sufficiently high temperatures both sets of modes behave classically and determine the activation energy. Finally, at intermediate temperatures the high-frequency modes are reorganized by a nuclear tunnel effect from a somewhat excited level and thus contribute to the activation energy to an extent which depends on the temperature. This effect should be manifested at room temperature if modes of frequencies around $k_B T/\hbar$ are reorganized during the process.

(3) The free energy relationship (energy gap law) must now be reformulated to incorporate the specific role of both low- and high-frequency modes. The deviations from the Gaussian dependence is thus expected to be more pronounced the more endothermic or exothermic the process, since reaction from or to excited intramolecular vibrational states, respectively, dominate the process in these free energy ranges (90,101).

We shall now proceed to a derivation of the rate expressions for simple electron transfer processes with coupling to both kinds of modes. Such expressions were firstly derived by Dogonadze and Kuznetsov (29,89), but lateron obtained in the same context by several others. We shall follow the commonly adopted approach that the intramolecular and medium modes constitute two independent sets of normal modes in the total reacting systems. From chemical intuition this seems a valid approximation for many processes in which the reactants are not strongly bonded (e.g. hydrogen-bonded) to the nearest solvent molecules, and in view of the fact that the dominating solvent modes have much lower frequencies than the intramolecular modes, thus ensuring a small intersystem coupling (95). However, in other cases this coupling may be strong, e.g. for reactions involving metal aquo ions for which the first coordination sphere is strongly bonded to the metal ions and also expected to be hydrogen-bonded to solvent molecules in higher coordination spheres.

Although the assumption of independent medium and intramolecular modes give results which can reproduce all important features of elementary reactions in liquid solution, it can be relaxed if warranted by experimental data or by the chemical nature of the reactants. The theory is then usually reformulated in either of the following two ways:

(a) The coupling between intramolecular and medium modes may give rise to the appearance of terms of the form $a_{qQ}\, q_\kappa Q_c$ in addition to terms of the forms $a_{qq}\, q_\kappa^2$ and $a_{QQ}\, Q_c^2$ in the expressions for the potential energy surfaces, where the a's refer to coupling constants. However, this only means that a new set of normal modes being a linear combination of the separate intramolecular and medium modes, can be defined. All subsequent formulae can then be reexpressed in terms of these coordinates, but are otherwise formally identical to those corresponding to independent subsets.

(b) A double adiabatic approximation has been invoked firstly in the description of both thermal electron transfer and proton transfer(29,33,96,97), and lateron for the theory of relaxation of dipolar molecules in solids(192). This procedure rests on the view that the intramolecular nuclear motion is so fast compared with the solvent molecular motion that the latter is practically fixed in a given position during the transitions of the former. In the dipolar relaxation the role of the 'interstitial' and the phonon modes is believed to be inverted the latter representing the fast subsystem(192). The total fast subsystem, i.e. the electrons and the fast nuclei, can then be separated from the slow nuclei by a procedure analogous to the Born-Oppenheimer approximation outlined in chapter 3. Subsequently a similar separation of the electrons and the fast nuclei is performed. We shall return to a discussion of this approach in chapter 6.

4.2 Quantum Modes in Electron Transfer Reactions

The potential energy surfaces of the initial and final states, $U_i(q_\varkappa, Q_{ci})$ and $U_f(q_\varkappa, Q_{cf})$ can now be separated into additive contributions, from the intramolecular donor, Q^D_{ci}, Q^D_{cf} and acceptor quantum modes Q^A_{ci}, Q^A_{cf} and from the solvent, viz.

$$U_i(q_\varkappa, Q_c) = f^D_i(Q^D_{ci}) + f^A_i(Q_{ci}) + \frac{1}{2}\sum_\varkappa \hbar\omega_\varkappa q_\varkappa^2 \tag{4.1}$$

$$U_f(q_\varkappa, Q_c) = f^D_f(Q^D_{cf}) + f^A_f(Q_{cf}) + \frac{1}{2}\sum_\varkappa \hbar\omega_\varkappa (q_\varkappa - q_{\varkappa o})^2 + \Delta E \tag{4.2}$$

The corresponding nuclear wave functions take the form

$$\chi_{iv}(q_\varkappa; Q^D_{ci}, Q^A_{ci}) = \chi^D_i(Q^D_{ci}, \varepsilon^D_i)\chi^A_i(Q^A_{ci}, \varepsilon^A_i)\chi^S_i(q_\varkappa, \varepsilon^S_i) \tag{4.3}$$

$$\chi_{fw}(q_\varkappa; Q^D_{cf}, Q^A_{cf}) = \chi^D_f(Q_{cf}, \varepsilon^D_f)\chi^A_f(Q^A_{cf}, \varepsilon^A_f)\chi^S_f(q_\varkappa, \varepsilon^S_f) \tag{4.4}$$

where $\chi^D_{i,f}$, $\chi^A_{i,f}$, and $\chi^s_{i,f}$ represent the nuclear wave function of the donor, the acceptor, and the solvent, respectively. The vibrational energy levels of the donor, acceptor, and solvent are labeled by $\varepsilon^D_{i,f}$, $\varepsilon^A_{i,f}$, and $\varepsilon^s_{i,f}$, respectively. Following the procedure outlined in chapter 3, i.e. separating the modes and invoking the Born-Oppenheimer and Condon approximations, the nuclear contribution to the probability of electron transfer per unit time can now be written as (97,101)

$$A = (Z_D Z_A Z_S)^{-1} \sum_{\varepsilon^D_i} \sum_{\varepsilon^A_i} \sum_{\varepsilon^S_i} \sum_{\varepsilon^D_f} \sum_{\varepsilon^A_f} \sum_{\varepsilon^S_f} \exp[-\beta(\varepsilon^D_i + \varepsilon^A_i + \varepsilon^S_i)] \tag{4.5}$$

$$S_D(\varepsilon^D_i, \varepsilon^D_f) S_A(\varepsilon^A_i, \varepsilon^A_f) S_s(\varepsilon^S_i, \varepsilon^S_f)\ \delta(\varepsilon^D_i - \varepsilon^D_f + \varepsilon^A_i - \varepsilon^A_f + \varepsilon^S_i - \varepsilon^S_f - \Delta E)$$

where Z_D, Z_A, and Z_s are the partition functions, i.e.

$$Z_I = \sum_{\varepsilon_i^I} \exp(-\beta\varepsilon_i^I) \;:\; I = D, A, s \tag{4.6}$$

for the nuclear energy levels. The Franck Condon vibrational overlap integrals are

$$S_I(\varepsilon_i^I, \varepsilon_f^I) = |\langle\chi_f^I|\chi_i^I\rangle|^2 \tag{4.7}$$

Eq.(4.5) can be rewritten in the form (96,101)

$$A = \int_0^\infty dxF_Q(x)F_S(\Delta E - x) \tag{4.8}$$

where we have defined the auxiliary functions

$$F_Q(X) = (Z_D Z_A)^{-1}\sum_{\varepsilon_i^D}\sum_{\varepsilon_i^A}\sum_{\varepsilon_f^D}\sum_{\varepsilon_f^A} \exp[-\beta(\varepsilon_i^D + \varepsilon_i^A)] \tag{4.9}$$

$$S_D(\varepsilon_i^D, \varepsilon_f^D)S_A(\varepsilon_i^A, \varepsilon_f^A)\delta(\varepsilon_i^D - \varepsilon_f^D + \varepsilon_i^A - \varepsilon_f^A + x)$$

and

$$F(\Delta E \quad x) = Z_s^{-1}\sum_{\varepsilon_i^s}\sum_{\varepsilon_f^s} \exp(-\beta\varepsilon_i^s)S_s(\varepsilon_i^s, \varepsilon_f^s)\delta(\varepsilon_i^s - \varepsilon_f^s - \Delta E + x) \tag{4.10}$$

These two equations define the energy dependent transition probabilities for the quantum and classical modes separately.

In eq.(4.10) we recognize the transition probability expression for the low-frequency solvent modes only. By invoking the classical high-temperature limit for these modes (cf. eq.(3.63)) this contribution can therefore be written as

$$F_s(\Delta E - x) = (\pi/\hbar^2 k_B T E_r)^{\frac{1}{2}} \exp[-\beta(E_r + \Delta E + x)^2/4E_s] \quad (4.11)$$

and together with eq.(4.10) this provides the final form of the rate probability. i.e.

$$A = (\pi/\hbar^2 k_B T E_r)^{\frac{1}{2}} (Z_D Z_A)^{-1} \sum_{\varepsilon_i^D} \sum_{\varepsilon_i^A} \sum_{\varepsilon_f^D} \sum_{\varepsilon_f^A} \exp[-\beta(\varepsilon_i^D + \varepsilon_i^A)]$$

$$S_D(\varepsilon_i^D, \varepsilon_f^D) S_A(\varepsilon_i^A, \varepsilon_f^A) \exp[-\beta(E_r + \Delta E + \varepsilon_f^D - \varepsilon_i^D + \varepsilon_f^A - \varepsilon_i^A)^2/4E_r] \quad (4.12)$$

We see that x coincides with the contribution of the quantum modes to the energy gap of the reactions.

Eq.(4.12) together with eq.(3.63) now represents a rather general theory of nonadiabatic electron transfer processes. The classical medium modes are treated within the harmonic approximation, whereas the intramolecular modes are deconvoluted but otherwise of a general form. Therefore, both equilibrium coordinate and frequency changes, anharmonicity, mode mixing etc. in these modes can be accounted for by introducing the appropriate Franck Condon overlap factors.

Eq.(4.12) also offers the following convenient interpretation (cf. chapter 5): The vibrational energy levels of the deconvoluted high-frequency modes in the initial and final states constitute a set of discrete energy values, and for each particular set of v_j, w_j an initial and a final state potential energy surface with respect to the classical modes only is defined. The system is distributed on the manifold of initial state surfaces according to the Gibbs function, and it reacts to a particular final state surface with a probability which is determined by the activation energy and a Franck Condon overlap factor corresponding to the particular couple of high-frequency levels.

We shall now illustrate some of these effects by actually providing the overlap factors and showing the results of numerical calculations of the relation between the transition probability and the energy gap for the process. We shall thus consider explicitly the following cases:

4.2.1 Displaced Potential Surfaces.

The simplest model involves a system characterized by displaced harmonic potential surfaces for the quantum modes, where the frequencies for each mode are identical in the initial and final states. The minimum of the final state potential surface is then displaced by an amount ΔQ°_{cj} for the j'th mode with respect to the initial state surface, corresponding to a dimensionless displacement $\Delta_{cj} = (\mu_j \omega_{cj}/\hbar)^{\frac{1}{2}} \Delta Q^{\circ}_{cj}$, where μ_j is the reduced mass associated with the mode. The vibrational overlap functions are then products of overlap functions corresponding to each mode. Using the following relationship between Hermite ($H_n(x)$) and Laguerre polynomia ($L^{\alpha}_n(x)$) (155)

$$\int_{-\infty}^{\infty} e^{-x^2} H_m(x+y) H_n(x+z)\,dx = 2^n \pi^{\frac{1}{2}} m!\, z^{n-m} L_m^{n-m}(-2yz) \qquad (4.13)$$

and putting $y = -\Delta_{cj}/2$ and $z = \Delta_{cj}/2$, the Franck Condon factor can be written in the form

$$S_j^{D,A}(v_j, w_j) = e^{-\Delta_{cj}^2/2} \frac{v_j!}{w_j!} (\Delta_{cj}^2/2)^{|w_j - v_j|} [L_{w_j}^{|w_j - v_j|}(\Delta_{cj}^2/2)]^2$$

(4.14)

where v_j and w_j are the vibrational quantum numbers of the mode j in the initial and final state, respectively. In the low-temperature limit and for exothermic processes, $v_j = 0$, and eq.(4.14) reduces to the simpler form also derived in chapter 3, i.e.

$$S_j^{D,A}(o,w_j) = e^{-\Delta_{cj}^2/2}(\Delta_{cj}^2/2)^{w_j}/w_j! \tag{4.15}$$

(cf. eq.(3.81)). If we combine the procedure outlined in section 3.3 with eqs.(4.8)-(4.12) it is not difficult to derive the following result which is valid at all temperatures

$$A = (\pi/\hbar^2 k_B T E_r)^{\frac{1}{2}} \exp[-\frac{1}{2}\sum_{j=1}^{N}\Delta_{cj}^2 \tanh(\beta\hbar\omega_{cj})/2] \tag{4.16}$$

$$\sum_{l_1=o}\sum_{k_1=o}\cdots\cdots\sum_{l_N=o}\sum_{k_N=o} \exp\left\{-\beta[E_r+\Delta E+\sum_{j=1}(l_j-k_j)\hbar\omega_{cj}]^2/4E_r\right\}$$

$$\prod_{j=1}^{N}\left\{\frac{1}{l_j!k_j!}[\Delta_{cj}^2\exp(-\beta\hbar\omega_{cj}/2)/4\sinh(\beta\hbar\omega_{cj}/2)]^{l_j+k_j}(l_j\beta\hbar\omega_{cj}/2)\right\}$$

This equation has the following form in the low-temperature limit, where $\beta\hbar\omega_{cj} >> 1$ (cf. eqs.(4.12) and (4.15))

$$A = (\pi/\hbar^2 k_B T E_r)^{\frac{1}{2}} \exp[-\frac{1}{2}\sum_{j=1}^{N}\Delta_{cj}^2]\ \exp[-\beta\sum_{j=1}^{N}(w_j+|w_j|)\hbar\omega_{cj}/2)]$$

$$\tag{4.17}$$

$$\sum_{w_1=o}\cdots\cdots\sum_{w_N=o}\prod_{j=1}^{N}\frac{(\Delta_{cj}^2/2)^{|w_j|}}{|w_j|!}\exp[-\beta(E_r+\Delta E+\sum_{j=1}^{N}w_j h\omega_{cj}^2)\ /4E_r]$$

where N is the total number of intramolecular modes.

We now show some numerical calculations of the relationship between A and ΔE, as given by eqs(4.16) and its low-temperature limit, eq.(4.17). The frequencies of the high-frequency modes were selected in the range 200-2000 cm^{-1} covering a representative variety of metal-ligand and organic skeletal modes, and the reduced displacements were chosen in the region 1-10 corresponding to representative values estimated from structural data. Furthermore, the value E_r = 1 ev, the energy gap region $-\Delta E$ = 0-4 ev, and the temperature 300 °K were chosen. At larger negative energy gaps transitions to excited electronic states may interfere. Fig.(4.1) portrays the numerical results for a system possessing a single displaced harmonic quantum mode. The numerical calculations revealed that even for fairly high values of $\hbar\omega_c$ several vibrational levels yield comparable contributions to A in the strongly exothermic region (up to w $\approx$ 10 for $\hbar\omega_c \approx 2000$ cm^{-1} and $\Delta_c \approx 10$). Thus, the effect of the vibrational excitation of quantum modes in the electron transfer process is appreciable. The numerical data also reveal the following quantum effects(101):

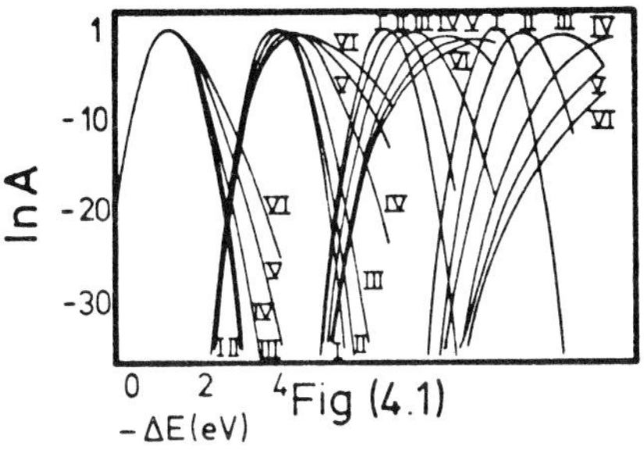

Fig (4.1)

Free energy plots for displacement of one high-frequency harmonic mode (eq. (4.16)). E_r = 1 ev, T = 298 °K. The four families of curves refer to (from left to right) Δ_c = 1, 2.5, 5, and 7.5, and the origins of the latter three are shifted by 3 ev, 6 ev, and 9 ev, respectively along the E axis. I: no quantum modes. II: ω_c = 206 cm^{-1}. III: ω_c = 514 cm^{-1}. ω_c = 1029 cm^{-1}. V: ω_c = 1543 cm^{-1}. VI: ω_c = 2058 cm^{-1}.

(1) In general, a maximum in the free energy relationship is exhibited. For a purely classical system this is located at $-\Delta E = E_r$, whereas quantum effects result in a shift of this maximum by roughly $\frac{1}{2}\hbar\omega_c\Delta_c^2$ towards higher $|\Delta E|$ values.

(2) The free energy relationships are asymmetric about the value of $\Delta E = \Delta E$, corresponding to the maximum value, A_{max}, of A. Incorporation of a quantum mode results in a slower decrease of lnA with increasing $-\Delta E$ for $|\Delta E| > |\Delta E_{max}|$, i.e. beyond the maximum.

(3) For high frequencies, i.e. $\hbar\omega_c \approx 2000$ cm^{-1}, and for intermediate values of $\Delta_c \approx$ 2-5 the free energy relationship at room temperature exhibits a broad flat maximum. For example for Δ = 2.5 and $\hbar\omega_c$ = 2000 cm^{-1} lnA varies weakly in the range $-\Delta E$ = 1.2-2.2 ev. An activationless region is not predicted for the simple model system which involves a single quantum mode. However, the dependence of A on ΔE is considerably weaker than for a purely classical system. This effect will be most pronounced when several high-frequency modes prevail such as in the case of electron transfer involving aromatic hydrocarbons and their radical ions, or if the high-frequency part of the medium is included(104).

(4) Interesting isotope effects are expected to be revealed when the role of quantum modes is important. For $|\Delta E| > |\Delta E_{max}|$ the electron transfer rate constant will increase with decreasing frequency of the quantum modes (i.e. an inverse isotope effect), whereas for $|\Delta E| < |\Delta E_{max}|$ a normal isotope effect will be exhi-

bited, i.e. the electron transfer rate constant will decrease with decreasing .

(5) If the temperature is changed from 300° K to 30° K (disregarding the effects of frequency dispersion) an oscillatory dependence of lnA on ΔE is revealed for high values of $\hbar\omega_c$, i.e. when $\hbar\omega_c > 2(E_r k_B T)^{\frac{1}{2}}$ (see eq.(4.17). This effect is smeared out when several high-frequency modes are present. At higher temperatures these oscillations which are analogous to vibrational structure of optical electronic transitions, are smeared out for the high values of E_r chosen, but such resonances would still be expected at room temperature for sufficiently weak medium coupling.

(6) Increasing the temperature for constant Δ_c and $\hbar\omega_c$ results in broadening of the ball-shaped curve which represents the free energy relationship. The broadening is roughly proportional to $T^{\frac{1}{2}}$, as in the situation for the purely classical case.

When two high-frequency modes are present - for example corresponding to the totally symmetric breathing frequencies of the donor and acceptor in electron transfer reactions involving octahedral metal complexes - the resulting free energy relationships are qualitatively similar to those obtained for a single displaced mode. However, the asymmetry is more pronounced with increasing number of quantum modes and with increasing $\hbar\omega_{cj}$ (j = 1, 2). This is borne out by more recent calculations on systems having many (10-20) displaced modes (193). As expected, the oscillatory behaviour revealed at low temperatures or weak medium coupling for a single quantum mode is largely eroded, when two or more displaced modes with different frequencies are involved.

4.2.2 Effects of Frequency Changes

In addition to equilibrium coordinate shifts, intramolecular modes are usually accompanied by changes of the vibration frequencies which are furthermore sometimes known experimentally. Expressions for the Franck Condon factors for a nuclear harmonic mode which undergoes both equilibrium coordinate and frequency shift are available (2). However, for the sake of simplicity we shall quote the simpler result which is obtained when the system is initially in its ground vibrational level. In this case (2,194)

$$S(0,w) = (1-\xi^2)^{\frac{1}{2}}(-\xi)^w \exp\left[-\Delta_c^2 \frac{\omega_{cf}}{\omega_{ci}+\omega_{cf}}\right]|H_w(ix)|^2/w! \qquad (4.18)$$

where

$$\xi = |\omega_{ci} - \omega_{cf}|/(\omega_{ci} + \omega_{cf}) \qquad (4.19)$$

$$\Delta_c = (\hbar/\mu_i\omega_{ci})^{\frac{1}{2}}\Delta Q_c^0 \qquad (4.20)$$

and

$$x = [\Delta_c^2(1-\xi)/4\xi]^{\frac{1}{2}} \qquad (4.21)$$

ω_{ci} and ω_{cf} are the vibration frequencies of the initial and final state, respectively, and the Hermite polynomial

$$H_w(ix) = (-1)^w w! \sum_{r=0}^{w/2} (-1)^r (2ix)^{w-2r}/(w-2r)!r! \qquad (4.22)$$

when $\xi \to 0$, eq(4.18) takes the form for pure displacement derived previously (eq.(4.15)). For pure distortion, i.e. $\Delta_c = o$ but $\xi \neq 0$, eq.(4.18) takes the following form

$$S(o,w) = (1 - \xi^2)^{1/2} \xi^w \frac{1.3.5.....(w-1)}{2 \cdot 4 \cdot 6 \cdots\cdots w} \quad \text{for w even} \qquad (4.23)$$

$$S(o,w) = o \quad \text{for w odd} \qquad (4.24)$$

This would correspond for example to skeletal deformation modes in electronic transitions of large organic molecules.

Free energy plots for a pure distortion display basically the same qualitative features as for pure displacement, in particular the asymmetry towards large $|\Delta E|$, and the larger ξ, the more pronounced are the effects (101).

4.2.3 Effects of Anharmonicity.

While the harmonic approximation is adequate for the medium modes, anharmonicity in the intramolecular modes is expected to prevail in systems which undergo substantial restructuration during the process. Both the absolute values of the transition probability and the form of the energy gap law may in fact be drastically affected by anharmonicity effects (195,196) as also revealed by experimental studies on both the nonradiative transition probability of transition metal ion impurities in crystals (121), and in atom group transfer processes (197).

We can illustrate this by giving the Franck Condon nuclear overlap factors corresponding to Morse potentials in the initial and final states, i.e. by choosing the following form of $f_i^{D,A}(Q_{ci}^{D,A})$ and $f_f^{D,A}(Q_{cf}^{D,A})$ in eq.(4.1)

$$f_i(Q_c) = D[1 - \exp(-a\,Q_c)]^2 \tag{4.25}$$

$$f_f(Q_c) = D\left\{1 - \exp[-a(Q_c - \Delta_c)]\right\}^2 \tag{4.26}$$

D is here the dissociation energy, and a the anharmonicity constant ($a = (\hbar\omega_c/2D)$). The total energy is

$$\mathcal{E}_w = \hbar\omega_c(w + \tfrac{1}{2}) - \tfrac{1}{2}\hbar\omega_c\, a^2(w + \tfrac{1}{2})^2 \tag{4.27}$$

The general form of the nuclear overlap integral between the initial ground state level and the w'th final state level is (101,121,198)

$$S(o,w) = [\cosh(\frac{a\Delta_c}{2})]^{-2p}(1 - 2w/p)\binom{p}{w} \tag{4.28}$$

$$\left\{\sum_{\sigma=o}^{w}(-1)^{\sigma}[-(1 + e^{-a\Delta_c})]^{\sigma}\binom{p-w}{r}\binom{p-1}{\sigma}^{-1}\binom{w}{\sigma}\right\}^2$$

where $p = 2a^{-2} - 1$. In particular, the overlap integral for the lowest levels take the following compact form

$$S(o,o) = [\cosh(\frac{a\Delta_c}{2})]^{2p} \tag{4.29}$$

$$S(o,1) = [\cosh(\frac{a\Delta_c}{2})]^{-2p}[-(p - 2)(1 - e^{-a\Delta_c})]^2 \tag{4.30}$$

Fig.4.2 shows several plots of lnA against ΔE for a system which contains a single displaced anharmonic high-frequency mode in addition to a continuous manifold of low-frequency harmonic modes, and for the sake of comparison the corresponding plots for harmonic potentials are also shown. The frequency range is 1000-2000 cm^{-1}, $a = 0.1$-0.2, $\Delta_c = 1$-10, and E_r and T are again 1 ev and 300° K, respectively. We should notice:

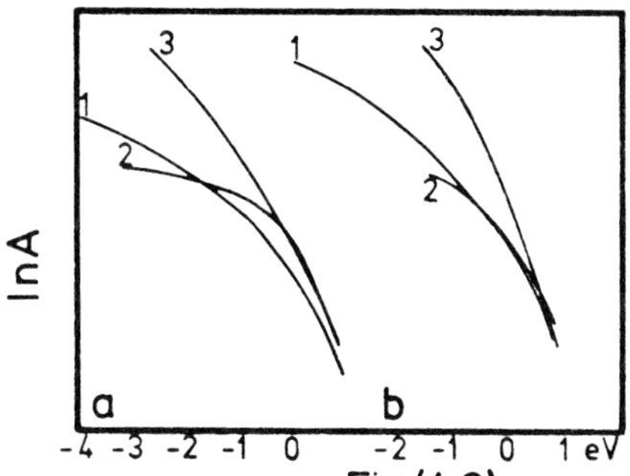

Fig (4.2)

Free energy plots for displacement in one anharmonic high-frequency mode. E_r = 1 ev, T = 298 ° K, $|\Delta_c|$ = 10. a: = 2058 cm^{-1}. b: ω_c : 1029 cm^{-1}. 1: harmonic mode; 2: Δ_c = 10, a = 0.2; 3: Δ_c = -10, a = 0.2.

(1) For weakly endothermic and exothermic processes and for high values of Δ_c and ω_c the transition probability is increased compared to the harmonic value for both positive and negative Δ_c. This effect increases with increasing ω_c and is due to the fact that the nuclear Franck Condon factor for transitions between the ground vibrational levels is always higher than for the corresponding harmonic potentials (equivalent to a lower barrier for nuclear tunnelling).

(2) When higher vibrational levels contribute to the rate, the latter is determined not only by the absolute value of Δ_c but also by its sign, and the deviations from harmonicity are reflected largely in the Franck Condon factors. This follows from eqs.(4.28)-(4.30) which show that for exothermic reactions a negative coordinate shift in a single quantum mode (corresponding to a bond stretching in the final state relative to the initial state) leads to an increased Franck Condon factor com-

pared with a harmonic potential, whereas a positive shift (corresponding to a bond compression) leads to the opposite effect. It is also obvious that this effect is larger, the higher the vibrational quantum number of the final state, and therefore the higher the exothermicity of the reaction. Fig.(4.3) and previous investigations (121,197) show that this effect may amount to many orders of magnitude for strongly exothermic processes.

(3) The free energy relationship is considerably flattened for strongly exothermic processes as compared with harmonic potentials and for $\Delta_c < 0$, whereas a more pronounced curvature is found for $\Delta_c > 0$. This reflects the larger and smaller values of the high-frequency Franck Condon factors, respectively, and the consequently larger (or smaller) contributions of several excited state levels.

Fig.(4.3) shows that this picture is changed when anharmonicity in the displacements of two modes are considered. This would be appropriate for homogeneous electron transfer reactions where commonly the symmetric stretching mode of both the donor and acceptor species are subject to equilibrium coordinate shifts. Of these two modes one would moreover be stretched (the acceptor) and the other one (the donor) compressed corresponding to Δ_c-values of opposite sign. Since we just noticed that positive and negative displacements have opposite anharmonicity effects, incorporation of both is expected to give a substantial cancellation of the anharmonicity effects.

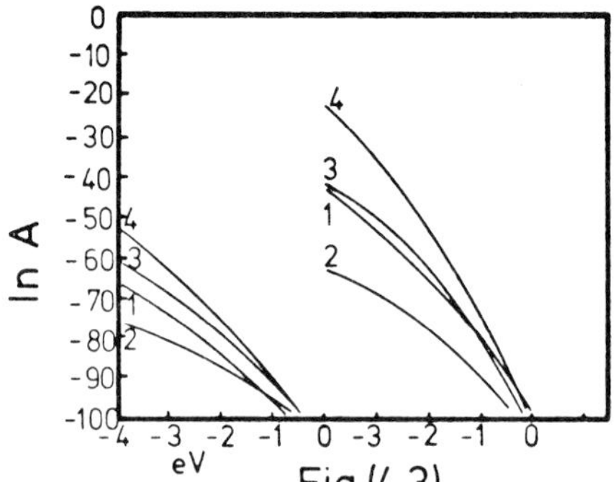

Fig (4.3)

Free energy plots for displacement in two anharmonic high-frequency modes. E_r = 1 ev, T = 298° K, $|\Delta_{c1}| = |\Delta_{c2}| = 10$. a: $\omega_{c1} = \omega_{c2} = 2058\ cm^{-1}$. b: $\omega_{c1} = \omega_{c2} = 1029\ cm^{-1}$. 1: harmonic modes. 2: $\Delta_{c1} = \Delta_{c2} > 0$, $a_1 = a_2 = 0.1$; 3: $\Delta_{c1} = -\Delta_{c2} < 0$, $a_1 = a_2 = 0.1$; 4: $\Delta_{c1} = \Delta_{c2} < 0$, $a_1 = a_2 = 0.1$.

This is borne out by the numerical results shown in fig.(4.3). When the two Δ_c-values have the same sign, the anharmonicity effects are reinforced in the appropriate direction, whereas a pronounced cancellation is observed when they have opposite signs. This represents an important difference from heterogeneous (electrochemical) processes of the same redox reactants. The latter processes involve only a single molecular reactant, and in contrast to homogeneous processes the anharmonicity effects are therefore expected to be fully manifested (for example by flat 'free energy plots (Tafel plots)).

4.3 Relation to Experimental Data

We have previously stated that studies of electron and atom group transfer systems involving reorganization of intramolecular modes have provided some rather unambiguous information as to the fundamental results of the general theory of multiphonon condensed phase chemical processes. This information refers to the quantum behaviour of intermediate or high-frequency discrete modes in different free energy and temperature regions. We have also seen that the quantum behaviour, i.e. the predicted behaviour which is not compatible with a purely classical formalism, is most clearly manifested at low temperatures and for strongly exothermic processes where the quantum modes are frozen, and increasingly vibrationally excited, respectively, giving rise to 'non-classical' effects. However, these ranges are difficult to investigate experimentally, and experimental data which fairly convincingly illustrate the predicted effects are few and have only recently begun to appear. An exception to this is the studies of kinetic isotope effects in proton transfer reactions for which quantum (tunnel) effects have been known for over 20 years (19). However, we shall treat these systems separately in chapter 6 in the context of atom group transfer processes in general.

In this section we shall mainly discuss free energy relations and isotope effects in the strongly exothermic ('abnormal') region. This is because the few studies of the temperature dependence of chemical processes at low temperatures which have been reported refer either to atom group transfer or to biological processes, and they are therefore more appropriately dealt with in the chapters on these processes.

Examples of the inverted isotope effects for strongly exothermic processes were recently reported (199). The processes were reactions between solvated electrons and various scavengers (aromatic and aliphatic hydrocarbons and their halide derivatives, acetone, dimethylsulphoxide etc.) in ethanol and O-deuterated ethanol. The rate constants are all much lower than corresponding to diffusion control, and they display values of the inverse isotope effect (k_{OD}/k_{OH}, where k_{OD} and k_{OH} are the rate constants of the process in the D- and H-substituted ethanol, respectively) in the region 1.2-3.2. These effects are understandable if (1) the electron is strongly coupled to the O-H and O-D solvent mode and (2) if the processes are so strongly exothermic that they correspond to the descending branch of the energy gap plots of figs.(4.1)-(4.3). The validity of the first condition is supported by the observed blue-shift of the optical transitions from the electronic ground state to excited bound states of trapped electrons in crystalline ice when going from hydrogen- to the deuteriumsubstituted ice (200). On the other hand, quantitative information about the electron affinities of the scavengers is not available, and the second condition can therefore not be checked, nor can proper energy gap laws be investigated.

We have seen that in contrast to the 'normal' free energy range, for which $|\Delta E| < E_r$, energy gap relationships are expected to reveal several interesting effects in the strongly exothermic region ($|\Delta E| > E_r$). In particular, the maximum and the decaying branch of the plot is a manifestation of nuclear quantum effects and sensitive to the nature and reorganization of the particular quantum modes.

The first studies of the 'abnormal' energy gap region were reported by Rehm and Weller (201). They measured the rate of fluorescence quenching of the excited states of several organic aromatic donor molecules in acetonitrile solution and in the presence of a variety of different acceptor molecules. This, and the estimate of energy gaps from spectroscopic and electrochemical data allowed to establish an energy gap relationship in the whole interval 10 kcal $> \Delta E >$ -60 kcal. While the Marcus relationship was found to be obeyed for $\Delta E >$ -15 kcal, the energy gap plot showed an activationless (ΔE-independent) region at lower values, corresponding to a rate constant of about 10^{10} M^{-1} s^{-1}. A qualitatively similar behaviour was observed for the electron transfer reactions between triplet duroquinone and several electron donors in water and ethanol solution (to form the duroquinone anion) (202), for the (both exothermic and endothermic) reactions of the superoxide ion with various quinones and quinone radicals, where Bronsted coefficients of approximately zero and unity, respectively, were reported (203), and in the reactions of excited triplet methylene blue with several ammines (204).

However, there are several indications that the energy gap plots may show an 'abnormal' behaviour for sufficiently exothermic processes, i.e. the rate constant decreases with increasingly negative ΔE. Thus, Schomburg, Staerk and Weller observed a decrease in the rate constant of the reactions between excited singlet pyrene and several organic donor and acceptor molecules, by a factor of about four, when the energy gap was changed from -0.77 ev to -2.41 ev. Such an effect was also observed by Van Duyne and Fischer (100,206) in the chemiluminescent diphenylanthracene anion-cation radical annihilation reaction in acetonitrile solution. This reaction may proceed to both excited singlet, triplet, and the ground state neutral molecules the energies of which decrease in the order given and which may in principle provide the basis of an energy gap law. Moreover, the photophysical parameters of this system are well known, and E_r could be estimated reasonably unambiguously from electron exchange reactions between organic molecules of similar size and their corresponding anions or cations. Using this value, the quantum yield could be calculated as a combination of the rate constants of the individual processes. However, in order to obtain agreement with the experimental quantum yield it was necessary to assume a slower decrease of the rate constant with increasing $-\Delta E$ in the abnormal region than predicted by the theory of Marcus, in line with the expectations of the general quantum theory. Finally, a very small decrease of the rate constant with decreasing ΔE for $-\Delta E > 1.6$ ev, following a maximum of about $3 \cdot 10^9$ M^{-1} s^{-1}, was obtained for the electron transfer reaction between the lowest excited state of $[Ru(bipy)_3]^{2+}$ (bipy is 2,2'-bipyridyl and its 4,4'-dimethyl derivative) and several polypyridine complexes of Os(III), Ru(III), and Cr(III) (207).

The processes considered so far are all very fast, and although a broad energy gap region of practically no dependence on this parameter is in some cases compatible with the theory, diffusion effects are likely to control the overall process in some cases. On the other hand, if the reactions were strongly electronically nonadiabatic, i.e. having a very small electronic coupling matrix element, they would proceed without diffusion control, even when the activation energy vanishes. This goal might be achieved if the electron transfer distance could somehow be 'artificially' increased thus giving a smaller electronic overlap between the donor and acceptor orbitals. Two techniques seem prospective in this respect. Thus, one of the reacting species may be trapped in a micelle, and the electron transferred between this species and an external reagent across the micelle double layer. Or, electron transfer between donor and acceptor species at fixed positions in solid solutions may be studied. The electron transfer distance may then be increased by diluting the solid solution. We shall in turn consider the results of some experimental studies belonging to each of these categories.

Henglein and his associates reported a free energy plot for the reactions between several organic molecules trapped in both cationic and anionic micelles and hydrated electrons in the external aqueous medium (208b). The free energies of reaction were calculated from gas phase electron affinities, solvation energies estimated from the Born formula using radii from the molar volume and an effective micelle dielectric constant of 2.5. The plot of the rate constants for reactants trapped in the anionic micelles against the free energy of reaction showed a maximum at about -1.0 ev, and corresponding to a bimolecular rate constant of $2 \cdot 10^{10}\ M^{-1}\ s^{-1}$ (which is smaller by a factor of three than the value estimated for a diffusion controlled reaction). The rate constant moreover decreased by a factor of 20 on the negative side of the maximum. In another study (208a) the rate constants of the electron transfer processes between the micelle-trapped ground and lowest triplet state pyrene and the anion radicals of formaldehyde, acetaldehyde, acetone, and carbon dioxide in the outer aqueous phase were reported. The data are shown in a free energy plot in fig.(4.4), which exhibits a form similar to the one predicted if $E_r \approx 1$ ev and a single or a few high-frequency modes are moderately strongly coupled to the electrons ($\Delta_c \approx 2\text{-}3$). The excitation energy of pyrene and the redox potentials of the carbonyl anions are known from spectroscopic data and polarographic half-wave potentials, respectively. Such data are not directly available for the redox couples involving CO_2 and its anions. Henglein and his associates suggested the values -2.0 ev and -1.0 ev for the CO_2/CO_2^- and CO_2^-/CO_2^{2-} couples, respectively. This estimate was based on comparison of the 'electron transfer capability of CO_2^- with several electron donors and acceptors' and the fact that the half-wave potential of the polarographic two-electron reduction of CO_2 is about -2.0 ev, while the cathodic and anodic branches of the CO_2^--polarogram show a single wave with a change from the cathodic to the anodic limiting current within a narrow potential region around approximately -1.0 v. In comparison, analysis of thermodynamic cycles gives the value -1.8 ev for the CO_2/CO_2^- couples (209).

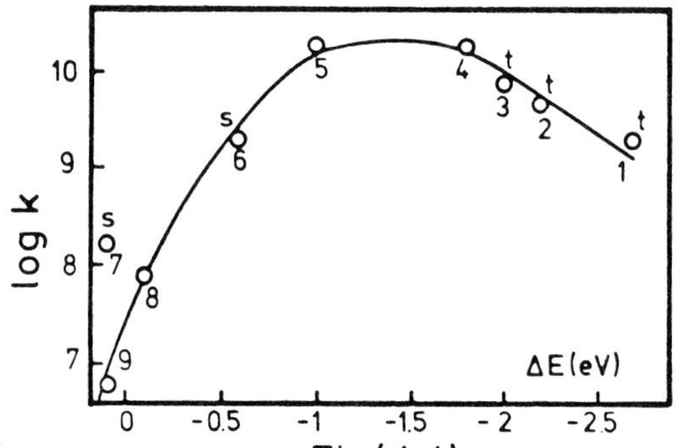

Fig(4.4)

Free energy plot for reactions between singlet (s) and triplet (t) micelle-trapped pyrene (P) and anionic radiacals. 1: $CH_3COCH_3^-$. 2: CO_2^-. 3: CH_3CHO^-. 4: CH_2O^-. 5: $P^-+CO_2^-$. 6: $CH_3COCH_3^-$. 7: CH_3CHO^-. 8: P^-+CH_3CHO. 9: P^-+CO_2.

Information about the role of high-frequency modes in strongly exothermic processes might conceivably be obtained from the study of reactions of electrons trapped in frozen aqueous and organic glasses at low temperatures (77° K) with different scavengers. After injection into these media electrons initially exist in a 'quasifree' ('dry') state the energy of which may be higher or lower than the vacuum state depending on the nature of the medium. For polar media the life time of this state is short and the electrons are subsequently trapped in localized states by motion of the medium molecules into new equilibrium positions under the influence of the polarization field of the electron. For water the relaxation time is a few picoseconds only (210), and the final state corresponds to an overall solvation energy of 1.7 ev (211). On the other hand, for polar media of rigid glasses involving the rotation of larger molecules the establishment of the nuclear equilibrium configuration around the trapped electron requires a much longer time (in the region 10^{-6} s-10^{-2} s) thus causing a continuous 'deepening' of the traps with time. This effect is reflected in a blue-shift of the optical absorption spectrum of the trapped electron with time in such glasses as ethanol and propanol (211-213).

Many molecular properties of the solvated electron in liquid phases are rationalized in terms of a semimolecular model according to which the trapped electron is viewed as a diffuse charge density located in a cavity constituted by a small number of solvent molecules (140). The interaction between the electronic charge density and this 'first solvation sphere' is described by electrostatic microscopic potentials, whereas the interaction with the more remote medium molecules is incorporated within a continuum model. One result of these calculations of particular importance in relation to electron transfer reactions is that the symmetric breathing motion of the first solvation sphere defines a strongly anharmonic potential energy surface the frequency of which is about 25-100 cm^{-1}. Under the temperature conditions (77-150 ° K) where most data for electron transfer of trapped electrons of interest in the present context have been obtained this mode thus represents a discrete classical anharmonic mode, in addition to the continuous manifold of outer medium modes.

The hydrated electron in aqueous solution is a strong reductant known to reduce many molecules and ions. Most of these reactions are almost diffusion controlled (211), and energy gap plots are not available. However, by trapping the electrons in solid matrices in such a way that the average distance to a potential scavenger molecule is sufficiently large (i.e. in sufficiently dilute solutions), the reaction may be expected to be slow even when the activation energy vanishes due to the small electronic overlap.

The evidence for long-range electron transfer from trapped electrons to the scavengers is based on several observations. Thus (212,213,214-216):

(1) The long life-times of the trapped electrons and the slow reactions (10^{-6} -10^{2} s) with scavengers which react with mobile hydrated electrons almost by diffusion control, are strongly indicative that trapping of the reactants does occur.

(2) The decay curves of the optical absorption of the trapped electrons in reactions with inorganic scavengers ($[Co(en)_3]^{3+}$, and BrO_3^-) in aqueous alkaline glasses are almost independent of temperature in the interval from 77 °K to 140°K (216). In contrast, at higher temperatures when the glass softens, the decay is faster. This is strongly suggestive of a tunnelling mechanism and a against 'trap hopping' by which the trap migrates by a diffusion mechanism.

(3) The phenomenology of the decay curves for reactions between species which are fixed at given relative positions is different from those between mobile species involving a collision complex. Thus, the electrons will firstly react with the nearest scavengers, subsequently with those located at more distant positions etc. For a random distribution of electrons and scavengers this will give rise to an exponential dependence of the optical absorption (concentration) of the trapped electrons on both time and scavenger concentration, in contrast to the behaviour of

second order processes involving mobile species. This is a typical feature for reactions of trapped electrons in aqueous and alcoholic glasses.

(4) The different electron acceptors investigated display a wide range (over a factor of 10^{10}) of reactivities towards the trapped electrons, even though the room temperature rate constants of the same scavengers towards the hydrated electron only vary by a factor of approximately 10^{2}. This certainly suggests some kind of 'direct' interaction between the two species, rather than trap-to-trap hopping which would be expected to be much less dependent on the nature of the scavenger.

These results are commonly interpreted in terms of (one-dimensional) long-range tunnelling (20-40 A) of the trapped electron to the nearest acceptor molecule (212,213,216). As the time passes, the electron has to tunnel to more and more remote acceptor molecules which thus gives a time dependent barrier width. With reference to the formalism outlined in chapters 3 and 4 this is equivalent to a representation of the electronic wave functions by quasiclassical wave functions (141) (chapter 6), and both this elaborate formalism and the interpretation based on electron tunnelling can thus account for the dependence of the decay rate on time and scavenger concentration.

Fig.(4.5) shows a number of experimental data referring to the reaction of trapped electrons with several classes of acceptors in a glass of 2-methyltetrahydrofuran (2-MTHF) at 77° K (212,213,217,218). The energy gap scale is basically a scale of electron affinity for the acceptors (due to uncertainties about solvation energies of the acceptor molecules). While the reactions of trapped electrons in polar media such as aqueous glasses would involve strong medium coupling, the situation is less clear for apolar media such as 2-MTHF. Thus, in addition to short-range interactions between the medium molecules and the localized electron, the cavity formation involves both surface and pressure-volume work of which the latter may cause a density distribution in the medium (chapter 2). These effects are commonly ignored and the data interpreted in terms of a single 'effective' Franck Condon factor representing intramolecular modes. In view of our lack of information about spectroscopic and structural data for the systems this may be an adequate procedure when the coupling to the medium is weak. On the other hand, for strong medium coupling the continuous medium spectrum provides energy gap laws with several features which are qualitatively different from systems with discrete modes only, even though an activationless region and a region of a decay rate exponentially decreasing with $|\Delta E|$ is still expected (104). With this in mind we then notice the following about fig.(4.5):

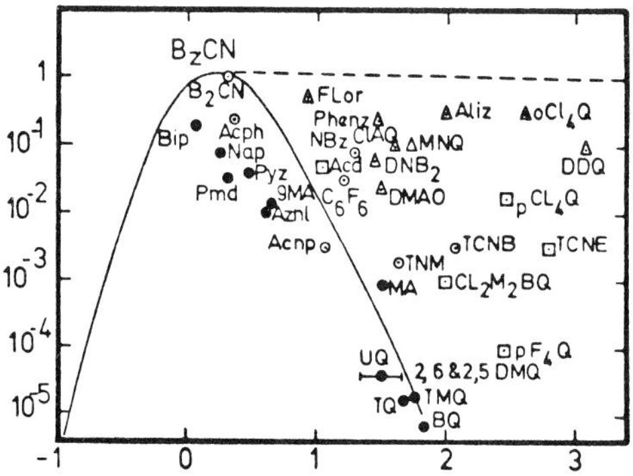

Fig (4.5)

Electron transfer rates (relative to benzonitrile (BzCN) for solvated electrons in glassy MTHF, T = 77 K. symbols ●, △, and ⊡ refer to different classes of compounds. Acd: Acridine. Acph:

Acenaphtylene. Acnp: Acetophenone. Aliz: 1,2-Dihydroxyanthraquinone. Azul: Azulene. Bip: Biphenyl. BQ: p-Benzoquinone. C_6H_6: Hexafluorobenzene. Cl_2M_2BQ: 2,6-Dichloro-3,5-dimethyl-p-Q. oCl_4Q: o-Chloranil. pCl_4Q: p-chloranil. ClAQ: 2-Chloroanthraquinone. DDQ: 2,3-Dichloro-5,6-dicyano-p-Q. DMAQ: 1,4-Dimethylanthraquinone. DMQ: 2,5- and 2,6-Dimethyl-p-Q. DNBz: p-Dinitrobenzene. Flor: Fluoranthene. pF_4Q: p-Fluoranil. MA: Maleic anhydride. 9MA: 9-Methylanthracene. MNQ: 2-Methyl-1,4-naphtoquinone. Nap: Naphtalene. NBz: Nitrobenzene. Phenz: Phenazine. Pmd: Pyrimidine. Pyz: Pyrazine. TCNB: 1,2,4,5,-Tetracyanobenzene. TCNE: Tetracyanoethylene. TMQ: Trimethyl-p-Q. TNM: Tetranitrometane. TQ: Methyl-p-Q. UQ: 2,3-Dimethoxy-5,6-dimethyl-p-Q.

(1) The points refer to several classes of compounds, and it is therefore not surprising that the overall picture displays a large amount of scatter.

(2) Some of the molecular anions are known (218) to possess low-lying (i.e. of lower energy than the exothermicity of the process) electronic states (the points indicated by the symbols '⚠' and '⊡'). According to the general theory electron transfer to these levels would be more facile than electron transfer to the ground state and cause a less rapid decrease of the rate with increasing $|\Delta E|$ in the exothermic region, in line with the experimental observations.

(3) The remaining points show a tendency that the relative rate decreases with increasing $|\Delta E|$ for sufficiently exothermic processes, in line with the predictions of the theory.

In addition to the effects of high-frequency modes discussed already, as manifested in the temperature dependence, the energy gap law in the 'abnormal' region, and the inverted isotope effect, we noticed still another effect of these modes, i.e. the 'vibrational structure' in the energy gap law. This is expected at low temperatures, small values of the reorganization energy of the low-frequency modes, and when only a single or a few high-frequency modes are displaced (101). No example of this effect is reported for chemical processes. However, a closely similar effect is observed in the current-voltage characteristics (which is analogous to an energy gap law) of the 'inelastic tunnelling' of electrons across metal-insulator-metal solid-state junctions in which the insulator is doped with impurities which can absorb or emit vibrational quanta. These systems will be further discussed in chapter 8.

5 SEMICLASSICAL APPROXIMATIONS

5.1 One-Dimensional Nuclear Motion

5.1.1 Classical Nuclear Motion

In chapters 3 and 4 we have been concerned with the calculation of the rate probability of chemical processes in terms of noninteracting ingoing and outgoing channel states. However, we have obtained closed rate expressions in the nonadiabatic limit only, i.e. we have assumed that the 'effective' coupling between the ingoing and outgoing states is sufficiently small that only first order terms in the expansion of the $\mathcal{T}$-operator need to be retained. This implies that V_{eff} is small, but in view of the composite nature of this operator the actual potential energy of interaction which appears in the exact Hamiltonian is not necessarily small.

At present no procedure for the summation of the complete expansion (eq.(3.13)) is available. If the 'effective' coupling between the initial and final states is not small (the adiabatic limit), it is therefore necessary to adopt alternative procedures. This analysis is most commonly performed within the semiclassical theoretical framework of Landau and Zener (142,219), the most important features of which we shall now briefly outline.

We consider at first two zero order potential energy surfaces, $\varepsilon_i(q)$ and $\varepsilon_f(q)$ characterized by a single nuclear coordinate q (fig.(5.1)) and the corresponding electronic states, ψ_i and ψ_f, which represent the isolated reactants and products.

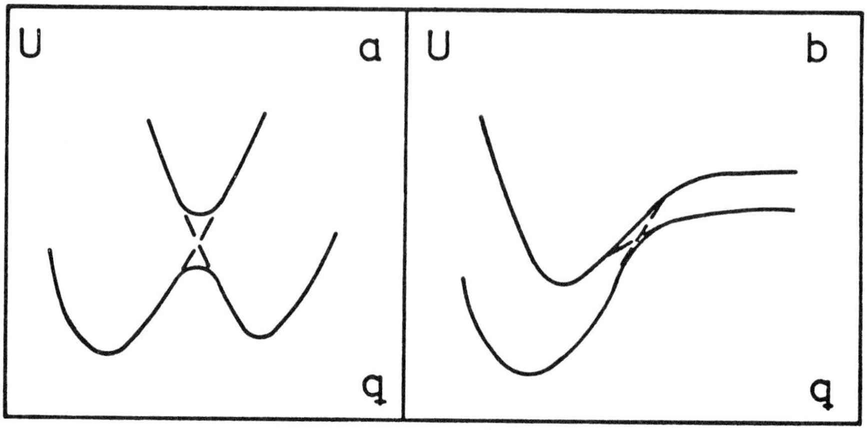

Fig (5.1)

In the crossing region, perturbation terms left out of the zero order Hamiltonians cause a 'mixing' of the zero order states, i.e. the actual surfaces are now no longer represented by $\varepsilon_i(q)$ and $\varepsilon_f(q)$, but by adiabatic surfaces $\varepsilon_-(q)$ and $\varepsilon_+(q)$, derived from a secular equation with respect to ψ_i and ψ_f (142). In the limit of small values of the overlap integral, $\langle\psi_f|\psi_i\rangle$, $\varepsilon_-(q)$ and $\varepsilon_+(q)$ take the following form

$$\varepsilon_\pm(q) = \frac{1}{2}[\varepsilon_i(q)+\varepsilon_f(q)+V_{ii}(q)+V_{ff}(q)] \mp \qquad (5.1)$$

$$+ \frac{1}{2}\left\{[\varepsilon_i(q)-\varepsilon_f(q)+V_{ii}(q)-V_{ff}(q)]^2+4[V_{fi}(q)]^2\right\}^{1/2}$$

where $V_{fi}(q) = \langle\psi_f|V_{i,f}|\psi_i\rangle$, and $V_{i,f} = H - H_{i,f}$ is the perturbation, i.e. the difference between the exact Hamiltonian and the zero order Hamiltonians.

The adiabatic surfaces thus correspond to a 'splitting' of the zero order surfaces, by an amount of $\Delta\varepsilon_{fi} \approx 2|V_{fi}(q)|$. Moreover, since both $V_{fi}(q)$ and the 'diagonal elements', $V_{ii}(q)$ and $V_{ff}(q)$

decrease rapidly when the system leaves the intersection region, the deviation from the zero order potential energy surfaces is appreciable only in this region. Thus, the lower adiabatic surface, $\varepsilon_-(q)$, coincides with the initial state zero order surface, $\varepsilon_i(q)$, sufficiently far to the left of this region, and with the final state surface $\varepsilon_f(q)$ sufficiently far to the right.

The electronic energies depend on the external parameter $q = q(t)$ which represents the classical nuclear motion. In regions outside the intersection region this corresponds to motion on the lowest potential energy surface, and no transition from a lower to a higher surface may occur here. Any such transition would thus be associated with a large change of the electronic energy which - in view of the Franck Condon principle - could not be compensated by a corresponding change of the nuclear energy. On the other hand, in the crossing region the electronic energies of the zero order surfaces coincide, and only in this region there may be an appreciable probability that the system changes its characteristics from those of the reactants (ψ_i) to those of the products (ψ_f). Whether this does occur depends on both the amount of splitting in the intersection region and the velocity, v, which characterizes the nuclear motion. For large v and sufficiently small splitting, the system may 'jump' from the lower to the higher surface. Since the higher surface corresponds to the reactants' characteristics, this implies that the system passes the intersection region without reacting. On the other hand, for large splitting or low velocities the system may perform a smooth passage over the intersection region from regions of q-space characteristic of the reactants to regions characteristic of the products. The semiclassical calculation of the rate probability is thus a calculation of the probability of transition from ψ_i to ψ_f by classical passage of the intersection region with a given velocity $v = \dot{q}$.

Following the procedure of Landau (142) and Zener (219) we seek the solution, $\psi(t)$, of the 'exact' time dependent Schrodinger equation near the intersection region

$$i\hbar\, \partial\psi(t)/\partial t = (H_{i,f} + V_{i,f})\psi(t) \tag{5.2}$$

in the form

$$\psi(t) = c_i(t)\exp(-i\varepsilon^* t/\hbar)\psi_i + c_f(t)\exp(-i\varepsilon^* t/\hbar)\psi_f \tag{5.3}$$

where $\varepsilon^* = \varepsilon_i(q^*) = \varepsilon_f(q^*)$ is the energy of the zero order potential energy surfaces at the intersection point of these surfaces, q^*, and $c_i(t)$ and $c_f(t)$ are time dependent coefficients to be determined subsequently. Insertion of eq.(5.3) in eq.(5.2), multiplication from the left by ψ_i^* and ψ_f^* (the complex conjugates of ψ_i and ψ_f), and integration with respect to the electronic coordinates gives the following two differential equations for $c_i(t)$ and $c_f(t)$

$$i\hbar\,\dot{c}_i = U_i(q)c_i + V_{if}\,c_f \tag{5.4}$$

$$i\hbar\,\dot{c}_f = U_f(q)c_f + V_{fi}\,c_i \tag{5.5}$$

where $U_i(q) = \varepsilon_i(q) + V_{ii}(q) - \varepsilon^*$, and $U_f(q) = \varepsilon_f(q) + V_{ff}(q) - \varepsilon^*$, and where we have considered the overlap integral negligible.

The following boundary conditions are now imposed on eqs.(5.4) and (5.5). For $t \to -\infty$ the lowest state, ψ, essentially coincides with ψ_i, whereas it essentially coincides with ψ_f for $t \to \infty$. ψ thus changes its character from the reactants' electronic configuration to that of the products. These features are expressed by the boundary conditions

$$c_i(t) \rightarrow 1 \quad ; \quad c_f(t) \rightarrow o \quad \text{for} \quad t \rightarrow -\infty \qquad (5.6)$$

We are moreover interested in the probability that the system remains on the lower surface after passage of the intersection region. This is expressed by specifying the system state in the 'infinitely remote' future, $t \rightarrow \infty$, when the electronic configuration of the system must be either that of the reactants or that of the products. With the representation of Ψ in the form of eq.(5.3) this means that $P = |c_f(\infty)|^2$ represents the probability that the system remains on the lower adiabatic potential energy surface, whereas $|c_i(\infty)| = 1 - |c_f(\infty)|^2$ is the probability that the system is located on the 'ascending' branch of the upper surface after passage of the intersection region(fig.(5.1)). $|c_f(\infty)|^2$ is thus the probability that a chemical reaction occurs during a single passage of the intersection region.

The transition probability from the initial to the final state is, however, not always determined by a single passage. Thus, if the potential energy surfaces are located as in fig.(5.1a) - which is representative for a chemical process - the system may be transformed from the initial to the final state also by multiple passages of the intersection region. The overall probability then becomes

$$P + (1-P)P(1-P) + (1-P)PPP(1-P) + ... = 2P/(1+P) \qquad (5.7)$$

For $P \rightarrow 1$, i.e. for the adiabatic limit, a single passage trajectory is seen to give the dominating contribution to the overall reaction probability, whereas for $P << 1$, i.e. the nonadiabatic limit, it is plausible that multiple passages are important. (cf.eq.(5.7)). We notice that for potential energy surfaces located as in fig.(5.1b) - which would correspond to electron transfer in gas phase processes - the transition requires a passage of the intersection region at least twice. Since this double passage can proceed in two ways, the simplest expression for the overall transition probability is in this case $P_{overall} = 2P(1-P)$ (142,219-222).

We can now proceed by giving the result for the semiclassical transition probability. Thus, in addition to classical nuclear motion we assume the following:

(a) $V_{fi}(q) \approx V_{fi}(q^*)$ is a constant in the intersection region. This is equivalent to the Condon approximation incorporated in the semiclassical framework.

(b) $q = v\,t$, where $v = \dot{q}$ is a constant velocity in the intersection region. This implies that

$$U_{i,f} = (\partial U_{i,f}/\partial q)_{q=q^*}(q-q^*) = F_{i,f}\,v\,t \tag{5.8}$$

where $F_{i,f} = (\partial U_{i,f}/\partial q)_{q=q^*}$ are the slopes of the potential energy surfaces at the intersection point, and $t = 0$ corresponds to the moment of passage of this point. Close to the intersection point the potential energy surfaces are thus replaced by straight lines.

With these assumptions the solution of eqs.(5.4) and (5.5) is (142,219-222)

$$P = |c_f(\infty)|^2 = 1-\exp(-2\pi\gamma)\ ;\ \gamma = |V_{fi}|^2/hv|\Delta F| \tag{5.9}$$

$$|\Delta F| = |F_f - F_i|$$

Before discussing the validity of this approach we shall briefly consider some implications of eq.(5.9) in the particular context of electron and atom group transfer processes. The transition probability is determined by the dimensionless quantity γ which depends on the coupling term, V_{fi}, the shape of the potential energy surfaces, ΔF, and the velocity of the nuclear motion v. The combination of these parameters emphasizes the role of both electronic and nuclear motion in the determination of P. Thus, a characteristic period for electronic transition between two levels separated by the quantity V_{fi}, is $\tau_e \approx \hbar/V_{fi}$. On the other hand, the time during which the system is located in a region,

Δq, around the intersection point, where the distance between the electronic levels is smallest, is $\tau_N \approx \Delta q/v \approx V_{fi}/v\Delta F$. We then notice that when $\tau_e >> \tau_N$, the probability of the electronic transition in the intersection region is small (33,93,94). This corresponds either to small coupling or to a large velocity, and with reference to eq.(5.9) the probability value is approximately $P \approx 2\pi\gamma$, i.e. it is given by the ratio τ_N/τ_e. If the opposite condition is valid, i.e. if $\tau_e << \tau_N$, which would mean that either V_{fi} is large or v low, then $\gamma >> 1$. The system then performs a smooth passage over the intersection region, and the chemical process occurs with a probability which approaches unity.

Secondly, in order to obtain the rate of the chemical process it is necessary to average P with respect to all q and v. In view of the classical nature of the nuclear motion the averaging is performed by means of the Maxwell-Boltzmann distribution function, $\Phi_{MB}(q,v)$

$$\Phi_{MB}(q,v)dqdv = (2\pi\hbar Z_i/m)^{-1} \exp\left\{-[U_i(q)-U_i(q_{io})]/k_BT - mv^2/2k_BT\right\} dqdv \tag{5.10}$$

where q_{io} is the equilibrium value of q in the initial state, m the mass associated with the nuclear motion, and Z the statistical sum of this state. The rate probability averaged with respect to q and v is then

$$W_{fi} = \int_{-\infty}^{\infty} dq \int_{-\infty}^{\infty} dv\, \Phi_{MB}(q,v)P(v)v \tag{5.11}$$

Since, in the one-dimensional semiclassical formulation $P \neq 0$ only for $U_i(q) \approx U_i(q^*)$, insertion of eqs.(5.9) and (5.10) in eq.(5.10) gives

$$W_{fi} = (m/2\pi \hbar Z_i) \exp \left\{ -[U_i(q^*) - U_i(q_{io})]/k_B T \right\} \int_{-\infty}^{\infty} dv\, P(v) v \exp(-mv^2/2k_B T) \tag{5.12}$$

The integrals with respect to v provide simple expressions for the limiting cases of strongly nonadiabatic ($P \approx 2\pi\gamma << 1$) and adiabatic processes ($p \approx 1$). In the former case $P(v) \cdot v$ is independent of v (cf.eq.(5.9)), whereas $P \cdot v \approx v$ in the latter case. In both cases these factors are thus slowly varying functions of v compared with the factor $\exp(-mv^2/2k_B T)$. Furthermore, the latter factor decreases rapidly with increasing v for $v > v_T = (2k_B T/m)^{\frac{1}{2}}$, i.e. the thermal velocity. We then obtain the following two expressions for the averaged rate:

(1) In the nonadiabatic limit $P \approx 2\pi\gamma_T << 1$ (where $\gamma = \gamma_T$ for $v = v_T$). Eq.(5.12) can then be rewritten in the following way which is formally analogous to the rate expression of absolute rate theory (45)

$$W_{fi} = \frac{k_B T}{h} \varkappa Z_i^{-1} \exp(-E_A/k_B T) \tag{5.13}$$

This expression contains an activation energy $E_A = U_i(q^*) - U_i(q_{io})$, equal to the difference between the energy of the intersection point and the initial state equilibrium energy, and a transmission coefficient, $\varkappa$, which determines the probability of electronic reorganization at the intersection point. $\varkappa$ is seen to have the explicit form

$$\varkappa = 4\pi^{3/2} \gamma_T = 4\pi^{3/2} |V_{fi}|^2 / hv_T |\Delta F| \tag{5.14}$$

For the particular case where both the initial and final state potential energy surfaces are given by harmonic oscillators of frequency ω, i.e. $U_{i,f} = \frac{1}{2}\hbar\omega(q - q_{i,fo}) + U_{i,fo}$, $\varkappa$ takes the form

$$\varkappa = |V_{fi}|^2 (4\pi^3/\hbar^2\omega^2 k_B T E_r)^{1/2} \tag{5.15}$$

where E_r is the reorganization energy along the mode q (cf. chapter 3). Moreover, in this case $Z \approx k_B T/\hbar\omega$ where we have noted that $\hbar\omega \ll k_B T$ for the classical mode. The rate expression for the reaction then becomes

$$W_{fi} = \varkappa \frac{\omega}{2\pi} \exp(-E_A/k_B T) \tag{5.16}$$

Eqs.(5.15) and (5.16) formally coincide with the result of the quantum mechanical approach (eq.(3.63)) in the limit of classical nuclear motion. However, the two approaches differ not only technically but also ideologically. In the quantum mechanical approach the $\mathcal{T}$-matrix and the perturbation expansion are evaluated in terms of isolated reactant and product (channel) states and potential energy surfaces, which constitute complete orthonormal sets inside each channel. On the other hand, in the semiclassical approach the electronic wave functions, potential energy surfaces etc. contain contributions from both reaction centres. However, by the particular form chosen for the adiabatic electronic wave functions it is subsequently possible to express these quantities in terms of the wave functions of the separated reactants and products.

We notice finally that for a classical harmonic oscillator eq.(5.13) also provides a formal expression for the activation entropy, i.e. $Z = [1-\exp(-\hbar\omega/k_B T)]^{-1} \approx k_B T/\hbar\omega$, and $-S_A \approx k_B \ln(k_B T/\hbar\omega)$.

(2) In the adiabatic limit $\gamma_T \gg 1$, and the reaction proceeds with a probability of approximately unity in the intersection region. Insertion of $P(v) \approx 1$ in eq.(5.12) and integration with respect to v then gives the following expression for the rate constant

$$W_{fi} = \frac{k_B T}{h} Z_i^{-1} \exp(-E_A/k_B T) \tag{5.17}$$

corresponding to a transmission coefficient of unity. Again, for harmonic potential energy surfaces eq.(5.17) has the form

$$W_{fi} = \frac{\omega}{2\pi} \exp(-E_A/k_B T) \tag{5.18}$$

(cf. eq.(5.16)). In this limit the integration with respect to v in eq.(5.12) is taken from zero to infinity. Integration from minus to plus infinity would correspond to passage of the intersection region in both directions. For the nonadiabatic limit this is appropriate since, with a large probability, the system passes this region without reacting, and eq.(5.13) is in a sense the rate probability during a whole period of the system motion. On the other hand, in the adiabatic limit this would correspond to a transition from the reactants to the products and back. This feature is characteristic of the one-dimensional motion in contrast to motion on many-dimensional surfaces.

In our formulation we have assumed that even though the splitting of the surfaces may be large for adiabatic processes, it still does not exceed approximately $k_B T$. This ensures that the energy of the intersection point is a good approximation for the activation energy (cf. the theory of Marcus). If the splitting is larger, the lower potential energy surface is deformed near the intersection region. This implies that also the activation energy is smaller than the value of the intersection point, by an amount determined by maximizing the energy of the lower potential energy surface. Thus, approximating $\varepsilon_i(q)+V_{ii}(q)$ and $\varepsilon_f(q)+V_{ff}(q)$ (eq. 5.1) by the linear functions, $F_i(q)$ and $-F_f q$, respectively (taking the intersection point as the origin) the maximum energy becomes $-2(F_i F_f)^{1/2} V_{fi}/(F_i+F_f)$ relative to the intersection point of the zero order surfaces. Inserting the Bronsted coefficient $\alpha = F_i/(F_i+F_f)$ as the derivative of the activation energy (for the corresponding nonadiabatic processes) with respect to the distance between the minima of the zero order potential energy surfaces the activation energy can be written as

$$E_A^{ad} = E_A^{na} - 2V_{if}(F_iF_f)^{1/2}/(F_i+F_f) = E^{na} - 2V_{if}[\alpha(1-\alpha)]^{1/2} \quad (5.19)$$

where the superscripts 'ad' and 'na' refer to adiabatic and nonadiabatic processes, respectively.

We shall finally give a semiquantitative estimate of the degree of coupling which would represent a 'critical' value for the limit between adiabatic and nonadiabatic processes. Eq.(5.15) provides the basis for this criterion, i.e. the conditions for which $\gamma \approx 1$. For E_r = 2 ev and a low 'effective' frequency (10^{11}-10^{12} s^{-1} or 10 - 10^2 cm^{-1}), such as provided by the medium frequencies of polar solvents, the criterion is $|V_{fi}| \approx 10^{-2}$ ev ($\approx$ 0.2 kcal $\approx$ 50 cm^{-1}). If the frequency is high, such as a proton stretching mode ($\hbar\omega \approx$ 0.3 ev) the nuclear motion is of quantum nature. Anticipating the discussion below, the adiabaticity criterion is now given by eq.(5.9) which can be rewritten in the approximate form $2\pi|V_{fi}|^2/\hbar|v||\Delta F| \approx 2\pi|V_{fi}|^2/\hbar|v|E_r$. Insertion of typical parameter values ($\omega \approx 3000$ cm^{-1}, E_r = 1 ev) shows that $|V_{fi}| \approx$ 0.5 ev ($\approx$13 kcal) now represents the critical value for electronic nonadiabaticity.

5.1.2 Nuclear Quantum Effects.

The semiclassical approach provides a compact result for the rate expression, formally covering the whole range of P from zero to unity. The fundamental assumptions of the theory - primarily that of the classical behaviour of the nuclei - have been investigated in the literature by comparison with more rigorous quantum mechanical calculations on simple model systems of which the crossing of two one-dimensional linear terms is most frequently applied. It is thus possible both to establish criteria for the classical behaviour of the nuclei and to provide modifications of the formalism when they are not met (33,221-223).

The formulae given above are thus valid for energies which are much higher than the energy of the intersection point and also much higher than the level distance in the potential well represented by the upper potential energy surface. If the system energy is lower than that of the intersection point, nuclear quantum effects are manifested. In the limit of small splitting P is then given by

$$P = 2\pi\gamma \exp(-\sigma) \tag{5.20}$$

where

$$\sigma = \frac{2}{\hbar} \int \left\{2m[U_{i,f}(q) - E]\right\}^{1/2} dq \tag{5.21}$$

is the Gamov factor of the probability of nuclear tunnelling through the barrier region (fig.(5.1)) which is classically forbidden for energies $E < U_{i,f}(q^*)$ (142). In this case the probability expression thus contains both a factor which expresses the probability of nuclear tunnelling through the barrier, and a factor which expresses the probability of electronic reorganization during the nuclear motion. The latter probability is identical to eq.(5.9) if v is put equal to $\{2[U_{i,f}(q)-E]/m\}^{1/2}$, i.e. the modulus of the imaginary velocity. For larger splitting but energies which are still lower than the energy of the intersection point the pre-exponential factor of eq.(5.20) is replaced by unity, and the appropriate barrier for nuclear tunnelling is now determined by the lower adiabatic potential energy surface. Finally, in the intermediate energy region when the system energy is approximately that of the intersection point, P displays an oscillatory dependence on E (155,221) which reflects the interference of the classical turning points with the intersection region.

5.2 Many-Dimensional Nuclear Motion

Condensed phase chemical systems are always characterized by a multitude of nuclear coordinates which may be of both quantum and classical nature, according to the criteria discussed in chapters 3 and 4. This has stimulated attempts to formulate a generalization of the semiclassical Landau-Zener formalism to incorporate the following effects (93,94):

(a) The generalized Landau-Zener formalism is applicable to systems which prossess an arbitrary number of classical nuclear modes. These modes define a many-dimensional classical potential energy surface.

(b) As for the one-dimensional motion, the potential functions representing the classical modes are not restricted to harmonic potentials but may take any physically reasonable form.

(c) The semiclassical approach can be extended to incorporate high-frequency quantum modes provided that the classical and quantum modes do not 'mix' during the reaction, i.e. provided that a given mode remains in the same class in both the initial and final states.

The extensions listed are conveniently incorporated in the theory in two stages (93,94). Firstly, with reference to chapter 3 and anticipating the discussion of chapter 6 the approximate criteria for classical and quantum nuclear motion for nuclear potential energy functions of rather general form are

$$\Delta\varepsilon_{cl} < k_B T \qquad (5.22)$$

$$\Delta\varepsilon_{qu} > k_B T \tag{5.23}$$

where $\Delta\varepsilon$ refers to the intervals between nuclear vibrational energy levels, and the subscripts 'cl' and 'qu' refer to 'classical' and 'quantum', respectively. This implies that the velocity of the motion of the quantum modes is much faster than that of the classical modes. It is therefore convenient to view the fast nuclei together with the electrons as an 'overall' fast quantum subsystem in a modified Born-Oppenheimer approximation, analogous to the one outlined in chapter 3. For each electronic state and each vibrational state of the fast nuclei (v and w in the initial and final electronic state, respectively) potential energy surfaces of the following form can therefore be introduced

$$U_{iv} = U_{io} + \varepsilon_v + f_i(\{q_k\}) \tag{5.24}$$

$$U_{fw} = U_{fo} + \varepsilon_w + f_f(\{q_k\}) + \Delta E \tag{5.25}$$

$\{q_k\}$ represents the total set of classical nuclear coordinates, f_i and f_f the potential energy surfaces with respect to these modes in the initial and final state, respectively, and ε_v and ε_w are the total energies of the fast nuclei in the initial and final states. The topology of these surfaces are thus solely determined by the classical nuclear modes $\{q_k\}$. Eqs.(5.24) and (5.25) represent families of potential energy surfaces where each member corresponds to a given set of vibrational quantum numbers (v,w) for each high-frequency nuclear mode.

During the chemical process the high-frequency vibrational states are generally subject to changes. The overall probability must therefore incorporate probabilities of transitions from all initial to all final state potential energy surfaces. If all these 'microprocesses' proceed in an independent fashion, the overall reaction probability takes the following form

$$W_{fi} = Z_{qu}^{-1} \sum_{v} \sum_{w} \exp(-\varepsilon_v / k_B T) W_{fi}(\varepsilon_v, \varepsilon_w) \tag{5.26}$$

where $W_{fi}(\varepsilon_v, \varepsilon_w)$ is the probability of transition between a given pair of surfaces (eqs.(5.24) and (5.25)) characterized by the nuclear quantum numbers v and w, and $Z_{qu} = \sum_v \exp(-\varepsilon_v/k_B T)$ is the statistical sum of these modes in the initial state. Eq.(5.26) thus reduces the transition probability to a calculation of the transition probability between individual classical potential energy surfaces and a knowledge of the 'spectrum' of the high-frequency nuclear modes. We also notice that eq.(5.26) is plausible as long as the adiabatic 'splitting' in the region of intersection between individual pairs of potential energy surfaces is small compared with the distance between vibrational energy levels of the high-frequency modes. If these two energy quantities are of comparable magnitude the assumption of independent transition probabilities may have to be modified. We shall return to this problem in chapter 8.

We shall now assume that incorporation of the high-frequency modes by eq. (5.26) is possible when the individual probabilities, $W_{fi}(\varepsilon_v, \varepsilon_w)$, and the vibrational spectrum, represented by ε_v and ε_w, are known and only consider a single initial and final state potential energy surface. The second stage is the calculation of W_{fi} extended to many-dimensional potential energy surfaces. The zero order surfaces now cross at a surface, S - which possesses one dimension less than the intersection surfaces - rather than at a point as for one-dimensional surfaces. The system may then pass from the initial to the final state across any point on the intersecting surface, and the overall transition probability is subsequently obtained by averaging over the entire intersection surface and over all velocities.

We shall invoke the same approximations as in our discussion of the one-dimensional curve crossing (93,94). We shall thus represent the total electronic wave function in the form of eq.(5.3).

As before, this leads to eqs.(5.4) and (5.5), where $V_{fi} \approx V_{fi}(\{q_k^*\})$ is assumed to be constant. However, close to the intersection surface $U_i(\{q_k\})$ and $U_f(\{q_k\})$ now take the form

$$U_i(\{q_k\}) \approx \sum_k (\partial U_i/\partial q_k)(q_k - q_k^*) \tag{5.27}$$

$$U_f(\{q_k\}) \approx \sum_k (\partial U_f/\partial q_k)(q_k - q_k^*) \tag{5.28}$$

q_k refers to each of the classical coordinates, q_k^* to its value at the intersection region, and $q_k - q_k^* = v_k t$, where $v_k = \dot{q}_k$ is the (constant) velocity component along the coordinate q_k. Insertion of this and eqs.(5.27) and (5.28) in eqs.(5.4) and (5.5) gives the following equations of motion for the coefficients $c_i(t)$ and $c_f(t)$

$$i\hbar\dot{c}_i = [\sum_k (\partial U_i/\partial q_k)^* \cdot v_k]\ t \cdot c_i + V_{if} c_f \tag{5.29}$$

$$i\hbar\dot{c}_f = [\sum_k (\partial U_f/\partial q_k)^* \cdot v_k]\ t \cdot c_f + V_{fi} c_i \tag{5.30}$$

where the superscript '*' refers to the value of the derivatives of U_i and U_f at the intersection surface. These equations are formally identical to eqs. (5.4) and (5.5), and we can then exploit our previous procedure and results to give the following expression for the transition probability (cf.eq.(5.9))

$$P = 1 - \exp(-2\pi\gamma) \tag{5.31}$$

The parameter γ is now

$$\gamma = |V_{fi}|^2/\hbar|\Delta F| \tag{5.32}$$

where

$$|\Delta F| = |\sum_k [\partial(U_i - U_f)/\partial q_k]^* \cdot v_k| = |\mathrm{grad}(U_i - U_f)^* \vec{v}| \quad (5.33)$$

As before, eqs.(5.31) - (5.33) represent the probability of transition at the intersection surface, and for a given velocity v. Subsequently we shall have to average with respect to all coordinates and velocities, i.e. over all possible paths of crossing the intersection surface. Since eq.(5.33) moreover shows that only velocity components perpendicular to the surface S in the intersection region contribute to P, the averaged transition probability must have the form (cf.eq. (5.11)

$$W_{fi} = \int \Phi_{MB}(\vec{q},\vec{v})\ P(\vec{v})[\sum_k v_{kn}(\{q_k^*\})]\ dS \prod_{k=1}^{N} dv_k \quad (5.34)$$

where v_{kn} are now the projections of the velocity components on a direction perpendicular to S at the points q_k^*, and N is the total number of classical modes.

In mass-weighted coordinates, $\xi_k = m^{\frac{1}{2}} q_k$, the classical distribution function, Φ_{MB} takes the form (93,94,149)

$$\Phi_{MB} = \Phi(\{\xi_k\},\{\dot{\xi}_k\}) = \exp\left\{-[\frac{1}{2}\sum_k \dot{\xi}_k^2 + U_i(\{\xi_k\})]/k_BT\right\}$$

$$[\int_{-\infty}^{\infty} \prod_{k=1}^{N} d\xi_k \int_0^{\infty} \prod_{k=1}^{N} d\dot{\xi}_k \exp\left\{-[\frac{1}{2}\sum_k \dot{\xi}_k^2 + U_i(\{\xi_k\})]/k_BT\right\}]^{-1} \quad (5.35)$$

Insertion of eq.(5.35) in eq.(5.34) then gives the averaged transition probability between a given pair of potential energy surfaces

$$W_{fi} = \int dS \int_0^{\infty} \prod_{k=1}^{N} d\dot{\xi}_k\ \dot{\xi}_{kn}(\{q_k^*\})P \exp[-E_i(\{\xi_k\}, \quad (5.36)$$

$$\{\dot{\xi}_k\})k_BT]\ \left[\int_{-\infty}^{\infty}\prod_{k=1}^{N}d\xi_k\int_{0}^{\infty}\prod_{k=1}^{N}d\dot{\xi}_k\exp[-E_i(\{\xi_k\},\{\dot{\xi}_k\})/k_BT]\right]^{-1}$$

where $E_i = \frac{1}{2}\sum_k \dot{\xi}_k^2 + U_i(\{\xi_k\})$ is the total energy of the system in the initial state at given $\{\xi_k\}$ and $\{\dot{\xi}_k\}$.

Eq.(5.36) is a general form of the transition probability when the potential energy terms are represented by many-dimensional surfaces. In our further discussion of this expression it is now convenient to consider the adiabatic and nonadiabatic limits separately.

(a) In the adiabatic limit $P \approx 1$, corresponding to a large 'splitting' of the adiabatic potential energy surfaces in the intersection region, or a low velocity in the passage of this region. We can then transform eq.(5.36) to a form which is formally identical to the rate expression of absolute rate theory (45) (cf. eqs.(5.13) and (5.17)), i.e.

$$W_{fi} = \frac{k_BT}{h}\exp(\ S_A/k_B)\ \exp(-E_A/k_BT) \qquad (5.37)$$

The activation energy, E_A, is here introduced as the distance from the minimum of the initial state potential energy surface to the point of lowest energy on the intersection surface. If the splitting, although large, is still smaller than k_BT this point is approximately the saddle point, $\{\xi_{ks}^*\}$, of the surface of intersection between the zero order surfaces, i.e. $E_A = U_i(\{\xi_{ks}^*\}) - U_i(\{\xi_{k0}\})$. This point is determined by the condition that the potential energy, $U_i(\{\xi_k\})$ is minimized with respect to all the coordinates ξ_k, subject to the additional condition that $U_i = U_f$ (cf. the theory of Marcus (chapter 1)). This is expressed by introduction of a Lagrange multiplier, α, to give

$$(1-\alpha)\,\partial U_i/\partial\xi_k + \alpha\,\partial U_f/\partial\xi_k = 0 \qquad (5.38)$$

for each classical coordinate ξ_k. α is subsequently found by noting that the saddle point, $\{\xi_{ks}^*\}$, determined from eq.(5.38) satisfies the equation $U_i(\{\xi_{ks}^*\}) = U_f(\{\xi_{ks}^*\})$.

The physical meaning of α can be illuminated by the following considerations. From eq.(5.25) we see that

$$\partial E_A/\partial\Delta E = U_i(\{\xi_{ks}^*\})/\partial\Delta E = \qquad (5.39)$$

$$\sum_k [\partial U_i(\{\xi_{ks}^*\})/\partial\xi_k](\partial\xi_k/\partial\Delta E)$$

Furthermore, from eq.(5.38) and since $U_i(\{\xi_{ks}^*\}) = U_f(\{\xi_{ks}^*\}) + \Delta E$,

$$\sum_k [\partial U_i(\{\xi_{ks}^*\})/\partial\xi_k](\partial\xi_k/\partial\Delta E) = \qquad (5.40)$$

$$\alpha\sum_k [\partial U_i(\{\xi_{ks}^*\})/\partial\xi_k - \partial U_f(\{\xi_{ks}^*\})/\partial\xi_k](\partial\xi_k/\partial\Delta E) = o$$

The Lagrange multiplier thus coincides with the Bronsted coefficient of the process, i.e. the derivative of the activation energy with respect to the distance between the minima of the initial and final state potential energy surfaces (cf. chapter 3).

If we had explicitly introduced the lower many-dimensional adiabatic surface in the form (cf.eq.(5.1))

$$U_- = \frac{1}{2}(U_i + U_f) - \frac{1}{2}[(U_i - U_f)^2 + V_{if}^2]^{1/2} \qquad (5.41)$$

and thus the finite value of V_{fi} (compared with k_BT), then the activation energy, E_A^{ad}, would be $E_A^{ad} = U_-(\{\xi_{ks}^*\})$, where ξ_{ks}^* is now the saddle point of the lower adiabatic surface. Introducing

$$\alpha^{ad} = \partial E_A^{ad}/\partial \Delta E = \partial U_-(\{\xi^*_{ks}\})/\partial \Delta E \qquad (5.42)$$

and noticing that

$$\partial U_-/\partial \xi_k = \frac{1}{2}\left(\frac{\partial U_i}{\partial \xi_k} + \frac{\partial U_f}{\partial \xi_k}\right) - \frac{1}{2}\,\frac{U_i - U_f}{[(U_i - U_f)^2 + V_{if}^2]^{1/2}}\left(\frac{\partial U_i}{\partial \xi_k} - \frac{\partial U_f}{\partial \xi_k}\right) \qquad (5.43)$$

and $U_i(\{\xi^*_{ks}\}) = U_f(\{\xi^*_{ks}\}) + \Delta E$, we find that

$$\alpha^{ad} = \sum_k [\partial U_-(\{\xi^*_{ks}\})/\partial \xi_k]\partial \xi_k/\partial \Delta E = \qquad (5.44)$$

$$= \frac{1}{2}\partial[U_i(\{\xi^*_{ks}\}) + U_f(\{\xi^*_{ks}\})]/\partial \Delta E + \frac{1}{2}[U_i(\{\xi^*_{ks}\}) - U_f(\{\xi^*_{ks}\})]/\left\{[U_i(\{\xi^*_{ks}\}) - U_f(\{\xi^*_{ks}\})] + V_{if}^2\right\}^{1/2} =$$

$$= \frac{1}{2} + \frac{1}{2}[U_i(\{\xi^*_{ks}\}) - U_f(\{\xi^*_{ks}\})]/\left\{[U_i(\{\xi^*_{ks}\}) - U_f(\{\xi^*_{ks}\})]^2 + V_{if}^2\right\}^{1/2}$$

Inserting this in eq.(5.43) gives

$$(1-\alpha^{ad})\partial U_i(\{\xi^*_{ks}\})/\partial \xi_k + \alpha^{ad}\partial U_f(\{\xi^*_{ks}\})/\partial \xi_k = 0 \qquad (5.45)$$

i.e. an equation formally identical to eq.(5.38). Eqs.(5.44) and (5.45) thus provide the saddle point coordinates and activation energy for the strongly adiabatic case (when $V_{fi} \gtrsim k_B T$).

While the activation energy is seen to have a quite transparent form, the apparent activation entropy, S_A, is more involved. By comparison of eqs.(5.36) and (5.37) we see that

$$\exp(S_A/k_B T) = \frac{h}{k_B T} \int dS \int_0^{\infty} \prod_{k=1}^{N} d\xi_k \dot{\xi}_{kn}(\{\xi_k^*\}) \exp\left\{-[\frac{1}{2}\sum_k \dot{\xi}_k^2 + U_i(\{\xi_k^*\}) - U_i(\{\xi_{ks}^*\})]/k_B T\right\}[\int_{-\infty}^{\infty} \prod_{k=1}^{N} d\xi_k \int_0^{\infty} \prod_{k=1}^{N} d\dot{\xi}_k \quad (5.46)$$

$$\exp\left\{-[\frac{1}{2}\sum_k \dot{\xi}_k^2 + U_i(\{\xi_k\})]/k_B T\right\}]^{-1}$$

The apparent activation entropy thus contains all information about the velocity distribution, in addition to the averaging over the potential energies of the whole intersection surface (as represented by the contribution $U_i(\{\xi_k^*\}) - U_i(\{\xi_{ks}^*\})$ in the exponent of the numerator of eq.(5.46).

These integrations can be performed when appropriate model potential energy surfaces are introduced (93,94). Thus, for one-dimensional harmonic potential energy surfaces of frequency ω, eq.(5.42) is seen to give $-S_A = k_B \ln(k_B T/\hbar\omega)$ (cf. eq.(5.19)) as also found previously. In the more general case of many-dimensional potential energy surfaces, S_A takes an analogous form where, however, ω is now replaced by ω_{eff} which is given by (93,94)

$$\omega_{eff}^2 = \sum_k \omega_k^2 E_{rk}/E_r \quad (5.47)$$

E_{rk} is the reorganization energy of the k'th mode and E_r the total reorganization energy of all modes. The rate expression for the adiabatic process is then

$$W_{fi} = \frac{\omega_{eff}}{2\pi} \exp(-E_A/k_BT) \qquad (5.48)$$

(b) In the limit of nonadiabatic processes $\gamma \ll 1$, and

$$P \approx 2\pi\gamma = 2\pi |V_{fi}|^2/\hbar|[\sum_k \partial(U_i-U_f)/\partial\xi_k]\dot{\xi}_k| \qquad (5.49)$$

With reference to eqs.(5.32) and (5.40), and if we replace the velocity by its 'thermal' value $(2k_BT/\pi)^{1/2}$, the adiabaticity criterion is equivalent to the condition

$$|V_{if}|^2 \ll \hbar|\sum_k \partial(U_i-U_f)/\partial\xi_k|(2k_BT/\pi)^{1/2}/2\pi =$$

$$= [(\hbar^2 k_BT/2\pi^3\alpha^2)\sum_k (\partial U_i/\partial\xi_k)^2]^{1/2} = |V_{cr}|^2 \qquad (5.50)$$

The rate expression can then be rewritten in the form (cf. eq.(5.16))

$$W_{fi} = \varkappa \frac{k_BT}{h} \exp(S_A/k_B) \exp(-E_A/k_BT) \qquad (5.51)$$

where S_A and E_A are determined by eq.(5.38) and (5.40) as for adiabatic processes, and the transmission coefficient, $\varkappa$, takes the form

$$\varkappa = 2|V_{if}|^2/|V_{cr}|^2 \qquad (5.52)$$

The factor 2 accounts for the fact that for nonadiabatic processes the intersection region is passed twice during a period of nuclear motion. When the nuclear motion is represented by harmonic oscillators of the frequencies ω_k, the 'critical' value of the 'splitting' of the surfaces, $|V_{cr}|$, becomes $(k_B T\hbar^2/\pi^3)^{\frac{1}{2}}$ $(\sum_k \omega_k^4 E_{rk})^{\frac{1}{4}}$, and

$$= (4\pi^3/k_B T\hbar^2\omega_{eff} E_r)^{1/2}|V_{if}|^2 \qquad (5.53)$$

where ω_{eff} and E_{rk} were both defined earlier. Eqs.(5.51) and (5.53) consequently give a rate expression which is formally identical to the one derived in both the semiclassical theory of one-dimensional motion and in the quantum theory of rate processes (chapter 3) in the limit of low-frequency modes subject to coordinate shifts only.

The semiclassical approach in the form outlined above represents a convenient formalism in the theory of condensed phase rate processes, in a sense complementary to the quantum mechanical approach. It is thus generally applicable, i.e. the nuclear potential energy functions can assume rather general forms and it covers in principle and in a compact form the whole range of transition probabilities from zero to unity. On the other hand, it is essential that at least one classsical nuclear mode is present and that the classical and quantum modes remain in these classes during the reaction, even though both frequency shift, coordinate shift, and mode mixing may occur inside each of the classes. Finally, we should emphasize that the activation energy and entropy emerge here in a different way than in the transition state theory, even though the rate expressions may appear identical. In the present formulation it is thus not necessary to invoke a concept of activated complex since the intersection region represents simply certain configurations of the initial and final electronic states, or certain regions on the lower adiabatic potential energy surface.

5.3 Relation to Experimental Data

We shall now proceed to show the applicability of the semiclassical theory to some chemical processes involving inorganic metal complexes in aqueous solution. The intramolecular modes undergoing major reorganization during the processes are largely metal-ligand stretching and bending modes the frequencies of which are sufficiently small (200-400 cm^{-1}) that a classical approach is adequate at room temperature (this is confirmed by the numerical calculations discussed in chapter 4 and in the literature (224)). We shall calculate the activation energies of these processes and show that for 'sufficiently closely related' processes a single set of parameters (e.g. solvent and intramolecular reorganization energies) can reproduce the experimental data. Also, in certain cases the activation energy expressions allow a destinction between alternative possible mechanisms. We consider in turn the following classes of processes:

5.3.1 Outer Sphere Electron Transfer Processes.

Simple outer sphere electron transfer processes are in principle subject to reorganization in both the solvent modes of the outer medium and the intramolecular metal-ligand stretching and bending modes. In some systems - for example the $MnO_4^{-/2-}$, $[Fe(CN)_6]^{3-/4-}$, $[W(CN)_8]^{3-/4-}$ and $[Mo(CN)_8]^{3-/4-}$ couples - the bond lengths in the oxidized and reduced forms of a reactant are practically identical and the intramolecular reorganization can be ignored. In other cases, however, involving for example aqua and ammine complexes, both subsystems contribute comparably to the activation energy. In addition, while the solvent modes are likely to undergo coordinate shift only, the intramolecular modes are commonly subject to frequency shifts as well (e.g. the vibration frequencies of the stretching modes are highest for the metal ions in the highest oxidation state) and possibly mixing of modes, i.e. the intramolecular normal coordinates in the final state differ from those of the initial state. We shall ignore the latter effect and furthermore assume that the intramolecular modes are adequately represented by harmonic potentials. This is plausible for electron transfer processes, both because of the relatively small coordinate shifts and because of the cancellation of anharmonicity effects when stretching in one mode occurs synchronously with compression in a different mode (chapter 4). The potential energy surfaces in the initial and final states then take the form

$$U_i = \frac{1}{2}\sum_k \omega_k^2 \xi_k^2 + \frac{1}{2}\sum_m \Omega_{Dm}^{i\,2} Q_{Dm}^2 + \frac{1}{2}\sum_n \Omega_{An}^{i\,2} Q_{An}^2 \qquad (5.54)$$

$$U_f = \frac{1}{2}\sum_k \omega_k^2(\xi_k - \xi_{ko})^2 + \frac{1}{2}\sum_m \Omega_{Dm}^{f\,2}(Q_{Dm} - Q_{Dm}^o)^2 + \qquad (5.55)$$

$$\frac{1}{2}\sum_n \Omega_{An}^{f\,2}(Q_{An} - Q_{An}^o)^2 + \Delta E$$

where ω_k and ξ_k are the frequencies and the (mass-weighted) coordinates of the medium, ξ_{ko} the shift of these coordinates in the final state relative to the initial state, and the second and third sums have the same meaning for the electron donor and acceptor, respectively. Finally, the difference between the minima of the surfaces ΔE, includes the work terms of bringing the reactants and products from an infinitely large distance to their positions in the collision complex (cf. the theory of Marcus).

Eq.(5.38) then gives for the coordinates of the saddle point (225)

$$\xi^{*}_{ks} = \alpha \xi_{ko} \tag{5.56}$$

$$Q^{*}_{Dm} = \alpha \Omega^{f\,2}_{Dm} Q^{o}_{Dm} / [(1-\alpha)\Omega^{i\,2}_{Dm} + \alpha \Omega^{f\,2}_{Dm}] \tag{5.57}$$

and a similar equation for Q^{*}_{An}. Since, at the saddle point $U_i = U_f$, insertion of eqs.(5.56) and (5.57) in eqs.(5.54) and (5.55) gives the following equation for the determination of α

$$(2\alpha-1)E_S + \sum_m E^{f}_{rDm} [\alpha^2 \Omega^{i\,2}_{Dm} - (1-\alpha)^2 \Omega^{f\,2}_{Dm}] / [(1-\alpha)^2 \Omega^{i\,2}_{Dm} +$$

$$+ \alpha \Omega^{f\,2}_{Dm}]^2 + \sum_n E^{f}_{rAn} [\alpha^2 \Omega^{i\,2}_{An} - (1-\alpha)^2 \Omega^{f\,2}_{An}] / [(1-\alpha)\Omega^{i\,2}_{Dn} +$$

$$+ \alpha \Omega^{f\,2}_{An}] - \Delta E = 0 \tag{5.58}$$

where $E_s = \frac{1}{2} \sum \omega_k^2 \xi_{ko}^2$ is the solvent reorganization energy, and $E^{f}_{rDm} = \frac{1}{2} \Omega^{f\,2}_{Dm} Q^{o\,2}_{Dm}$ and $E^{f}_{rAn} = \frac{1}{2} \Omega^{f\,2}_{An} Q^{o\,2}_{An}$ refer to the reorganization energies along the appropriate modes and with frequencies corresponding to the final state. Eq. (5.58) can be solved exactly in two cases. Firstly, if all frequencies retain their values when going from the initial to the final state, then

$$\alpha = \frac{1}{2} + \Delta E / 2(E_S + E_{rD} + E_{rA}) \tag{5.59}$$

where E_{rD} and E_{rA} are now the total reorganization energies of all modes in the donor and acceptor complex, respectively. Insertion of eqs.(5.56) and (5.59) in eq.(5.55) then gives for the activation energy

$$E_A = (E_S + E_{rD} + E_{rA} + \Delta E)^2 / 4(E_S + E_{rD} + E_{rA}) \tag{5.60}$$

which is identical to the expression derived by Marcus for the same model system. Secondly, the conditions $\Delta E = 0, \Omega^i_{Dm} = \Omega^f_{An} = \Omega_{Ox}$, and $\Omega^f_{Dm} = \Omega^i_{An} = \Omega_{Red}$, but $\Omega^i_{Dm} \neq \Omega^f_{Dm}$ and $\Omega^i_{An} \neq \Omega^f_{An}$ are valid for electron exchange reactions between the oxidized and reduced forms of the same ion. Then, from eq.(5.58), α = 0.5, and

$$E_A = \frac{1}{4}[E_S + 4\Omega^4_{Red} E^i_{rD}/(\Omega^2_{Red} + \Omega^2_{ox})^2 + 4\Omega^4_{ox} E^i_{rA}/(\Omega^2_{Red} + \Omega^2_{ox})^2] \tag{5.61}$$

In other limiting cases approximate solutions to eq.(5.58) can also be provided (225). For example if $\alpha = 0.5 + \Delta\alpha$, where $|\Delta\alpha| \ll 0.5$, the equation can be solved for general frequency changes, and for small frequency changes it can be solved for all values of α (99). In addition, if large frequency changes occur, the solution of eq.(5.58) takes a relatively simple form derived in the next section. Except for the latter case the resulting expressions are rather voluminous, however, and numerical solution is usually the most convenient way of achieving the α-values in any concrete case.

The expression for the rate constant, k_r, can now be written formally as

$$k_r = \frac{k_B T}{h} \exp(-G_A/k_B T) = \frac{k_B T}{h} \exp\left\{-[E_A + w_r - k_B T \ln(\kappa \frac{\hbar\omega_{eff}}{k_B T} \Delta\vec{R})]/k_B T\right\} \tag{5.62}$$

where we recall that $\Delta\vec{R}$ is an effective reaction volume, and w_r the work in bringing the reactants from infinity to their contact distance. In order to compare this expression directly with experimental data we invoke - for the sake of simplicity - the following assumptions: (a) E_A is given by eq.(5.61) and E_S by eq.(1.28) which we write as $E_S = Q_S(\varepsilon_o^{-1} - \varepsilon_s^{-1})$. (b) w_r and w_p are represented by coulomb terms, i.e. $w_{r,p} = Q_{r,p}/\varepsilon_s R = z_1 z_2/\varepsilon_s R$, where the z's refer to the ionic charges and R to the distance between the ionic centres which is furthermore assumed to be approximately constant for all the metal complexes. Since ε_s depends on the temperature, the apparent activation entropy (i.e. an experimentally observable quantity) is then

$$S_A = -\partial G_A/\partial T = k_B \ln(\kappa \hbar\omega_{eff} \Delta\vec{R}/k_B T) - \tag{5.63}$$

$$- k_B + \alpha(1-\alpha)Q_S\gamma + \alpha\Delta S_o - Q_r\gamma - \alpha\gamma(Q_r - Q_p)$$

where $\gamma = \partial\varepsilon_s^{-1}/\partial T$ (= $6\cdot10^{-5}$ K^{-1} for water (226)), the standard entropy of reaction $\Delta S_o = -\partial\Delta E/\partial T$, and the other quantities have been defined previously. For electron exchange processes ΔS_o = 0, $\alpha = 0.5$, and $Q_r = Q_p$ reducing eq.(5.63) to

$$S_A \approx k_B \ln(\kappa\hbar\omega_{eff} \Delta\vec{R}/k_B T) - Q_r\gamma \tag{5.64}$$

A plot of the experimental activation entropies against the charge product of the reactants should then give a straight line of intercept $k_B \ln(\kappa\hbar\omega_{eff}\Delta\vec{R}/k_B T)$, and a slope from which the interreactant distance R can be estimated. This relationship was found to be well obeyed for the electron exchange processes involving the $MnO_4^{-/2-}$, $Co^{3+/2+}$, $Fe^{3+/2+}$, $V^{3+/2+}$, and $Ce^{4+/3+}$ couples giving R ≈ 7 A (226). Taking this as the linear dimension of the 'effective' reaction zone gives $\Delta\vec{R} \approx 1$ $dm^3 mol^{-1}$. This value together with the value $k_B \ln(\kappa\hbar\omega_{eff}\Delta\vec{R}/k_B T) = -6\cdot10^{-3}$ kcal mol^{-1} K^{-1} read from the intercept give $\omega_{eff} = 3\cdot10^{12}$ s^{-1} for $\kappa = 1$. On the other hand, if the upper limit of the effective classical frequency is taken as $k_B T/h = 4\cdot10^{13}$ s^{-1} at room temperature, then $\kappa \approx 0.1$ which thus represents a lower limit of for these processes and within this simple model.

We now adopt this value of R in a subsequent calculation of the free energy of activation for several electron transfer processes involving metal complexes of the same kind as those used in the estimate of R, i.e. aqua and ammine complexes. For the calculation of E_s the ionic radii were taken as the sum of the distances from the central atom to a hydrogen atom and the covalent radius of a hydrogen atom (0.35 A). The distances were taken from a literature collection of X-ray data (227,228), subsequently inserted in Marcus' equation for E_s (eq.(1.28)) and corrected for frequency dispersion effects by multiplication with a factor of 0.82 (cf. chapter 3). (This was not done in the orginal analysis (225)). In the calculation of E_{rD} and E_{rA} it is also necessary to know the force constants of the metal-ligand stretching modes (which are the only ones subject to reorganization) of both the oxidized and reduced forms. Unfortunately, accurate data are here only available for $[Co(NH_3)_6]^{3+}$ and $[Co(NH_3)_6]^{2+}$ for which $\Omega_{ox} \approx 500$ cm^{-1} and $\Omega_{Red} \approx 320$ cm^{-1}(229) The original analysis of these rate data (225) were therefore based on metal-ligand stretching force constants determined from model calculations in which the metal-ligand bond was represented as an overlap of an attractive electrostatic ion-dipole potential and a core repulsion potential (230). However, infrared spectroscopic data on solid aqua complexes of transition metal ions (Cr^{3+}, Mn^{2+}, Fe^{2+}) suggest (231,232) that representative values for the metal-ligand stretching frequencies of divalent and trivalent aqua ions are 390 cm^{-1} and 490 cm^{-1}, respectively. These values which we shall adopt give somewhat higher E_{rD} and E_{rA} than the earlier calculation.

Tabel 5.1

System	a	b	r_{m-L}^{ox}	r_{m-L}^{red}	w_r	E_r	E_s	G_A^{theor}	G_A^{exp}
$[Co(NH_3)_6]^{3+/2+}$	2.73	2.89	1.96	2.12	3.6	25.0	34.8	20.4	22-30
$[Co(H_2O)_6]^{3+/2+}$	3.02	3.16	1.95	2.09	3.6	20.5	29.2	17.8	16.5
$[Fe(H_2O)_6]^{3+/2+}$	3.03	3.19	1.97	2.14	3.6	30.2	29.2	20.2	16.8
$[Mn(H_2O)_6]^{3+/2+}$	3.06	3.25	2.0	2.18	3.6	33.9	28.4	20.9	18
$[V(H_2O)_6]^{3+/2+}$	3.00	3.20	1.95	2.15	3.6	41.8	29.2	23.1	20.1

The ionic radii, a and b, and the metal - ligand bond in A, energy quantities in kcal.

Table 5.1 shows the various contributions to the free energy of activation estimated in the way described above and the experimental values of the total free energy of activation (16,230). From this comparison we notice: (a) within the semiclassical theory and by invoking quite simple models it is possible to perform a straightforward and consistent calculation for a given class of outer sphere electron transfer process from which the important parameters, such as reorganization energies can be determined; (b) the kinetic parameters, calculated from structural and spectroscopic data generally show a quite good agreement with the same parameters estimated kinetically. There is a tendency, however, that the theoretical values for the aqua complexes are higher than the experimental values. This discrepancy would be diminished by taking into account the space dispersion effects discussed in chapter 2; (c) in one case, i.e. for the $[Co(NH_3)_6]^{3+/2+}$ couple the discrepancy between the theory and the experimental data is in the opposite direction. This is a well know observation for electron transfer reactions involving Co(III) in moderately strong ligand fields (16,30), and is likely to be associated with a markedly smaller transmission coefficient for this system than for the others. The effect originates from the poor electronic coupling, V_{eff}, associated with the change from a low-spin state to a high-spin state when Co(III) is converted to Co(II).

We can illustrate this by comparing this couple and the $[Ru(NH_3)_6]^{3+/2+}$ couple where the spin change effect is absent. The rate constant for the cobalt couple is reported to be smaller than 10^{-8} $dm^3 mol^{-1} s^{-1}$ while the rate constants of a OH^-- and a Cl^--catalyzed outer sphere path are $5 \cdot 10^{-3}$ $dm^3 mol^{-1} s^{-1}$ (G_A = 22 kcal) and $7 \cdot 10^{-4}$ $dm^3 mol^{-1} s^{-1}$ (G_A = 23 kcal), respectively (233) (at 337° K). The rate constant for the electron exchange in the $[Ru(NH_3)_6]^{3+/2+}$ couple at 298° K is $8.2 \cdot 10^2$ $dm^3 mol^{-1} s^{-1}$ and G_A = 13.6 kcal (E_A = 10.3 kcal, S_A = -11 cal K^{-1}), i.e. higher by 6-11 orders of magnitude. The Ru-N bond lengths are 2.14 A and 2.10 A for $[Ru(NH_3)_6]^{2+}$ and $[Ru(NH_3)_6]^{3+}$ (234), respectivly giving ionic radii of 2.75 A and 2.71 A and E_s = 36 kcal if R is taken as 7 A as before. The vibration frequencies of the Ru-N symmetric stretching mode is 474 cm^{-1} for $[Ru(NH_3)_6]^{3+}$ and probably only slightly less for $[Ru(NH_3)_6]^{2+}$ (189,235). If we take the value 437 cm^{-1} (235), then E_r = 3 kcal. From this we see that E_s is practically identical for the two couples, whereas differences in E_r can explain a rate difference involving a factor of 10^2-10^3. The remaining difference, amounting to a factor of 10^4-10^8, is then most likely located in the pre-exponential factor, i.e. the transmission coefficient. This is compatible with the observation of 'abnormally' large negative activation entropies in other electron transfer processes involving Co(III) complexes (16,230b) and the order of magnitude of this effect furthermore suggests that spin-orbit coupling is the electronic interaction which induces the process (cf. chapter 6).

5.3.2 Nucleophilic Substitution Reactions.

A similar analysis can in principle be performed for inner sphere electron transfer processes (18). However, several practical difficulties arise in these systems: (a) the processes are of a composite nature, and the electron transfer step is both preceded and succeeded by ligand substitution; (b) the electron transfer occurs in a binuclear complex for which thermodynamic and spectroscopic parameters cannot be measured, even though estimates can be made; (c) it is seldomly possible to vary any of these parameters in a systematic way (e.g. to provide a free energy relationship). (d) the mechanism of the actual electron transfer step may involve direct or coupled elementary processes, or it may rather be viewed as an atom group transfer process (cf. chapter 1 and 6); (e) there are seldomly extensive activation parameter data available, and in the few cases where they are avaible it is difficult to identify the elementary process to which they refer.

For these reasons we shall rather consider briefly the applicability of the semiclassical theory to a different class of processes, i.e. ligand substitution in inorganic metal complexes (98a,98c,236). These processes are really atom group transfer processes, which raises some important questions about the modifications of the general theory in the form which we have applied so far. This will be discussed further in chapter 6, and we shall at present assume that the outcome of the discussion provides sufficient justification for invoking the semiclassical theory for ligand substitution processes as well.

Both the acid hydrolysis of aqua and ammine complexes of cobalt(III), chromium(III) and rhodium(III) (98a), the substitution in square planar palladium(II) complexes (236), and the nucleophilic substitution in alkyl halides (98b) have been subject to theoretical analysis in terms of the semiclassical theorv. The analysis of each system raises several questions specific to their individual nature. We shall therefore choose a particular system, i.e. substitution in square planar palladium(II) complexes to illustrate the method. This system furthermore possesses the advantage that the mechanism of substitution in square planar d^8 metal complexes is generally well established experimentally (236,237) to be of the associative (A) or interchange (I_a) type, i.e. both breaking of the bond to the outgoing ligand and bond formation to the incoming ligand are important in the rate determining step.

The process to be considered is

$$[Pd(H_2O)_4]^{2+} + H_2NC_6H_4X \underset{k_{-1}}{\overset{k_1}{\rightleftharpoons}} [Pd(H_2O)_3(H_2NC_6H_4X)]^{2+} + H_2O \qquad (5.65)$$

where X is a substituent in the incoming ligand (aniline). This reaction can be followed spectrophotometrically, and both the forward (k_1) and back (k_{-1}) rate constants and activation energies determined. The forward rate constant was found to increase and the activation energy to decrease with increasing equilibrium constant of eq.(5.65), whereas the inverse relationship was found for the back reaction. With reference to the general possible reaction patterns of ligand substitution in square planar complexes (237) the data are conveniently interpreted in terms of the following three possible molecular mechanisms:

(a) An associative (A) mechanism with a rate determining formation of a pentacoordinated intermediate followed by a rapid release of a water molecule.

(b) An analogous associative mechanism where, however, the release of the outgoing H_2O molecule is rate determining, i.e. this step is preceded by a rapid pre-equilibrium involving the pentacoordinated intermediate.

(c) A synchronous incorporation of the incoming ligand and release of the outgoing ligand (an I_a mechanism (237)).

Rather than invoking a detailed kinematic pattern we shall assume that the motion of both the incoming and outgoing ligands can be represented by a single classical coordinate (the metal-ligand distance). Representing eq.(5.65) schematically as

$$MX + Y \rightleftarrows MY + X \tag{5.66}$$

the potential energy surfaces in the initial and final states are

$$U_i = \frac{1}{2}\sum_k \omega_k^2 \xi_k^2 + u_i(r_{MX}) + v_i(r_{MY}) \tag{5.67}$$

$$U_f = \frac{1}{2}\sum_k \omega_k^2(\xi_k - \xi_{ko})^2 + u_f(r_{MY}) + v_f(r_{MX}) + \Delta E \tag{5.68}$$

where r_{MX} and r_{MY} are the distances between the metal atom and the incoming and outgoing ligand, respectively. $u_i(r_{MX})$ and $u_f(r_{MY})$ are the potential energies of the M-X and M-Y bonds in the initial and final states, respectively, while $v_i(r_{MY})$ and $v_f(r_{MX})$ represent the intermolecular interaction between a molecular complex and a ligand group separated from the complex but still located in the 'collision complex'. We have not made any particular assumptions about the form of these potentials, except that $u_i(r_{MX})$ and $u_f(r_{MY})$ correspond to intramolecular motion, whereas $v_i(r_{MY})$ and $v_f(r_{MX})$ correspond to intermolecular motion. The characteristic velocity of the latter is expected to be close to that of the Debye motion in the solvent, which means that v_i and v_f are slowly varying functions compared with u_i and u_f. This assumption implies that the potential energy dependence on the coordinates r_{MX} and r_{MY} takes the form qualitatively illustrated in fig.5.2, in other words the characteristic frequency of the motion along r_{MX} decreases drastically during the process, whereas that of the motion along r_{MY} similarly increases.

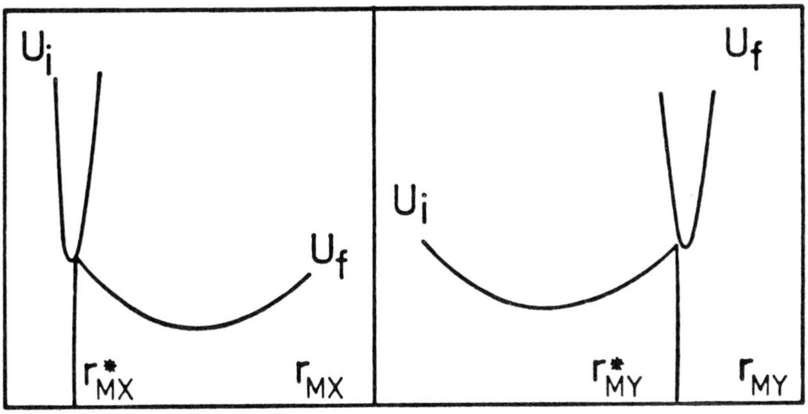

Fig (5.2)

For the sake of simplicity we shall assume that the activation energy can be determined by the saddle point of the intersection surface of the zero order surfaces given by eqs.(5.67) and (5.68), i.e. the effect of the electronic splitting is ignored. Following the procedure outlined in the previous section the saddle points are then given by the equations

$$\xi^*_{ks} = \alpha \xi_{ko} \tag{5.69}$$

$$(1-\alpha)\partial U_i/\partial r_{MX} + \alpha\partial U_f/\partial r_{MX} = 0 \tag{5.70}$$

$$U_i(\{\xi^*_{ks}\}, r^*_{MX}, r^*_{MY}) = U_f(\{\xi^*_{ks}\}, r^*_{MX}, r^*_{MY}) \tag{5.71}$$

and an equation similar to eq.(5.70) for r_{MY}. With our previous assumptions about the nature of $u_{i,f}$ and $v_{i,f}$ we can now see that provided neither $1-\alpha$ nor α are close to zero, the solutions to eqs.(5.69) and (5.70) are approximately

$$r^{*}_{MX} \approx r^{oi}_{MX} \quad \text{and} \quad r^{*}_{MY} \approx r^{of}_{MY} \tag{5.72}$$

where the superscripts refer to the appropriate equilibrium coordinate values in the initial and final states. In the 'transition state' the outgoing ligand is thus practically located at its initial state equilibrium configuration and does not contribute to the activation energy, whereas the incoming ligand is transferred practically to its final state equilibrium configuration. This is a quite general result, solely arising from the assumption of the large frequency differences of the two kinds of modes.

The activation energy is now $E_A = U_i(\{\xi^{*}_{ks}\}, r^{*}_{MX}, r^{*}_{MY})$. Insertion of eqs.(5.69)-(5.71) in eq.(5.67) gives

$$\alpha = \frac{1}{2} + (E^{X}_{r} - E^{Y}_{r} + E_{s})/2E_{s} \tag{5.73}$$

and

$$E_{A} = E^{Y}_{r} + (E_{s} + E^{X}_{r} - E^{Y}_{r} + \Delta E)^{2}/4E_{s} \tag{5.74}$$

where we have introduced the notations $E^{Y}_{r} = v_i(r^{of}_{MY})$ and $E^{X}_{r} = v_f(r^{oi}_{MX})$, i.e. the reorganization energy for bringing the incoming ligand from its initial state equilibrium position to its equilibrium position as a ligand in the complex, and in bringing the outgoing ligand from its position in the complex to its final state equilibrium position separated from the complex. E_s is the solvent reorganization energy which may also include reorganization energy terms from the modes of other ligands which are not substituted.

We shall now exploit these results in examining the three possible mechanisms of the Pd(II)/ aniline reactions. Recalling that the activation energies for both the forward and back reactions were determined experimentally we notice that for mechanism (a)

$$E_{Af}^{(a)} = E_r^{an} + (E_s + \Delta E_2 - E_r^{an})^2/4E_s \qquad (5.75)$$

$$E_{Ab}^{(a)} = \Delta E_1 + (E_s - \Delta E_2 + E_r^{an})^2/4E_s \qquad (5.76)$$

where the subscripts 'f' and 'b' refer to the forward and back reaction, respectively, ΔE_1 is the reaction enthalpy for association of the water ligand to form the intermediate (the distance between the minima of the corresponding potential energy surface), and ΔE_2 has the same meaning when aniline is the incoming ligand ($\Delta E_2 - \Delta E_1 = \Delta E$, where ΔE is the overall enthalpy of reaction for the process). Similarly, for mechanism (b)

$$E_{Af}^{(b)} = \Delta E_2 + (E_s - \Delta E_1 + E_r^{H_2O})^2/4E_s \qquad (5.77)$$

$$E_{Ab}^{(b)} = E_r^{H_2O} + (E_s + \Delta E_1 - E_r^{H_2O})^2/4E_s \qquad (5.78)$$

The experimental values take the form (E_A in kcal)

$$E_{Af}^{ex} = 8.5 + (0.195\,\Delta E + 2.75)^2 \qquad (5.79)$$

$$E_{Ab}^{ex} = 16.1 + (0.195\,\Delta E - 0.181)^2 \qquad (5.80)$$

In both cases this requires that the fast step of formation of the intermediate is strongly endothermic (by 8.5 kcal and 16.1 kcal) corresponding to a much higher activation energy than the slow step. This seems very unlikely, and furthermore, a fit of either of the theoretical activation energy expressions, eqs.(5.79) and (5.80), gives a negative value of at least one of the reorganization energy terms.

In contrast, for the interchange mechanism (mechanism (c)) eq.(5.74) takes the form

$$E_A^{(c)} = E_r^{an} + (E_s + \Delta E + E_r^{H_2O} - E_r^{an})^2/4E_s \qquad (5.81)$$

and the following physically reasonable parameter values provide a good fit to the experimental data: E_s = 6.6 kcal, E_r^{An} = 8.5 kcal, and $E_r^{H_2O}$ = 16.1 kcal.

This formulation of the semiclassical theory thus allows us not only to estimate the important kinetic parameters by comparison with experimental data but also to distinguish between various possible mechanisms.

6 ATOM GROUP TRANSFER PROCESSES

6.1 General Features of Nuclear Motion

Atom group transfer (AT) processes represent a special class of vibrational and electronic relaxation phenomena in which large reorganization occurs in certain intramolecular modes, i.e. those associated with the motion of the transferring atom group. This kind of process is widely represented in chemical reactions such as nucleophilic substitution (32,237,238), inner sphere electron transfer processes (13,24,32), and in proton and hydrogen atom transfer reactions (11,19). However, AT between different molecular fragments also occurs in several other - largely solid-state - processes, among which we notice in particular:

(a) In amorphous solids atom groups may be located in two different neighbouring positions separated by an energy barrier. At low temperatures this gives rise to the formation of 'tunnelling states', i.e. a splitting of the vibrational energy levels in the neighbouring potential wells, associated with the finite probability of atom group tunnelling through the barrier (6-8). Since the barrier widths and heights in the amorphous solids are almost continuously distributed this gives rise at low temperature to an additional specific heat (and thermal conductivity) from the phonon energy absorption associated with the transition between the tunnelling levels. Both experimental data and a closer theoretical analysis show that this 'anomalous' specific heat is proportional to the absolute temperature (as opposed to the T^3 dependence expected for crystalline materials at low temperatures).

(b) Rotational and translational motion of lattice impurities in crystalline solids, e.g. rotation of CN^- and OH^- or hindered translation of Li^+ in alkali halide lattices (239). These effects are revealed by fine-structure in the infrared absorption spectra and by the temperature dependence of the thermal conductivity at low temperatures (a few K). At low temperatures the 'impurities' are localized in individual librational wells but interaction between neighbouring wells leads to a splitting, most frequently of the range up to a few cm^{-1}. The appearance of these new 'tunnelling levels' are the origin of the effects listed.

(c) The dynamics of hydrogen-bonded systems (e.g. the vibrational properties of water, ice, carboxylic acids etc, proton mobility) is physically closely related to proton transfer reactions in chemical systems (240). These systems are for example subject to both single and correlated proton tunnelling giving rise to a fine-structure in the infrared absorption peaks. Moreover, the theory of local defect dynamics is commonly handled within a modified Born-Oppenheimer approximation - in which the proton stretching modes constitute the fast subsystem and are strongly coupled to slower local modes or to the lattice phonon modes constituting the slow subsystem - to give optical line shape functions analogous to the ones derived in chapter 3.

(d) Diffusion of 'interstitials' or lattice defects through crystal lattices also involves essentially atom group transfer between neighbouring local equilibrium positions. Examples are the diffusion of excess cations or anions in alkali and earthalkaline halides (5,128,241) and diffusion of light atoms (in particular hydrogen) in metals (242).

(e) We should finally notice that intramolecular conversions in 'fluxional' molecules bears a resemblance to the kind of AT processes to be discussed in the following. The classical example is the inversion of the ammonia molecule which exists in two equilibrium configurations corresponding to inversion of the pyramidal structure. The molecules can thus be represented by a symmetric double-well potential energy surface in which the 'zero order' levels in each well are split due to finite probability of localization of the proton in either well. Direct evidence for the proton tunnelling between the two configurations was first obtained by the microwave spectrum of gaseous ammonia which shows a characteristic peak at about 1 cm^{-1} corresponding to the magnitude of the level splitting (243).

There are now several examples of fluxional molecules in which rapid interconversion between different configurations of equal energy occurs (244). However, the ammonia molecule is of particular interest being the first example of application of quantum mechanics to the problem of particle penetration through a potential energy barrier (245) preceding both the first study of field-induced electron emission from metals (246) and the radioactive decay of atomic nuclei (247).

In the theory of chemical AT processes it is often adequate to assume that the molecular nuclear framework is deformed to some configuration intermediate between those of the initial and final states, prior to the electronic reorganization, and that the energy of this state (the activation energy) can be calculated by classical mechanics (cf. chapter 5). However, it was suggested early (248-251) that this nuclear transition may also proceed by a quantum mechanical tunnel effect. We shall discuss this by considering nuclear motion along a particular nuclear coordinate, Q, between two bound states separated by a barrier, V(Q) (fig. 6.1). The quantum statistical probability that Q assumes a particular value, Q_o, is then (33,34,145)

$$I(Q_o) = Z^{-1}\sum_n |\chi_n(Q_o)|^2 \exp(-E_n/k_BT) \qquad (6.1)$$

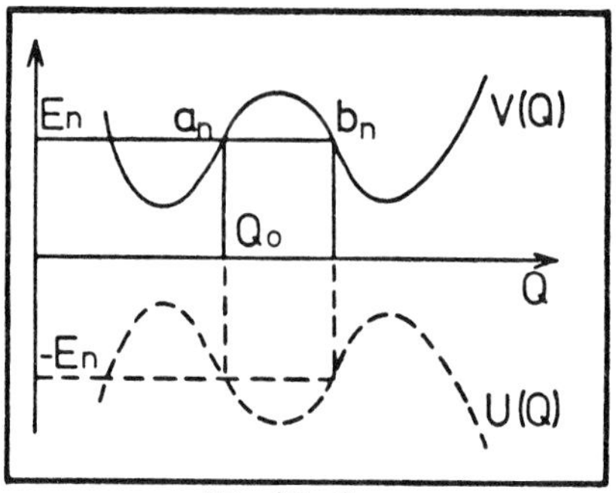

Fig(6.1)

where $\chi_n(Q)$ is the nuclear wave function associated with the mode, E_n the energy of the level n, and Z the statistical sum of the appropriate mode. For sufficiently low energies Q is located in the classically forbidden region, and Q_o can only be reached by tunnelling. In this case $|\chi_n(Q_o)|$ increases with increasing values of n due to the lowering of the barrier height for nuclear penetration. On the other hand, the Gibbs factor decreases with increasing n, and as a result of these two opposite effects a group of levels around a given number n^*, gives the dominating contribution to $\Phi(Q_o)$. If the barrier potential rises sharply with Q, the Gibbs factor will decrease more rapidly with increasing E_n than $|\chi_n(Q)|$ increases. The system therefore passes through the barrier by nuclear tunnelling from the ground vibrational level, i.e. $n^* = 0$, and

$$\Phi(Q_o) \approx |\chi_o(Q_o)|^2 \tag{6.2}$$

If, on the other hand, V(Q) is a sufficiently slowly varying function of Q, then the nuclei have to penetrate the classically forbidden region over a large distance, and levels near the barrier peak now give the dominating contribution to $\Phi(Q)$, which can therefore be approximated by a classical Boltzmann distribution function, i.e. (33,34,145)

$$\Phi(Q) = A\exp[-V(Q)/k_BT] \qquad (6.3)$$

where A is a normalization constant. For a harmonic oscillator with the frequency ω and mass μ, $\Phi(Q)$ takes the form (145)

$$\Phi(Q) = \left(\frac{\mu\omega}{\pi\hbar}\,\mathrm{th}\frac{\hbar\omega}{2k_BT}\right)^{\frac{1}{2}}\exp\left[-\frac{\mu\omega Q^2}{\hbar}\,\mathrm{th}\frac{\hbar\omega}{2k_BT}\right] \qquad (6.4)$$

and the limits of quantum and classical motion corresponding to eqs.(6.2) and (6.3) are obtained from the high- and low-temperature limits of the hyperbolic tangent

$$\Phi_{qu}(Q) = \left(\frac{\mu\omega}{\pi\hbar}\right)^{\frac{1}{2}}\exp(-\mu\omega^2Q^2/\hbar) = |\chi_o(Q)|^2 \qquad (6.5)$$

and

$$\Phi_{cl} = (\mu\omega^2/2\pi k_BT)^{\frac{1}{2}}\exp(-\mu\omega^2Q^2/2k_BT) \qquad (6.6)$$

respectively. The activation energy can thus be calculated from the statistical distribution functions corresponding to the classical low-frequency modes only ($\omega < k_BT/\hbar$). Reorganization of modes for which $\omega > k_BT/\hbar$ does not primarily contribute to the activation energy but is reflected in the pre-exponential factor providing apparently negative activation entropy contributions.

A semiquantitative criterion for the nature of nuclear motion across a potential energy barrier can be given by means of the quasiclassical approximation for the nuclear wave function. Thus, in the classically forbidden region (142)

$$|\chi_n(Q)|^2 \propto \exp\left\{-\frac{2}{\hbar}\int_{a_n}^{Q}\{2\mu[V(Q)-E_n]\}^{\frac{1}{2}}dQ\right\} \qquad (6.7)$$

where a_n is the left-hand limit of the classically forbidden region. The characteristic energy interval over which the Gibbs factor decreases is of the order of $k_B T$. The corresponding interval over which $|\chi_n(Q)|$ increases is found by noting that in the potential well which results by reflection of the barrier of fig. (6.1) in the Q axis the action integral is equal to the exponent of the barrier penetration probability, in other words (33,34,145)

$$\frac{2}{\hbar}\int_{a_n}^{b_n}\{2\mu[V(Q) - E_n]\}^{\frac{1}{2}}dQ = \frac{2}{\hbar}\int_{a_n}^{b_n}\{2\mu[-E_n - U(Q)]\}^{\frac{1}{2}}dQ \qquad (6.8)$$

where U(Q) is the reflected potential, and b_n the right-hand turning point. The latter factor is by the Bohr-Sommerfeld quantization rule (142)

$$\frac{2}{\hbar}\int_{a_n}^{b_n}\{2\mu[-E_n - U(Q)]\}^{\frac{1}{2}}dQ = 2\pi(n + \frac{1}{2}) \qquad (6.9)$$

A convenient estimate of the characteristic energy interval over which $|\chi_n(Q)|$ - and therefore the barrier penetration probability - changes, is thus given by the energy level difference $\Delta E_n = E_{n+1} - E_n$ in the inverted potential. If we only consider barriers for which ΔE decreases monotonically with increasing n (e.g. a triangular or an Eckart barrier) we can conclude:

(1) If

$$\Delta E_o < k_B T \qquad (6.10)$$

the system passes the barrier classically.

(2) If

$$\Delta E_{N-1} > k_B T \qquad (6.11)$$

where N is the number of levels in the inverted potential, then the corresponding nuclear mode tunnels from the initial state ground vibrational level, and $\Phi(Q) \approx |\chi_0(Q)|^2$.

(3) If, finally, the inverse of both inequalities are valid ($\Delta E_{N-1} < \Delta E_0$) the mode is excited to some 'effective' level from which the transition proceeds by a tunnel effect, and the number of which, n^* ($0 < n^* < N$), is determined by maximizing eq.(6.1) with respect to E, i.e. from the condition $2\pi k_B T \approx E_{n^*}$.

Since most real systems possess frequencies of such values that they are weakly thermally excited at room temperature, we can also conclude:

(a) The classical or quantum behaviour of a particular nuclear mode depends critically on the temperature. For temperatures higher than some value, T_{cl}, all modes are classical, and the reaction displays an Arrhenius temperature dependence. On the other hand, below a certain value, $T_{qu} < T_{cl}$ all modes satisfy the condition of eq.(6.11) and undergo a tunnel transition with an apparently vanishing activation energy. Finally, in the intermediate temperature range $T_{qu} < T < T_{cl}$ thermal excitation to an intermediate level followed by tunnelling from this level occurs. In this temperature region both the apparent activation energy and the pre-exponential factor depend on the temperature.

(b) The apparent temperature dependence is modified when several modes are reorganized during the process. These effects are of two kinds:

(1) If two modes of sufficiently widely separated characteristic frequencies are present, such as solvent and intramolecular

modes, then all modes are again classical at sufficiently high temperatures and display tunnelling behaviour at sufficiently low temperatures. However, in an intermediate temperature region the low-frequency (medium) modes are classical and the high-frequency modes of quantum nature. The presence of several modes may therefore give rise to a 'structure in the Arrhenius relationship.

(2) For reactions in condensed media the reaction centre is always coupled to a continuous manifold of medium nuclear modes, characterized by a certain frequency dispersion. At sufficiently low temperatures the number of thermally excitable modes is therefore an increasing function of the temperature. This may conceal a finite activation energy in a phenomenological Arrhenius relationship (cf. section 3).

6.2 Semiclassical Approaches to Atom Group Transfer

In the present section we shall only discuss the simplest semiclassical models commonly adopted in the description of condensed phase AT processes. In the context of chemical processes, proton transfer has received by far the greatest attention in the literature, and a vast amount of data relating to a variety of systems in this category is available (19,238,252-254). Since among room temperature processes in the 'normal' free energy range quantum effects are also most clearly manifested here, we shall refer our discussion to this important class in particular, even though many of the conclusions are valid for other AT processes as well.

The major aims of the attempts to formulate a theory for proton transfer reactions have been to rationalize the following phenomena:

(a) The Bronsted relationship between kinetic and thermodynamic parameters. For a series of 'sufficiently closely related' reactions this relationship is usually expressed as (11,19)

$$\log k_r = \text{const} + \alpha \Delta pK \tag{6.12}$$

where k_r is the rate constant, ΔpK the difference in the pK values of the proton donor (acid) and the proton acceptor (base), and α a coefficient between zero and unity which is constant over sufficiently narrow intervals of ΔpK.

(b) The qualitative difference between Bronsted plots for reactions in which the proton donor and acceptor atoms are oxygen or nitrogen and those in which it is carbon. In the former group a transition between a diffusion controlled activationless region ($\alpha \approx 0$) and a barrierless region where $\alpha \approx 1$ occurs within a few pK units (a few kcals) (255), whereas in the latter group this transition occurs over a much wider interval.

(c) The absolute values of deuterium and tritium kinetic isotope effects.

(d) The variation of the kinetic isotope effects with the free energy of reaction. Typically a maximum value of the isotope effect is observed for $\Delta pK = 0$.

According to the semiclassical view of proton transfer reactions in solution the proton transfer between the donor (D) and acceptor (A), proceed in a collision complex

$$D - H \cdots\cdots A \rightarrow D \cdots\cdots H - A \tag{6.13}$$

When the transferring group is a light particle, such as a proton, the D and A residues are considered to be stationary, and (for the sake of simplicity) the motion of H described by a single coordinate, i.e. the stretching coordinate of H. When the mass of the transferring group approaches that of D or A several intramolecular coordinates are required, and a change of normal modes moreover occurs when going from the initial to the final state. In the simplest formulation the AT can therefore be described by a one-dimensional double-minimum potential energy surface where the minima correspond to the equilibrium positions of the transferring group at the donor and acceptor.

Within this simple picture the Bronsted coefficient provides a measure of the transition state structure. Thus, a substituent in one of the reactants sufficiently far from the AT centre will induce a vertical shift in the relative position of the two wells without deforming the potentials or changing the AT distance. α is therefore a measure of the 'extent' of proton transfer in the transition state, and it is given by the equation

$$\alpha = dE_A/d\Delta E \qquad (6.14)$$

where E_A is the activation energy and ΔE the distance between the two minima. When $\alpha \to 0$ the final state surface passes through the minimum of the initial state, and the transition state then coincides with the initial state. Similarly, when $\alpha \to 1$, the transition state coincides with the final state.

The simple one-dimensional semiclassical picture is thus capable of explaining the experimentally observed smooth Bronsted relationships and the fact that $0 < \alpha < 1$. It can also account for the qualitative difference in Bronsted relationships for systems involving different donor and acceptor centres. The free energy range over which the reaction shifts from activationless to barrierless, is determined by the vibration frequencies and proton transfer distances. The proton transfer distance is likely to be larger in processes involving C-acids than in those involving O- or N-acids due to the less effective hydrogen-bonding in the former systems, and is seen to provide a larger free energy interval over which α changes fron zero to unity, or a less curved Bronsted plot.

In the semiclassical picture the role of the solvent in proton transfer reactions is usually considered to be threefold: (1) solvent molecules may constitute part of the transition state complex providing a special reaction path for the proton transfer (255,256); (b) a certain amount of desolvation may have to occur before the collision complex is formed, corresponding to finite work terms of nonelectrostatic origin (254) (cf. chapters 1 and 5); (c) the solvation is different in the initial, final, and transition states. This is, however, a static effect rather than the dynamic effect ascribed to the solvent in the quantum mechanical formulation of ET and AT processes. However, in molecular terms, the explicit role of the solvent is in fact most commonly ignored and only indirectly considered as determining the geometry of the collision complex, the free energy of reaction, the proton transfer distance, etc.

The last effect of major concern is the kinetic isotope effect, expressed quantitatively by the ratio of the rate constant of a proton transfer reaction, k_H, and the corresponding deuteron (k_D) or triton transfer (k_T), and quantum effects in chemical processes at room temperature are probably revealed more clearly by kinetic isotope effects in proton transfer reactions than in any other kinetic phenomena.

According to a theory originally suggested by Westheimer (257) the kinetic isotope effect arises in the following way: The potential wells for the bound proton and heavier isotope are practically identical. Initially this gives rise to a lower vibrational zero-point energy for the heavier isotope due to the lower frequencies ($\hbar\Omega_D < \hbar\Omega_H$). If the transition state is viewed as a trinuclear centre [D.....H.....A], then the two bending and one stretching modes of the proton and the mode of intermolecular motion of DH and A are converted to two doubly degenerate bending, one symmetric, and one asymmetric stretching mode. Both bending modes involve hydrogen motion, i.e. they are isotope sensitive and since the hydrogen is much lighter than the donor and acceptor fragments, the zero point energies of these two modes are roughly equal to those of the bending modes in the initial state. The asymmetric stretch is associated largely with 'translational' motion of the proton and this velocity varies approximately as the inverse of the square root of the isotope mass. Finally, the symmetric stretch involves a substantial motion of the heavier fragments and is therefore expected to have a low frequency and to be less isotope dependent. For symmetric transition states this mode involves a stationary transfer atom and is then insensitive to isotope substitution.

Within this simple model the isotope effect is predicted to have a maximum value for reactions which possess a symmetric transition state (ΔpK = 0). The effect primarily originates from the loss of zero-point energy in one vibrational stretching mode when going from the initial to the transition state, i.e. the maximum value is determined by the difference in activation energy arising from the zero-point energy difference between modes of the two isotopes. This effect is therefore given by $\exp[(\varepsilon_0^H - \varepsilon_0^D)/k_B T] \approx 6\text{-}10$, where ε_0^H and ε_0^D are the zero-point energies of the proton and deuteron modes, respectively. Reactions which are not thermoneutral have asymmetric transition states, i.e. there is now some zero-point energy involving isotope motion in the symmetric stretch mode in the transition state. The loss of zero-point energy is then smaller than $\varepsilon_0^H - \varepsilon_0^D$ giving also a smaller isotope effect.

Experimentally observed deuterium isotope effects are in fact often located in the range 3-10 and show maximum values at approximately ΔpK = 0. However, attempts to go beyond the qualitative picture have raised a number of difficulties (19). Thus, the variation of k_H/k_D with ΔpK must be related to the variation of the force constants in the activated complex. However, these parameters cannot usually be associated with any observable characteristics of the processes. Also, model calculations give force constant values which generally vary very little with ΔpK giving correspondingly slow variations in the isotope effects. In addition, experimental values of kinetic isotope effects frequently display the following features which are quite incompatible with this approach:

(a) The isotope effects may be 'unusually' large, i.e. much larger than predicted from the zero-point energy loss.

(b) The isotope effect increases with decreasing temperature with an 'activation energy' which is larger than $\varepsilon_0^H - \varepsilon_0^D$, i.e. the difference in activation energy for the proton and deuteron transfer is greater than this amount.

(c) The Arrhenius relationship may be nonlinear corresponding to a decreasing activation energy with decreasing temperature.

(d) The pre-exponential Arrhenius factor may be lower for the lighter isotope than for the heavier one by orders of magnitude. This ratio is predicted to be close to unity on the basis of the semiclassical theory.

These effects may be very pronounced for proton or hydrogen atom transfer processes in solid glasses at low temperatures. For example, in the hydrogen atom transfer reactions between acetonitrile or methyl isocyanide and methyl radicals in solid glasses, recently studied by Williams and associates (258), the apparent activation energy increases from 1.4 to 4.5 kcal/mol when the temperature is increased from 77°K to 125°K, and the gas phase activation energy in the temperature region 373-573°K is about 10 kcal/mol. Moreover, the pre-exponential factors for the solid-state reactions are lower than that of the gas phase reaction by many orders of magnitude, and a lower limit of the deuterium isotope effect was estimated to be about 10^3.

These effects are usually explained by quantum mechanical tunnelling of the proton, whereas the heavier isotopes are assumed to pass over the barrier by thermal activation.

Together with the assignment of spectroscopic doublets (as indicated above) these effects in fact provide some of the most accurate and direct ways of detecting quantum mechanical nuclear tunnelling. The tunnelling is usually viewed as a one-dimensional stationary barrier problem in which the proton stretching motion is identified with the reaction coordinate. The basis of further analysis is then eq.(6.15)

$$W_{fi} = Z^{-1} \sum_{n} \Gamma(E_n) \exp(-E_n / k_B T) \tag{6.15}$$

where $\Gamma(E_n)$ is identified by the Gamov tunnelling probability for penetration through the barrier, i.e. (142,247)

$$\Gamma(E_n) = \exp \left\{ - \frac{2}{\hbar} \int_{a(E_n)}^{b(E_n)} \left\{ 2m[U(Q) - E_n] \right\}^{1/2} dQ \right\} \tag{6.16}$$

and E_n is a particular energy level, m the mass of the proton, and U(Q) the barrier potential along the proton stretching mode Q.

In subsequent steps U(Q) is approximated by various potentials (Eckart or parabolic barriers) and the summation over E_n in eq.(6.15) is replaced by an integration assuming essentially classical proton motion.

Proton tunnelling can explain the effects (a) to (d) above (19,252). We notice in particular two effects here. Firstly, by analysis of experimental data on temperature effects in terms of the tunnelling correction formulas it is possible to obtain effective barrier heights and widths and the energy values which give the dominating contribution to the summation in eq.(6.15). One interesting conclusion is that even in cases where tunnelling corrections are substantial for proton transfer, it is much smaller for deuterium or tritium transfer due to their larger masses. Passage of the barrier by thermal activation is thus favoured relative to tunnelling for these particles. This can explain the temperature effects listed above. Secondly, since the reaction coordinate is identical to the coordinate along which the proton tunnels, the tunnelling corrections should have maximum values for thermoneutral reactions and vanish for both activationless and barrierless processes where the Bronsted coefficient approaches zero and unity, respectively. While the former effect is borne out by experimental data, the latter seems to be in conflict with recent experimental evidence for the electrochemical hydrogen evolution reaction (259). We shall discuss these experiments further in chapter 8.

We conclude this section by noting that when the motion of the AT group is separable from other intramolecular and solvent modes the literature on semiclassical methods in gas phase kinetics is in principle also available for condensed phase processes. This was exploited by Marcus (260) who applied the bond-energy-bond-order (BEBO) method (261) to proton transfer reactions in solution. The BEBO method focuses on the motion of the AT group and is therefore particularly adequate when the medium is weakly coupled to the reaction centre (e.g. apolar solvents). The potential energy of the AT group is represented by an empirical relationship between the order (n) of the bond to be broken or formed and the bond energy (V), of the form

$$V = -V_o n^p \qquad (6.17)$$

where V_o is the constant potential energy of the initial equilibrium state, and p an empirical constant of the order of unity. The activation energy for motion along this mode is found by maximizing the potential energy of the atom transfer centre with respect to n under the restriction that the sum of the orders of the bonds which are being broken and formed, remains constant.

The BEBO and other semiclassical methods can be combined with an account of the intramolecular and solvent modes to provide a general theory applicable to a wide variety of condensed phase processes and represent in this version an appropriate complement to the quantum mechanical approach. However, it is inadequate for example in relation to several quantum effects of nuclear motion manifested in low-temperature AT processes. For a proper account of these and other effects it is therefore necessary to reformulate the AT theory in quantum mechanical terms.

6.3 Quantum Mechanical Formulation of Atom Group Transfer

6.3.1 Nuclear Tunnelling between Bound States

In our attempts to reformulate the theory of condensed phase AT processes we notice at first that in these processes the reaction centre is typically coupled to two sets of nuclear modes: (a) the high-frequency molecular modes, which include in particular those associated with the group to be transferred; (b) the low-frequency medium modes. When the AT process involves the transfer of electrically charged groups the coupling to the medium modes is expected to be strong, while in cases where the transferring group is electrically neutral, the role of the medium coupling is primarily to ensure vibrational relaxation of the modes associated with the transferring group subsequent to the AT step. In any event, the AT processes are most typically viewed as a special class of multiphonon radiationless processes and therefore expected to exhibit a number of features characteristic for this general class.

However, the role of the molecular group transfer modes and the medium modes also differ in a way which is analogous to ET processes. We can see this by considering again the single AT mode of fig.(6.2), corresponding to the motion of the molecular group on a one-dimensional double-well potential surface. Considering at first a symmetrical double-well potential we notice that this represents localization of identical nuclei at different equilibrium sites. However, the localization of the system in either the left- or right-hand well does not represent a stationary state of the system.

In order to find the stationary states it is necessary to solve the time-independent Schrodinger equation using an appropriate approximation for the double-well potential (245,253,262-265). The nature of these solutions can, however, be illustrated by expanding the wave functions in terms of 'basis' functions corresponding to localization of the molecular group in either of the two wells separately, i.e. in terms of the solution to the Schrodinger equations of each well if no interaction between the wells occurred. This could correspond to a representation of the real potential by a superposition of two single-well potentials the wave functions of which would then constitute the basis set. For the symmetric double-well potential the nuclear wave functions could then be represented as $\Psi_+ = 2^{-\frac{1}{2}}(\phi_l + \phi_r)$ and $\Psi_- = 2^{-\frac{1}{2}}(\phi_l - \phi_r)$, i.e. the symmetric and antisymmetric combinations of the wave functions of the left- and right-hand well separately, ϕ_l and ϕ_r. In terms of this formulation, there is now a finite probability of finding the molecular group in the barrier region due to the finite value of the wave function in this region. This is commonly expressed as tunnelling of the molecular group from one well to the other.

The single-well wave functions ϕ_l and ϕ_r are associated with a set of discrete (degenerate) energy levels also indicated in fig.(6.2). However, if the barrier width is sufficiently small, the finite overlap between ϕ_l and ϕ_r causes a splitting of the zero-order levels. For the ammonia inversion and several of the solid-state rotational systems the splitting is commonly a few wave numbers and can thus in principle, and in some cases in practice, be detected by fine-structure in the IR spectra or by microwave spectroscopy.

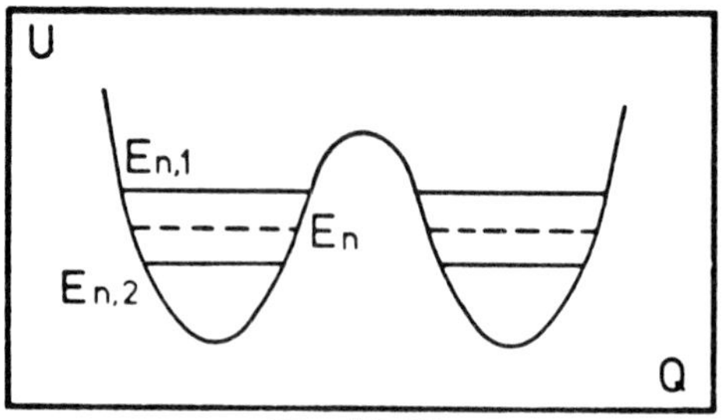

Fig (6.2)

The splitting of the zero-order levels thus define the new stationary energy levels in the double-well potential. The energy differences, $E_{n,2} - E_{n,1}$ where $E_{n,m}$ is the energy of the level which corresponds to the n'th zero-order level and m (= 1,2) refers to the components into which it is split, is commonly related to a frequency of tunnelling, τ_T^{-1} , by the (uncertainty) relation $\Delta \mathcal{E} = E_{n,2} - E_{n,1} = \hbar/\tau_T$. This means that if the system is 'prepared' in the left-hand well at a given time t = 0, then at t = τ_T it has been transferred to the righ-hand well by tunnelling, at t = $2\tau_T$ it is back in the left-hand well etc. This also points to the conditions under which the nonstationary states could possibly be detected. Quite generally, the measurement of some property, R, requires a certain interval, Δt, corresponding to an energy uncertainty $\Delta \mathcal{E}$. If $\Delta \mathcal{E}$ is large compared with the separation of the nonstationary states we may observe an 'average' value of R appropriate to mixtures of stationary states. Since the states ϕ_l and ϕ_r correspond to such mixtures this means that a sufficiently fast measuring device may detect the nonstationary states, i.e. the system 'caught' in one of the wells. On the other hand, a large Δt corresponds to a small $\Delta \mathcal{E}$. If $\Delta \mathcal{E}$ is smaller than the level splitting the stationary states can now be detected. This is the situation prevailing in the 'slow' techniques of micro- and radiowave spectroscopy.

Asymmetric double-well potentials correspond to a representation of ψ_+ and ψ_- in the form $\psi_\pm = N(\phi_l \pm \lambda \phi_r)$, where N is a normalization constant and λ a 'mixing' coefficient. The most striking result is now that even for a very slight asymmetry the ground state wave function is strongly located in the lower well, and only thermal excitation to a higher level may induce tunnelling.

We notice finally that the considerations above implicitly have assumed that the AT processes are electronically adiabatic, i.e. we have only considered a single nuclear potential energy surface and presupposed that the electrons adiabatically follow the nuclear motion.

We now turn to an application of these principles to condensed phase AT processes. In the limit of high temperatures the AT mode and the medium modes are equivalent and define together a many-dimensional classical potential energy surface from which the activation energy and rate probability can be determined in principle. At present we shall, however, assume that the AT modes have sufficiently high frequencies that they exhibit nuclear quantum effects. This is expected for the important class of proton transfer reactions at room temperature and otherwise for AT processes in general at sufficiently low temperatures. In addition to ensuring vibrational relaxation the role of the medium modes in AT processes is then three-fold. Firstly, when the medium modes are subject to appreciable coordinate displacement, they constitute an additional set of 'accepting' modes, i.e. modes which provide and accept the thermal activation energy and the energy arising from the energy gap between the initial and final state. Secondly, in the limit of nonadiabatic processes the vibrational levels of the molecular group in the initial and final states only coincide approximately for certain relative positions of the initial and final state surfaces. In all other cases the probability of nuclear tunnelling vanishes due to the gap between the vibrational donor and acceptor levels. However, fluctuations in the medium coordinates, of the kind discussed in chapters 1, and 2, may induce this degeneracy and the reaction probability is then determined by the characteristic times of electronic and (medium) nuclear motion as discussed in chapter 5. Thirdly, in the limit of electronically adiabatic processes where only a single potential energy surface is considered the role of the medium is important by defining the characteristics of the AT mode. Thus, if we consider the nuclear potential energy as a function of this coordinate, then the medium determines the topology of the resulting surface.

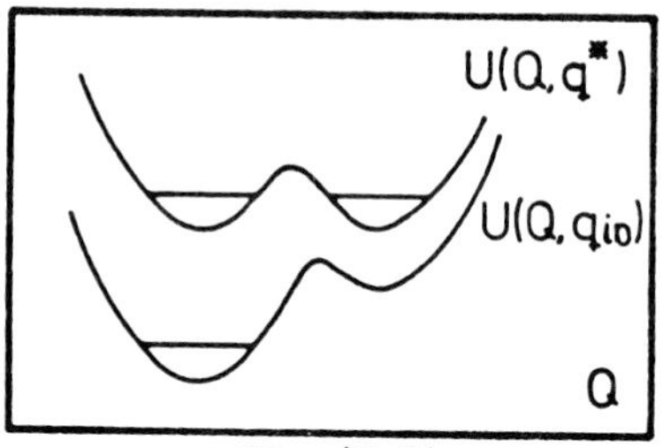

Fig (6.3)

With reference to fig.(6.3) which we may interprete as a representative for a proton transfer reaction, we notice for example that when the solvent configuration is that of the initial state (q_{io}) in which the proton is localized at the donor, the energy of this state is lowest and there is no possibility of proton tunnelling. Fluctuations in the solvent coordinates to a value $q^* \neq q_{io}$ provide, however, a state in which the proton vibrational levels in the two wells coincide. This gives rise to a splitting of the zero-order levels, or a finite probability of nuclear tunnelling. If the splitting is large or the tunnelling time low compared with the time during which the coordinates of the solvent system are close to q^*, there is also a large probability of proton transfer in this region. Finally, subsequent to the AT step the solvent system now relaxes to its final state equilibrium value q_{fo}.

6.3.2 Adiabatic and Nonadiabatic AT

When attempting to formulate a quantum mechanical theory for AT processes we should recognize that even though progress towards this aim has certainly been made (95,96,267), the theory is still far less comprehensive than the theory of simple nonadiabatic ET processes. This is due to several causes:

(a) Nonadiabatic ET processes involve electronic transitions between recognizable zero-order electronic states which correspond to the localization of the electron on either the donor or the acceptor. In contrast, AT processes in which chemical bonds are broken and formed, are commonly expected to be electronically adiabatic. However, proton transfer reactions and certain heavy atom group transfer processes involving spin multiplicity changes may correspond to the nonadiabatic limit (cf. section 6.4).

(b) AT processes are subject to large displacement in the stretching mode of the transferring group. In this way AT processes are analogous to such intramolecular electronic processes as photoinduced cis-trans isomerization of excited organic molecules (268) in which the torsional motion of the molecular 'halves' around the double bond is analogous to the AT mode. Unless the mass of the transferring group is small compared with the molecular donor and acceptor fragments (e.g. in proton transfer reactions) the motion of this group cannot be separated from the intramolecular motion of other parts of the reaction centre. This would lead to a change of the normal modes when going from the initial to the final state and may be to a change in the character of some modes (e.g. from quantum to classical, or from vibrational to rotational motion), and incorporation of these effects requires an elaborate analysis (33,90,269).

With these reservations we shall now proceed to a formulation of the theoretical results for AT processes by considering in turn the nonadiabatic and the adiabatic limits.

<u>(a) Nonadiabatic AT Processes.</u> This limit prevails when the residual electronic interaction which couples the zero-order states is sufficiently small compared with the vibrational level spacing of the AT mode. Our approach towards an AT theory then essentially follows the lines for ET processes as outlined in chapters 3 and 4 and can be summarized as follows:

(1) The entire system can be characterized by two distinct zero-order electronic states. These states correspond to the isolated reactants but equilibrium bond lengths etc. may be modified by the 'diagonal' part of the residual interaction (cf. eqs.(3.43) and (3.46)).

(2) For each of these states we can construct many-dimensional Born-Oppenheimer potential energy surfaces determined by the nuclear displacements of the entire system.

(3) Two sets of vibronic levels for the nuclear potential surfaces can subsequently be found. These two sets constitute the quantum mechanical initial and final (zero-order) states of the system.

(4) A microscopic rate constant is derived by considering the system to be initially present in a vibronic level of the initial state potential energy surface. Residual interactions which were not incorporated in the zero-order Hamiltonians then couple the initial state vibronic level to a manifold of final state vibronic levels, degenerate with the initial state level. The initial state level is then metastable and undergoes a decay process, and when the manifold of the final state levels is dense, such as for condensed phase systems, the decay is irreversible (2,3,167).

(5) When the residual coupling is weak relative to the vibrational frequencies of the AT modes, all the microscopic decay processes can be described in terms of first order time dependent perturbation theory. This is the basic feature of the nonadiabatic description of rate processes.

(6) The time dependent perturbation theory gives microscopic rate constants determined by the Franck Condon nuclear overlap integrals which can be handled by the theory outlined above.

(7) The macroscopic nonadiabatic rate is finally expressed in terms of a thermal average of the microscopic rates, the averaging being taken over the manifold of initial vibronic states (cf. eqs.(3.19), (3.47), and (4.5)).

Following this scheme and the procedure discussed in chapters 3 and 4 we can write the macroscopic nonadiabatic AT rate constant as

$$W = \frac{2\pi}{\hbar}|V_{eff}|^2 Z^{-1} \sum_v \sum_w \exp(\beta E^o_{iv}) |\langle X^o_{fw}|X^o_{iv}\rangle|^2 \delta(E^o_{fw}-E^o_{iv}) \tag{6.18}$$

(cf.eq.(3.48)) where all the symbols have been defined previously and where the nuclear wave functions, energies etc. now refer to both the AT modes and to all other intramolecular and medium modes. In eq.(6.18) we have also invoked the Condon approximation separating the integrations over the electronic and nuclear coordinates. This is adequate for electronic operators such as the two-centre one-electron exchange integrals encountered in simple ET and proton transfer reactions, and for spin-orbit coupling operators which couple electronic states of different spin multiplicities. However, if the nuclear momentum operator is of importance, non-Condon effects are expected to be manifested (153), and modifications of eq.(6.18) have to be introduced.

The rate expression for nonadiabatic AT is formally identical to that for nonadiabatic ET processes. With reference to eq.(6.18) the important analogies between ET and AT in the nonadiabatic limit, can thus be summarized in the following way emphasizing several of the general features of multiphonon processes:

(1) Both nonadiabatic ET and AT rate constants are expressed as products of an electronic coupling term and a thermally averaged nuclear Franck Condon factor.

(2) At low temperatures the rate expressions in both cases exhibit temperature independent nuclear tunnelling. This limit prevails when the level spacings between the lowest initial vibrational level, E^{0}_{io} (v = 0), and all other states in the initial manifold are larger than the thermal energy, i.e. $|E^{0}_{iv} - E^{0}_{io}| > k_B T$ and only exothermic processes may then occur, from the lowest zero-point energy state of the initial nuclear configuration to the final vibronic states degenerate with this level, i.e. W_{fi} now takes the form

$$W_{fi} = \frac{2\pi}{\hbar} Z^{-1} |V_{eff}|^2 \sum_{w} |\langle X^{o}_{fw} | X^{o}_{io} \rangle|^2 \, \delta(E^{o}_{fw} - E^{o}_{io}) \qquad (6.19)$$

(3) In both cases the high-temperature activated rate involves thermal activation to the lowest intersection point of the nuclear Born-Oppenheimer potential energy surfaces. In this limit and for nuclear potential energy surfaces of rather general form, $U_i(q)$ and $U_f(q)$, the rate constant becomes (119) (cf. chapter 5)

$$W = \frac{2\pi}{\hbar} Z_{cl}^{-1} |V_{eff}|^2 \int d\vec{q} \, \exp[-U_i(\vec{q})/k_B T] \, \delta[U_f(\vec{q}) - U_i(\vec{q})] \qquad (6.20)$$

with the classical partition function

$$Z_{cl} = \int d\vec{q} \, \exp[-U_i(\vec{q})/k_B T] \qquad (6.21)$$

For potential surfaces characterzed by a single nuclear coordinate, q, this becomes(119)

$$W = \frac{2\pi}{\hbar} Z_{cl}^{-1} |V_{eff}|^2 \, |\partial[U_f(q) - U_i(q)]/\partial q|^{-1} \exp[-U_i(q^*)/k_B T] \qquad (6.22)$$

where q^* is the intersection point of the one-dimensional zero-order potential energy surfaces.

(4) By invoking the harmonic approximation for the nuclear modes we can exploit a comprehensive literature on multiphonon processes (cf. chapters 3 and 4). Thus, if we consider only a single mode characterized by two harmonic potentials of frequency ω and with the energy gap ΔE, then the configurational change is specified by the reduced displacement $\Delta = d(\mu\omega/\hbar)^{\frac{1}{2}}$, where d is the coordinate distance between the minima of the potential energy surfaces and μ the mass associated with the nuclear motion. The 'coupling strength' is then $S = \Delta^2/2$, and the vibrational reorganization energy $E_r = S\hbar\omega$. The single-mode rate expression, eq.(6.18) becomes (51,108) (cf. section 3.3.1)

$$W_{fi} = A \exp[-S(2\bar{v} + 1)]\, I_p \left\{ 2S[\bar{v}(\bar{v} + 1)]^{\frac{1}{2}} \right\} [(\bar{v} + 1)/\bar{v}]^{p/2} \tag{6.23}$$

where $A = 2\pi\, |V_{eff}|^2/\hbar^2\omega$, $p = |\Delta E|/\hbar\omega$ is the normalized energy gap, I_p the modified Bessel function of order p, and $\bar{v} = [\exp(h\omega/k_B T)-1]^{-1}$ is the Bose occupation number which reflects the temperature dependence of the rate. This expression exhibits in particular a continuous transition from a low-temperature tunnelling expression

$$W = A \exp(-S)S^p/p! \tag{6.24}$$

to a high-temperature activated rate expression

$$W = A\hbar/(k_B T E_r/\pi)^{\frac{1}{2}} \exp[-(E_r + \Delta E)^2/4 E_r k_B T] \tag{6.25}$$

formally identical to eq.(3.81) and (3.63), respectively.

If the nonadiabatic AT processes involve coupling to both a high-frequency AT mode and a multitude of low-frequency medium modes, the procedure of chapter 4 finally gives a rate expression formally identical to eq.(4.12), i.e.

$$W_{fi} = (\pi/\hbar^2 \omega_{eff} k_B T E_s) |V_{eff}|^2 Z_{qu}^{-1} \sum_v \sum_w \exp(-\beta \varepsilon_v) \qquad (6.26)$$

$$\exp[-(E_s + \Delta E + \varepsilon_w - \varepsilon_v)^2 / 4E_s k_B T]$$

where $S_{v,w}$ is the Franck Condon overlap factor for the AT mode, ε_v and ε_w the vibrational energy levels of this mode in the initial and final state, respectively, and Z_{qu} the partition function for this mode in the initial state.

We shall conclude this section by noticing the conditions under which a nonadiabatic approach to AT processes is at all appropriate. This requires firstly that the nuclear Franck Condon factors are appreciable in order that the relaxation process is efficient and is ensured by the intramolecular nuclear rearrangement in the reaction centre and by coupling to the medium modes. Secondly, it requires that two electronic states can be distinguished, and that the residual coupling between the states (V_{eff}) is 'small'. This condition is expected to be generally much more restrictive for AT processes than for ET processes (section 5.1). Apart from AT processes involving spin-orbit coupling and some proton transfer reactions most AT processes are thus expected to proceed on a single potential surface and therefore to correspond to the opposite limit of adiabatic AT processes. Most attempts to formulate a theory for adiabatic AT so far rest on a double (nuclear) adiabatic approximation and semiclassical rate theory (5,95,96,267) which we shall now briefly discuss.

(b) Partially and Totally Adiabatic AT Processes.

Eq.(6.26) can be interpreted in the following way. The reaction can be viewed as proceeding by motion of the system on potential energy surfaces defined by classical (medium) coordinates only. Fluctuations in these coordinates may create a certain configuration corresponding to the saddle point of the intersection surface between the initial and final state surfaces, reflected by the activation factor which is thus solely determined by these modes. At the intersection region the high-frequency coordinates (e.g. the stretching and bending proton modes in proton transfer reactions) proceed from their initial equilibrium to their final equilibrium values by nuclear tunnelling, with a probability given by the Franck Condon overlap factor of these modes. The electronic redistribution also occurs in this region which is reflected in the electronic coupling term and leads to a breaking of the bond between the transferring group and the donor fragment and the formation of a bond to the acceptor fragment at some optimal value, $\vec{Q}^*$, of the coordinates of the transferring group. The overall rate expression is subsequently obtained by averaging and summation over all vibrational states of the high-frequency modes.

These considerations are the basis of an alternative approach to the formulation of a theory of AT processes which is appropriate when the electronic coupling term is not sufficiently small that the purely nonadiabatic approach is adequate. According to this double adiabatic approximation (267,270) the view is taken that since the high-frequency modes are much faster than the low-frequency medium modes, they are conveniently viewed together with the electrons as the total fast system and the medium as the slow system in a modified Born-Oppenheimer scheme. Thus, the characteristic frequency for the proton stretching modes is about 3000 cm^{-1} , whereas those of the solvent are commonly taken as 1-10 cm^{-1} . However, in reality the solvent spectrum is continuous and include also much higher frequencies (100-500 cm^{-1}), and for this reason the separation of the AT modes and the solvent modes is in fact less justified than for the separation of electronic and nuclear motion.

The separation of the combined (i.e. electrons and fast nuclei) quantum system from the medium modes would correspond to the introduction of total modified channel wave functions of the form

$$\Psi(\vec{r},\vec{Q},\vec{q}) = \psi(\vec{r},\vec{Q}\ ;\ \vec{q})\ \chi(\vec{q}) \qquad (6.27)$$

where $\vec{r}$, $\vec{Q}$, and $\vec{q}$ refer to the coordinates of the electrons, the fast nuclei, and the slow nuclei, respectively. The zero-order functions $(\vec{r},\ \vec{Q}\ ;\ \vec{q})$ are accordingly determined by stationary Schrodinger equations of the form

$$[T(\vec{r})+T(\vec{Q})+V(\vec{r},\vec{Q}\ ;\ \vec{q})]\psi(\vec{r},\vec{Q}\ ;\ \vec{q}) = \varepsilon(\vec{q})\,\psi(\vec{r},\vec{Q}\ ;\ \vec{q}) \qquad (6.28)$$

and the wave functions of the slow system by (cf. eqs.(3.21) and (3.27))

$$[T(\vec{q}) + \varepsilon(\vec{q})]\ \chi(\vec{q}) = E\ \chi(\vec{q}) \qquad (6.29)$$

where $T(\vec{r})$, $T(\vec{Q})$, and $T(\vec{q})$ are the kinetic energies of the three subsystems, and $V(\vec{r},\vec{Q}\ ;\ \vec{q})$ the potential energy of interaction between the total fast subsystem and the medium. The coupling between the slow and the fast systems is thus incorporated by the parametric dependence of $\psi(\vec{r},\vec{Q}\ ;\ \vec{q})$ and $\varepsilon(\vec{q})$ on the coordinates of the slow subsystem. $\varepsilon(\vec{q})$ defines the total potential energies of the slow subsystem. A pair of potential energy surfaces is thus determined for each value of both electronic quantum numbers and the vibrational quantum numbers of the high-frequency modes.

The motion of the electrons and the fast nuclear subsystems can subsequently be separated by a 'conventional' Born-Oppenheimer approximation, i.e. by representing $\psi(\vec{r},\vec{Q}\ ;\ \vec{q})$ in the form

$$\psi(\vec{r},\vec{Q}\ ;\ \vec{q}) = \phi(\vec{r}\ ;\ \vec{Q}\ ;\ \vec{q})\ \Phi\ (\vec{Q}\ ;\ \vec{q}) \qquad (6.30)$$

where ϕ and Φ now represent the electronic wave function and the wave function of the fast nuclear subsystem, respectively. This double separation then gives rise to the following three distinguishable cases;

(a) Totally Nonadiabatic Processes. This implies that the process is electronically nonadiabatic with respect to the fast nuclei and furthermore the total fast system, electrons and fast nuclei, nonadiabatic with respect to the slow subsystem. We can give conditions for this in terms of the semiclassical rate expression as discussed in chapter 5. Thus, if we can view the slow subsystem classically, the semiclassical rate expression can be written

$$W_{fi} = \varkappa_{ep} \frac{\omega_{eff}}{2\pi} \exp(-E_A^{na}/k_B T) \qquad (6.31)$$

where ω_{eff} is the effective frequency of the classical modes, E_A^{na} the (nonadiabatic) activation energy determined by these modes, and $\varkappa_{ep}$ the transmission coefficient with respect to the total quantum system. Two conditions must then be valid for the process to be totally nonadiabatic. Firstly, with reference to eq.(5.9), the process is electronically nonadiabatic if

$$2\pi(\Delta\varepsilon^e/2)^2/\hbar|v_p|\,|F_i^p - F_f^p| \ll 1 \qquad (6.32)$$

$\Delta\varepsilon^e$ is here the 'splitting' of the zero-order electronic potential surfaces, $U_i(\vec{Q},\vec{q})$ and $U_f(\vec{Q},\vec{q})$ with respect to the coordinates of both the fast and slow nuclei in the intersection region of lowest energy, F_i^p and F_f^p the slopes of these surfaces with respect to the coordinates of the fast nuclear subsystem, i.e. $F_{i,f}^p = \partial U_{i,f}(\vec{Q},\vec{q})/\partial\vec{Q}$, and $|v_p|$ the numerical value of the velocity of the fast nuclear system in the intersection region (v_p is here imaginary corresponding to tunnelling of the fast nuclear system). We saw in chapter 5 that the inequatity (6.32) can be expected to be valid for some proton transfer reactions, if the splitting, $\Delta\varepsilon^e$, is approximated by the interaction between donor and acceptor in a hydrogen bond.

In addition to eq.(6.32) the inequality

$$\varkappa_{ep} = 2\pi(\Delta\varepsilon^{ep}/2)^2/\hbar v_T|F_i^m - F_f^m| \ll 1 \qquad (6.33)$$

must be valid for the system to be totally nonadiabatic. $\Delta\varepsilon^{ep}$ is now the splitting of the potential energy surfaces with respect to the coordinates of the slow nuclear subsystem only, F_i^m and F_f^m the slopes of these surfaces with respect to the coordinates of this subsystem ($F_{i,f}^m = \partial\varepsilon_{i,f}(\vec{q})/\partial\vec{q}|_{\vec{q}=\vec{q}^*}$), and v_T the thermal velocity of the slow (medium) nuclei (provided that the latter can be represented as a classical system). Eqs.(6.31)-(6.33) can be rearranged to the same form as the one derived above on the basis of first order perturbation theory. Thus, if we can invoke the Condon approximation, then

$$\Delta\varepsilon^{ep}/2 = (\langle\varphi_f(\vec{r};\vec{Q};\vec{q}^*)\ \Phi_f(\vec{Q};\vec{q}^*)|V\ |\varphi_i(\vec{r};\vec{Q};\vec{q}^*)\ \Phi_i(\vec{Q};\vec{q}^*)\rangle) \approx$$

$$\langle\varphi_f(\vec{r};\vec{Q}\ ;\vec{q}^*)|V\ |\varphi_i(\vec{r};\vec{Q}\ ;\vec{q}^*)\rangle(\Phi_f(\vec{Q};\vec{q}^*)|\Phi_i(\vec{Q};\vec{q}^*)) \tag{6.34}$$

where we have distinguished between the integrations with respect to the electronic coordinates and the coordinates of the fast nuclear subsystem by the symbols < > and ( ), respectively. V refers to the residual coupling between the initial and final states, and $\vec{Q}^*$ to the value of $\vec{Q}$, for which the electronic coupling is maximum. Insertion of eqs.(6.33) and (6.34) in eq.(6.31) then gives an expression identical to eq.(6.26).

(b) Partially Adiabatic Processes. Within the double adiabatic approximation, even if the inverse inequality of eq.(6.32) is valid, the overall process may still be nonadiabatic, when eq.(6.33) remains valid. This is associated with a small overlap of the nuclear wave functions of the AT modes, and this limit is conveniently named the partially adiabatic limit, i.e. the process is now electronically adiabatic but nonadiabatic with respect to the total fast subsystem.

The electronic factor can now be disregarded and the rate probability expressed with reference to adiabatic potential energy surfaces of the form (cf.eq.(5.1))

$$\varepsilon_{\pm}^{ep}(\vec{q}) = \frac{1}{2}[\varepsilon_i^{ep}(\vec{q})+\varepsilon_f^{ep}(\vec{q})] \pm \left\{[\varepsilon_i^{ep}(\vec{q})-\varepsilon_f^{ep}(\vec{q})]^2+(\Delta\varepsilon_{ad}^{ep})^2\right\}^{\frac{1}{2}} \tag{6.35}$$

where $\varepsilon_i^{ep}(\bar{q})$ and $\varepsilon_f^{ep}(\bar{q})$ are the zero-order potential energy surfaces (i.e. solutions of eq.(6.28)) and $\Delta\varepsilon_{ad}^{ep}$ the splitting of these surfaces in the intersection region due to the interaction between the wave functions of the total fast subsystem. The rate constant of the partially adiabatic processes is

$$W = \varkappa_{ep}^{ad} \frac{\omega_{eff}}{2\pi} \exp(-\tilde{E}_A/k_B T) \tag{6.36}$$

where $\tilde{E}_A$ is now the energy of the saddle point of the potential energy surface of the coordinates of the slow system only, and the transmission coefficient

$$\varkappa_{ep}^{ad} = (\Delta\varepsilon_{ad}^{ep}/2)^2 (4\pi^3/\hbar^2 \omega_{eff}^2 k_B T E_r)^{\frac{1}{2}} \tag{6.37}$$

(cf. eq.(5.53)). In complete analogy with the analysis of electronic adiabaticity, the splitting of the vibrational levels of the fast nuclear subsystem, $\Delta\varepsilon_{ad}^{ep}$, depends on the instantaneous coordinate values of the slow system, q, and takes a minimum value at the intersection point of the classical potential energy surfaces. This point also corresponds to a maximum probability of nuclear tunnelling, whereas the transferring molecular group is localized either at the donor or at the acceptor molecular fragment for other values of the slow nuclear system coordinates. In the quasiclassical approximation this amount of splitting is approximately (263,267)

$$\Delta\varepsilon_{ad}^{ep} \propto \hbar\omega \exp(-\sigma/2) \tag{6.38}$$

where σ is the Gamov factor (eq.(6.16)), ω the frequency of the transferring group, and the proportionality factor depends on the nature of the nuclear potential. Even if the electronic coupling term is large, the condition of overall nonadiabaticity may thus be met if the probability of nuclear tunnelling is small. Partial nonadiabaticity is thus likely to be generally of major importance for AT processes at sufficiently low temperatures.

<u>(c) Totally Adiabatic Processes.</u> If both the electronic coupling and the nuclear tunnelling splitting are large so that the inverse inequalities of both eq.(6.32) and (6.33) are valid, then $\kappa_{ep}^{ad} \approx 1$. The rate expression then becomes

$$W = \frac{\omega_{eff}}{2\pi} \exp(- E_A^{ad} / k_B T) \qquad (6.39)$$

(cf. eq.(5.48)), where E_A^{ad} is now lower than the activation energy for the corresponding nonadiabatic process by an amount determined by the splitting of the classical potential energy surfaces in the intersection region (cf. eq.(5.19)). This equation corresponds to the totally adiabatic limit in which both the electrons and the fast nuclei have sufficient time to be reorganized as the slow subsystem passes the intersection region of the classical potential energy surfaces.

6.3.3 Relation to the Gamov Tunnelling Factor

Before we proceed to a discussion of some experimental data we shall make few observations concerning the relationship between the tunnelling corrections in the semiclassical formalism and the quantum mechanical formulation of AT reactions. We noticed how the Gamov tunnelling factor energes from the Condon approximation when quasiclasical nuclear wave functions are used for the transferring molecular group. We can illustrate this correspondence further by considering molecular group motion along a single displaced harmonic mode of frequency ω. The potential energy of this system in the initial and final states, respectively, is (cf. eqs.(6.23) - (6.25))

$$U_i = \frac{1}{2}\mu\omega^2 x^2 \qquad (6.40)$$

$$U_f = \frac{1}{2}\mu\omega^2 (x - d)^2 - |\Delta E| \qquad (6.41)$$

The reaction probability in the low-temperature limit is then given by eq.(6.24). On the other hand, the nuclear contribution to the Gamov tunnelling probability would be

$$\Gamma = \exp(-\sigma) \qquad (6.42)$$

where

$$\sigma = \frac{2}{\hbar}\int_0^{x^*} [2\mu(\frac{1}{2}\mu\omega^2 x^2)]^{\frac{1}{2}} dx + \frac{2}{\hbar}\int_{x^*}^{x^{**}} 2\mu[\frac{1}{2}\mu\omega^2 (x-d)^2 - |\Delta E|]^{\frac{1}{2}} dx \qquad (6.43)$$

$x^* = d/2 - |\Delta E|/\mu\omega^2 d$ is the intersection point of the one-dimensional potential surfaces ($U_i(x^*) = U_f(x^*)$), and $x^{**} = d-(2|\Delta E|/\mu\omega^2)$ is the coordinate value for which U_f coincides with the zero-point energy of the initial state potential energy surface ($U_f(x^{**}) = 0$). Integration of eq.(6.43) and application of Stirling's approximation for $(|\Delta E|/\hbar\omega)! = p!$, i.e. $p! = \exp[p(\ln p - 1)]$ subsequently gives an expression identical to eq.(6.24). This result is useful but not surprising, as it shows that the semiclassical WKB approximation (and the representation of the nuclear wave functions by the quasiclassical approximation (142)) is applicable for the calculation of Franck Condon factors not only between a bound and a continuum state such as in the original Gamov tunnelling formula (247) but also between two bound states.

6.4 Relation to Experimental Data

We shall now complete our discussion of AT processes with a brief analysis of some experimental data. We have already (chapter 5) discussed how the semiclassical AT theory can be used to estimate several important kinetic parameters and to distinguish between various molecular mechanisms for ligand substitution processes. We shall now discuss two condensed phase chemical AT systems for which the theory is both particularly illustrative and for which the semiclassical theories appear to be inadequate. The two systems are

(a) The low-temperature recombination reaction between CO and hemoglobin (hb) subunits, with CO produced by photodissociation of its complex with the latter. This is in fact a biological system but we discuss it here rather in chapter 9, as it appears to be particularly illuminating with respect to many of the results of nonadiabatic AT theory.

(b) Homogeneous proton transfer reactions.

Before proceeding we notice that a similar analysis can be performed for the low-temperature γ-radiation-induced formaldehyde polymerization as studied by Gol'danskij, Frank-Kemenetskij and Barkalov (271,272). At temperatures higher than about 80 °K the rate of this process (eq. (6.44)) follows an Arrhenius law with an activation energy of 0.1 ev, whereas no temperature dependence is

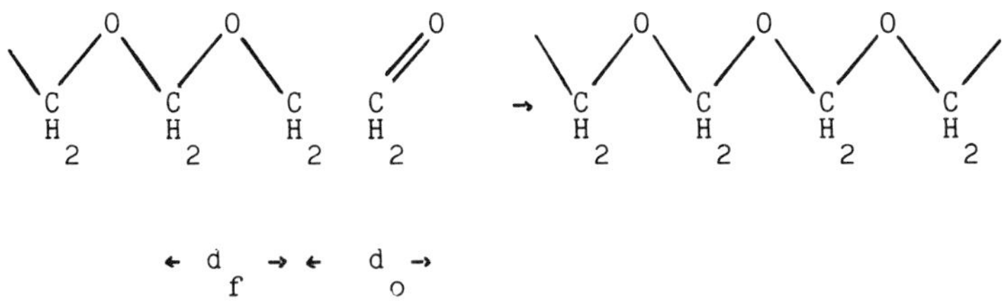

(6.44)

observed for temperatures lower than about 10° K. This process is of course in itself of great interest by documenting the rapid ($W_{fi} \approx 10\ s^{-1}$ as $T \rightarrow 0$) spontaneous high-yield polymer chain growth close to the absolute zero temperature ('astrophysical' conditions). However, we shall refrain from a closer analysis of these data (cf.ref. 271 and 272), as a large number of parameters necessarily have to be introduced for this system. Thus, at least three local modes are displaced, i.e. the motion of a formaldehyde monomer as a whole, the change of a C=O double bond to a single bond, and at least one deformational mode in the polymer chain. In addition, a value of ΔE and at least two parameters to characterize the medium coupling, i.e. altogether nine parameters, characterize the system and are correspondingly ambiguously determined by the data.

(a) The CO-Hemoglobin Recombination. During the last decade several elementary biological processes were investigated over a broad temperature range from about 2° K up to room temperature (see further chapter 9). One notable example is the recombination of CO and other small ligand molecules produced by photodissociation of their complexes with β-hemoglobin subunits with their parent deoxy form (273). At high temperatures CO has to pass several barriers on its way back to the heme group which is located in a 'hydrophobic pocket' formed by the protein residue (fig.(6.4)). However, at temperatures lower than about 180° K the pocket is sealed off thus trapping the CO molecule, and the system only overcomes a single barrier corresponding to a shift in the position of the CO molecule, probably from a site close to the 'distal' histidine to its bound position at the iron atom. The experimental kinetic studies provided the following information (273):

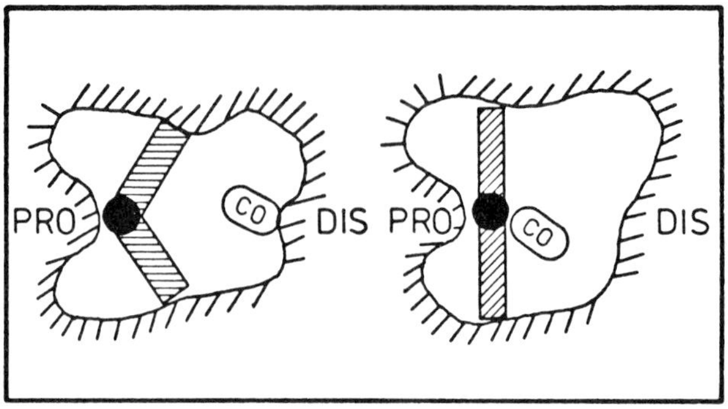

Fig (6.4)

Schematic view of the initial (left) and final (right) state of the hb/CO recombination. The heme group is seen from the edge and 'proximal' and 'distal' histidines indicated.

(1) The rebinding process does not show an exponential decay of the concentration of the deoxy form with time, but rather follows a power law. This effect originates from the energetic spread of the barrier heights due to the freezing of different conformational states. (2) The average half-times (or rather $\tau_{0.75}$ which refers to the time when the deoxy-hb concentration has dropped to 75 % of its initial value) is practically temperature independent in the range 2-10 °K. (3) A transition from a tunnelling region (2-10 ° K) to a temperature-activated region occurs in the range 10-20 ° K. Above this region $\tau_{0.75}^{-1}$ is temperature dependent corresponding to an apparent activation energy of 0.045 ev.

Crystallographic data originally suggested that in the CO-free five-coordinated state the Fe atom is located 0.75 A out of the mean puckered heme plane (275). This value has been questioned by other data which suggest a value of 0.3-0.4 A (276) similar to myoglobin (277), whereas certain model compounds show a somewhat larger value of 0.55 A (278). The heme group is otherwise linked to the protein residue via its peripheri and via axial coordination of a 'proximal' histidine to the fifth coordination site of Fe, while the sixth site is vacant. In the bound CO-hb state the iron atom is shifted into the heme plane, while the CO molecule moves into its bound state at the sixth coordination site of the Fe atom. The geometry of this bond is not known with certainty but recent structural data show that for related model compounds the Fe-C-O unit is linear (279). Finally, the CO-free five-coordinated heme group is in the high-spin state ($S = 2$), whereas the heme group is in the low-spin state ($S = 0$) in the bound hb-CO complex (S is the total spin quantum number).

This system was recently subject to an analysis within the theoretical framework outlined in the present chapter (280). The fundamental equation was here eq.(6.18) associated with a Hamiltonian of the entire system of the form

$$H(\vec{r},\vec{q}) = T(\vec{q}) + H_e(\vec{r},\vec{q}) + H_{so}(\vec{r},\vec{q}) \qquad (6.45)$$

where T is the kinetic energy, H_e the total electronic Hamiltonian at fixed nuclear configuration, H_{so} the spin orbit coupling operator which has to be introduced as we shall consider transitions between states of different spin multiplicities, and $\vec{r}$ and $\vec{q}$ refer to all the electronic and nuclear coordinates, respectively. Following the procedure of the present chapter and of chapter 3 we introduce zero-order Born-Oppenheimer states and potential energy surfaces, i.e. eigenfunctions and eigenvalues of the operator H_e. Considering furthermore only the two lowest electronic states, Ψ_i and Ψ_f, the electronic coupling term, V_{fi}, in eq.(6.18) takes the form

$$V_{if} = \langle \Psi_f \mid R \mid \Psi_i \rangle \qquad (6.46)$$

where R is the overall transition operator. R is expanded in a perturbation series (cf. section 3.1)

$$R = (L + H_{so}) + \sum_{d \neq i,f} \left\{ \frac{(L+H_{so})|\Psi_d(\vec{r},\vec{q}_{io})\rangle\langle\Psi_d(\vec{r},\vec{q}_{io})|(L+H_{so})}{\varepsilon_i(\vec{q}_{io}) - \varepsilon_d(\vec{q}_{io})} + \right.$$

$$\left. + \frac{(L+H_{so})|\Psi_d(\vec{r},\vec{q}_{fo})\rangle\langle\Psi_d(\vec{r},\vec{q}_{fo})|(L+H_{so})}{\varepsilon_i(\vec{q}_{fo}) - \varepsilon_d(\vec{q}_{fo})} \right\} + \ldots\ldots \qquad (6.47)$$

where L is the 'nonadiabaticily operator' (chapter 3), and the summation over all the electronic states d involves only excited electronic configurations of the system (cf. the discussion in section 3.1). q_{io}, q_{fo}, and q_{do} refer to the appropriate equilibrium nuclear configurations. The operator L can, however, only couple the same spin states, whereas the spin-orbit coupling operator couples states the spins of which differ by unity. Since the initial and final spin states of the CO/hb system differ by 2, we can therefore conclude that the lowest term of finite value in the expansion of eq.(6.46) is that of second order, that the intermediate states must correspond to a spin quantum number S = 1 and that H_{so} must be the operator which induces the reaction. The resulting electronic coupling appropriate to the system is then $V_{fi} \approx \langle H_{so}\rangle_{di} \langle H_{so}\rangle_{fd} / \Delta\varepsilon$ where $\langle H_{so}\rangle_{\alpha\beta}$ are the integrals of spin-orbit coupling between the electronic states α and β (α,β = i,f,d) and $\Delta\varepsilon$ the energy gap between the i or f states and the state d. Typical values of the spin-orbit coupling terms for first-row transition metal ions are $\langle H_{so}\rangle_{\alpha\beta} \approx$ 100 cm^{-1} which together with $\Delta\varepsilon \approx 10^4$ cm^{-1} gives a crude a priori estimate of $V_{fi} \approx$ 1 cm^{-1} (121). We notice that such an analysis would be more involved for the O_2/hb recombination as the hb-O_2 complex, unlike the hb-CO complex, is characterized by several low-lying electronic states (281).

In addition to the nuclear mode which represents the motion of the Fe atom into the heme plane it is necessary to consider two other modes of the reaction centre. Firstly, the motion of the heme group as a whole relative to the (distal) histidine and protein residue. This motion, however, essentially mimics the medium modes. It is expected to exhibit a small configurational change only and a very small frequency and is therefore reasonably incorporated with other medium modes which can be assumed to act essentially as 'hidden' variables ensuring vibrational relaxation subsequent to the AT step. All these modes could, however, be incorporated explicitly in the analysis if necessary. Secondly, the CO-histidine motion will also be disregarded. This is partly because no structural information about this mode is available, but it is also plausible in view of the presumably much smaller frequency of this mode compared to the Fe motion. The rate equation, eq.(6.18) therefore only incorporates a single nuclear mode, i.e. that of the Fe-heme motion which is similar to a metal-ligand bending mode. The analysis is then completed by the following scheme:

(1) A given value for the energy gap, $-\Delta E$, is chosen (in the range 0.05-0.5 ev). From the high-temperature activation energy (eq.(6.25)) this provides a corresponding value of the vibrational reorganization energy $E_r = S\hbar\omega$.

(2) The general expression (the harmonic approximation)(eq.(6.23)) is subsequently plotted numerically for various E_r and ΔE. Examples of this are shown in fig. (6.5) and show that the main features of the experimental data, in particular the transition from an activationless to an activated temperature region is reproduced well only for certain values of ΔE and E_r. Moreover, the transition region occurs at $k_B T/\hbar\omega \approx 0.1-0.2$ which gives $\omega \approx 100\ cm^{-1}$. This gives again the best values of S and p (= $|\Delta E|/\hbar\omega$) of about 20 and 2-5, respectively, or $E_r \approx 0.25$ ev and $-\Delta E \approx 0.05$ ev.

Using these values and the low-temperature expression for the Franck Condon factor (eq.(6.24)) we can finally calculate a value for the electronic coupling term of 0.1-1 cm^{-1}, and for the shift distance of the Fe atom of 0.4-0.5 A.

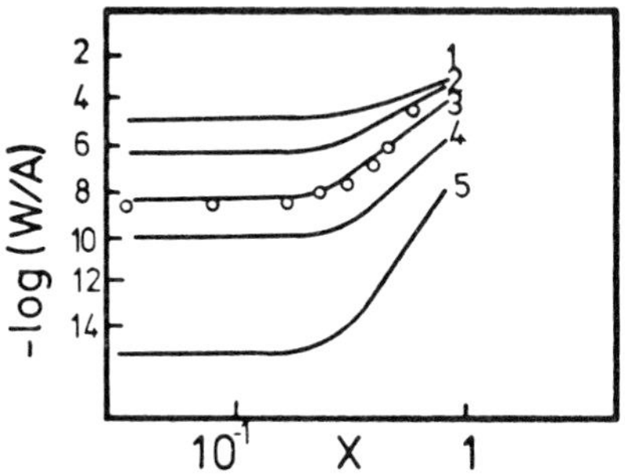

Fig (6.5)

Theoretical plots of log(W/A) against $X = k_B T/\hbar\omega$. 1: S = 60, p = 30. 2: S = 30, p = 8. 3: S = 25, p =2. 4: S = 80, p = 20. 5: S = 80, p = 20. The circles show the experimental $(\tau_{0.75})^{-1}$ with values at T → 0 matched with curve 3.

This analysis on the basis of the single-mode model thus provides a satisfactory reproduction of all the features of the CO/hb reaction compatible with those known from independent experimental investigations. The low value of V_{fi} furthermore justifies the nonadiabatic approach. The analysis could be improved for example by introducing also the medium modes explicitly, which would, however, introduce at least two additional parameters. Also, we have given no explicit attention to the distribution over the frozen conformational states but taken $\tau_{0.75}^{-1}$ at the peak of this distribution as the rate constant. This is justified in view of the narrow distribution (273,280) and by analysis of rate constants referring to given barrier heights, properly disentangled from the overall rate constant (280).

(b) We shall finally notice a few points of special relevance to proton transfer reactions in solution (19,238,252,254).

It is appropriate to investigate here the same features as the ones we considered at the beginning of the chapter in terms of a simple semiclassical theory. We shall thus now view the proton transfer as a multiphonon radiationless process which can be described within the general theory of these processes adapted to AT as outlined above. More specifically, we shall assume that the proton transfer can be viewed as a totally nonadiabatic process which can be described by means of eq.(6.26), but our fundamental coclusions would not be strongly affected if the process were partially adiabatic. We shall also assume that the proton transfer can be described by means of a single mode which is separated from all other intramolecular and medium modes. Furthermore, with reference to the general criteria about the quantum or classical behaviour of a given nuclear mode, we shall now view the proton mode as a quantum mode. Thus, at room temperature the level spacings of the proton stretching and bending modes are about 15 $k_B T$ and 7 $k_B T$, respectively, which implies that quantum effects of these modes are expected to be strongly manifested. Finally, calculations on the kinetic isotope effect show that experimental data are occasionally equally well or better reproduced by the theory if the proton is assumed to move along a bending mode or a 'mixed' mode rather than along a stretching mode. This is not surprising as the Franck Condon factors are generally larger the smaller the appropriate frequency. If steric effects and repulsion forces between the reactants could be ignored, transfer along bending modes would therefore be more likely than transfer along stretching modes. We shall then consider the following points specifically:

(1) From eq.(6.26) we can see that a smooth Bronsted relationship is expected for a series of proton transfer reactions when the Franck Condon factors of the proton, the solvent reorganization energy, and the reorganization energy of all other modes

are constant throughout the series. This implies that the geometry, charge distribution etc. of the reactants and products should also be constant. Moreover, for proton transfer reactions which are not strongly exothermic or endothermic only the ground vibrational levels of the proton contribute ($v = w = 0$) in eq.(6.26), and the Bronsted relationship is then quadratic with the Bronsted coefficient varying from zero to unity as ΔE varies from $-E_r$ to E_r. The Bronsted relationship is less curved if the excited vibrational states also contribute.

A variety of proton transfer reactions are known to exhibit this behaviour, even though the whole expected range of the Bronsted coefficient is seldomly observed (19,282). We notice thus in particular simple acid-base reactions between O- and N-fragments; the base-catalyzed enolization of ketones (cf. eq.(1.4)), and acid- and base-catalyzed hydration of carbonyl compounds.

(2) From eq.(6.26) we see that for $v = w = 0$ the curvature of the Bronsted plot is determined solely by the reorganization energy of the classical modes. This important kinetic quantity can therefore be estimated from experimental Bronsted plots by fitting a quadratic free energy relationship to the experimental data. Analysis along such lines has shown (282,283) that E_r is often quite small e.g. 0.2-1 ev for proton transfer between O- and N-fragments (i.e. the Bronsted plots display a large curvature), whereas E_r may be 2-3 ev for proton transfer involving C-acids. This is understandable from the following: (i) the proton transfer distance is generally much smaller (0.3-0.7 A) than the electron transfer distance in ET reactions (6-10 A). This would give lower values of the medium reorganization energies (for ET reactions E_s is commonly 2-4 ev). (ii) For O- and N-acids the proton charge is largely localized at the reaction centre, i.e. only the proton itself and the donor and acceptor atoms are involved in the charge redistribution. For C-acids the charge is delocalized over several atoms giving a larger 'effective' charge transfer distance and therefore a larger E_s

(cf.eq.(1.28)) and a smaller curvature. (iii) In contrast to O- and N-acids, carbon acids are often subject to substantial reorganzation in intramolecular modes of moderately high frequencies which are slightly thermally excited at room temperature (e.g. carbon skeletal modes of frequencies in the range 1000-1500 cm^{-1}). This would be reflected in both a higher reorganization energy term and lower intramolecular Franck Condon factors than for O- and N-acids, and both effects would be displayed as lower rate constants and less curved Bronsted plots than for processes involving O- and N-acids.

(3) We shall finally analyze a few experimental data on the primary kinetic deuterium isotope effect. While Bronsted plots could be explained by classical nuclear motion in a double-well potential, essentially with no specification of the nature of the nuclear mode (e.g. proton or solvent modes), the kinetic isotope effect is inevitably a reflection of the quantum behaviour of the proton. As a quantitative measure of the isotope effect we shall take k_H/k_D (section 6.2) where k_H and k_D are given by (cf. chapter 1)

$$k_{H,D} \propto \int \exp[-U(Q_{H,D})/k_BT]W(Q_{H,D})dQ_{H,D} \approx \qquad (6.48)$$

$$\exp[-U(Q^*_{H,D})/k_BT]W(Q^*_{H,D})\,\Delta V$$

$Q^*_{H,D}$ is the 'effective' value of the proton or deuteron coordinate, $Q_{H,D}$, in the collision complex, ΔV the reaction volume, $U(Q_{H,D})$ the interaction potential between the reactants (i.e. the work required to bring the reactants from infinity to their positions in the collision complex), and $W(Q_{H,D})$ is given by eq.(6.26) where the dependence of W on $Q_{H,D}$ is reflected largely in the proton or deuteron Franck Condon factors and in the electronic coupling term.

$Q^*(=Q^*_{H,D})$ is determined by maximizing the integral of eq.(6.48), i.e. from the equation

$$U'(Q^*) = k_B T[W(Q^*)]^{-1} dW(Q)/dQ \,|_{Q=Q^*} \qquad (6.49)$$

In order to see the physical meaning of this we insert a parabolic repulsion potential of the form $U(Q) = \frac{1}{2} b(Q-Q_0)^2$ and the Franck Condon factors for harmonic potentials for the ground vibrational states only, i.e. $S_{0,0} = \exp(-m\Omega Q^{*2}/\hbar)$ where Ω is the proton or deuteron frequency, and m the corresponding mass. b characterizes the curvature and Q_0 the range of the repulsion potential. Eq.(6.49) then becomes

$$Q^* = Q_0 \,/\, (1 + m\Omega/\hbar b) \qquad (6.50)$$

We then notice (283) that for a sufficiently rapidly rising repulsion potential (large b), Q^* is independent of the isotope mass, in other words the proton and heavier isotopes are transferred over the same distance. When the potential does not rise sufficiently rapidly the tunnelling distance is smaller for the heavier isotope. This would also imply that the activation energy is larger for the heavier isotope due to the closer contact distance (i.e. $U(Q_D^*) > U(Q_H^*)$) (283-285)

We shall here assume that the proton and heavier isotope transfer distances are identical. However, in order to rationalize large activation energy differences for proton, deuteron, and triton transfer, it is necessary in certain cases to incorporate this effect (196). We shall then approximate $k_H/k_D \approx W_H/W_D$ and directly use eq.(6.26) for both W_H and W_D. The isotope effect now primarily arises from differences in $S_{v,w}$, ε_v, and ε_w when the proton is substituted by heavier isotopes, whereas we shall ignore differences in E_r and ΔE (284). Thus, $S_{v,w}^H > S_{v,w}^D$ due to the smaller mass of the proton, and $\varepsilon_v^H, \varepsilon_w^H > \varepsilon_v^D, \varepsilon_w^D$. Fig. 6.6 shows representative plots of W_H/W_D against ΔE for various values of E_r and the proton transfer distance. The proton frequencies were those typical of proton stretching (≈ 3000 cm^{-1}) and bending (≈ 1500 cm^{-1}) modes, and the corresponding proton motion was represented by Morse potentials and square hyperbolic tangent potentials, representatively. Leaving out details of the calculations (196) we draw the following conclusions from fig. 6.6 (196,283,284):

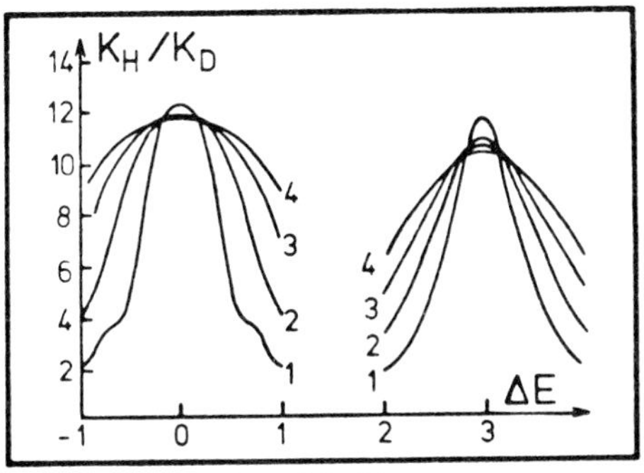

Fig (6.6)

Examples of plots of k_H/k_D against ΔE(ev). Left: stretching mode (3000 cm^{-1}) and proton transfer distance 0.37 A. Right: 1500 cm^{-1} and 0.62 A (shifted 3 ev along ΔE axis).

(1) A maximum is always observed. This maximum is located at $\Delta E = 0$ when the proton frequencies are identical in the initial and final states, whereas it is shifted to positive and negative ΔE when the frequency decreases or increases, respectively during the reaction.

(2) W_H/W_D increases approximately exponentially with Δ^2 (the reduced displacement). For E_r = 1 ev, and Ω_H = 3000 cm^{-1} the absolute value thus increases from 4 to 56, i.e. the range of most experimental values, when the proton transfer distance increases from 0.27 A to 0.47 A.

(3) The proton transfer distance for 'typical' maximum values of W_H/W_D of 10 is about 0.35 A and 0.60 A for transfer along a stretching and a bending mode, respectively.

(4) While the absolute values of W_H/W_D is determined largely by Δ and Ω, the width of the maximum depends strongly on E_r. This

provides a possibility of obtaining this important parameter in cases where it cannot be calculated from the curvature of a Bronsted plot.

(5) The isotope effects are commonly smaller when anharmonic potentials for the proton are used than in the harmonic approximation. Thus, a Morse potential of dissociation energy 4 ev gives a proton transfer distance of 0.4 A for a maximum isotope effect of 12 and a frequency of 3000 cm^{-1}. The corresponding value for a harmonic potential is 0.35 A, and this difference originates from the smaller barrier for proton tunnelling (larger Franck Condon factors) for Morse potentials than for harmonic potentials.

We notice also that the apparent activation energy $d\ln W/d\beta$ is higher for deuteron transfer than for proton transfer, whereas the pre-exponential Arrhenius factor is typically lower. Thus, $E_A^D - E_A^H \simeq 0.3$ kcal and 1.7 kcal, while $A_H/A_D = 0.13$ and 0.69 for stretching and bending modes, respectively, (when $(W_H/W_D)_{max} = 10$ and $E_r = 1$ ev). In the present formalism these effects are due to thermal excitation of the deuteron mode, and as expected, the effet is larger, the larger $|\Delta E|$, the lower the frequencies, and the more pronounced the anharmonicity.

Extensive investigations of the dependence of W_H/W_D on both ΔE and the temperature are available in the literature (19,196). As examples we show the data for the deprotonation of phenylnitromethane and substituted phenylnitroethanes by several N-bases (286) and the best theoretical fits in fig. 6.7.

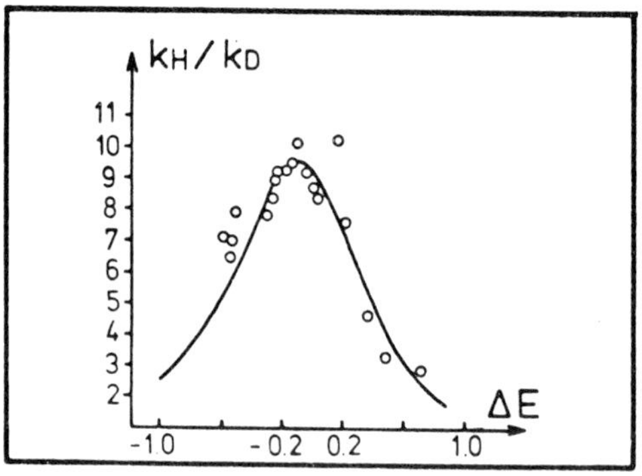

Fig (6.7)

Plot of k_H/k_D against ΔE(ev) for the ionization of phenylnitromethane (ref. 286). Theoretical fit: E_r = 0.7 ev, proton transfer distance 0.61 A and transfer along bending modes.

We can now summarize the results of the application of the multiphonon rate theory to proton transfer reactions and the differences from the semiclassical theory as follows:

(1) The semiclassical theory views the proton transfer as classical motion over a potential barrier determined by the proton stretching mode. The topology of this potential energy surface determines the Bronsted relationship and the different behaviour of proton donor and acceptor molecules having different donor atoms. In contrast, the formalism outlined in the present chapter is based on the view that the proton is present in discrete vibrational states the level spacing of which at room temperature is much higher than the thermal energy. With reference to the general criteria given at the beginning of this chapter proton transfer by quantum mechanical tunnelling is therefore the most likely general transfer path whereas the activation energy is provided by thermal excitation in other nuclear modes, primarily those of the medium.

(2) The role of the solvent is indirectly incorporated in the semiclassical theory. In the quantum mechanical theory the effect of the solvent modes is specified not only as an inert heat bath but as an additional set of low-frequency nuclear modes which contribute to the rate probability on equal footing with other low-frequency modes.

(3) The multiphonon rate theory contains parameters of the initial and final states only, i.e. information which is principle experimentally available. On the other hand, the semiclassical theory requires that estimates of transition state properties can be made.

(4) In the semiclassical theory the kinetic isotope effect is ascribed to the loss of different zero-point energy contribu-

tions when going from the initial to the transition state. In some cases tunnelling through the proton transfer potential barrier also has to be invoked. In terms of multiphonon theory and on the basis of general criteria given at the beginning of the present chapter, the proton is transferred by tunnelling through a barrier which differs from the one which determines the activation energy. The isotope effect then arises from the fact that tunnelling of the heavier isotope is a more difficult process, that the heavier isotope tunnels from a higher 'effective' vibrational level and is in some cases possibly also transferred over a smaller distance.

As most experimental data on chemical proton transfer reactions apparently can be rationalized in terms of both the semiclassical and the quantum mechanical rate theories, the design of new types of experiment which could distinguish between the predictions of the two theories would obviously be a challenging goal. Important progress in this direction was recently achieved by studies on the electrochemical hydrogen evolution reaction. We shall discuss these interesting new data obtained by Krishtalik and his associates in chapter 8 after a general description of the adaption of the multiphonon rate theory to electrochemical processes.

7 HIGHER ORDER PROCESSES

We have considered two complementary approaches to the formulation of a rather complete theory for ET and AT reactions. In the first approach the time evolution of the system is described as an infinite perturbation series in eigenstates of the individual reactants (eq.3.13). However, in practice, in most cases all terms higher than first order in the perturbation series have to be omitted, which leads to the rate probability in the nonadiabatic limit.

On the other hand, if the interaction between the initial and final states is not small the system may pass from the reactants' to the products' configuration in such a way that the electronic subsystem follows the nuclear motion during the reaction. A formal description of such a process so far rests on an alternative approach based on semiclassical trajectory considerations and generalizations of the Landau-Zener formalism. This approach gives both the adiabatic and nonadiabatic limits, the latter in a form identical to that of first order perturbation theory. However, the semiclassical approach differs from the perturbation approach by defining the potential energy surfaces as adiabatic surfaces. i.e. eigenstates of a stationary Schrodinger equation for the total, two-centre, system including the interaction between the centres.

The total reaction probability expressed as the time evolution in the two eigenstates of the separate reactants (eq.(3.13)) can be given the following formal interpretation (143). The first term corresponds to a direct transition from the initial to the final state induced by the perturbation at the saddle point of the reaction hypersurface. The second induces a transition from the initial to the final, back to the initial and eventually to the final state. The third term represents a transition involving still an additional pair of transitions back and forth, and so on. When all these reaction paths are summed the exact result for the adiabatic ET is obtained. This interpretation is useful but of course it does not correspond to really occurring processes. Each of the terms in the perturbation series which is formally interpreted as a sequence of ET's are the result of a special way of decomposing the overall expression for the reaction probability, namely, in terms of certain eigenfunctions, i.e. those of the separated reactants. The terms may be calculated separately but only the final summation result corresponds to a physically occurring process.

In general several states constitute the total electronic basis set for the reacting system. These states may belong to the ingoing or outgoing channels (as the states involved in spin-orbit coupling discussed in section 6.4), or they may belong to other channels different from the ingoing and outgoing channels, which would correspond to the localization of the electron on intermediate molecular species and with Hamiltonians which differ from those of the initial and final states. This would add new manifolds of ET steps via intermediate states in the interpretation of the rate probability, and in some cases these processes may furthermore correspond to physical reality.

We shall discuss this further by considering the second order terms in particular. When the first order term in the perturbation expansion vanishes, this term acquires a special importance being the lowest one of finite value. This may happen if a first order transition is symmetry forbidden, as in the CO-hemoglobin system discussed in section 6.4, or if the donor and acceptor in an ET process are so widely separated that the first order electronic coupling terms are vanishingly small. In this case electronic orbitals located on intermediate species and belonging to an 'intermediate' channel may provide a better coupling. The perturbation term of lowest order then becomes (cf.eq.(3.16))

$$V_{fi}^{(2)} = \sum_{d} \frac{\langle c'f|V_{c''}|c''d\rangle\langle c''d|V_c|ci\rangle}{E_{ci} - E_{c''d}} \qquad (7.1)$$

where c, c", and c' refer to the ingoing, the intermediate (which may be identical to either the ingoing or the outgoing channel), and the outgoing channel, respectively, and i, d, and f to the total set of states in these channels. This leads to the following expression for the averaged rate probability per unit time

$$W^{(2)} = \frac{2\pi}{\hbar} \sum_{i,f} \exp(-\beta E^{o}_{ci}) \left| \sum_{d} \frac{\langle c'f|V_{c''}|c''d\rangle\langle c''d|V_{c}|ci\rangle}{E_{ci} - E_{c''d}} \right|^{2}$$

$$\delta(E_{ci} - E_{c'f}) \tag{7.2}$$

if all contributions to the perturbation series other than those of second order can be ignored.

This expression is usually interpreted by saying that even if no direct transition from the initial to the final state can occur, the transition may nevertheless proceed through an intermediate state belonging to the channel c". The transition probability is furthermore summed over all such possible states for which $\langle c'f|V_{c''}|c'd\rangle$, $\langle c'd|V_{c}|ci\rangle \neq 0$. The intermediate states may have energies different from those of the initial and final states, but since they are only temporarily occupied no energy conservation rule is violated. For this reason the states are called virtual (142,143). It should again be emphasized that usually it is not right to say that the system actually passes through some intermediate state, but the electronic contributions to the transition matrix element contain contributions from such states. This is again related to our desire of expressing the reaction probability in terms of eigenfunctions of the separated reactants. However, in certain cases, when the intermediate states have sufficiently low energies, the description of the second order process as proceeding through an intermediate state has a special physical content, namely, it corresponds to chemical reactions through non-relaxing intermediate states. Such processes have also been named concerted, or quantum dynamic processes (287,288) (as opposed to quantum statistical processes (288)), and analysis and application of these concepts to various kinds of ET and AT processes will form the topic of the present section. At first we notice, however, that interpretation of electronic processes in terms of higher order effects is also commonly invoked in other contexts than chemical processes. In fact, this concept was introduced as a microscopic chemical reaction mechanism by the apparent conceptual analogy between certain chemical processes (primarily inner sphere ET) and these other processes (24). Examples of higher order 'physical' processes are

(1) Electronic spin coupling between paramagnetic transition metal ions in cubic lattices of MnO, MnSe, MnTe, and other materials where the transition metal ions are separated by non-magnetic ions. The antiferromagnetism of these materials was first explained by Kramers (21) by the assumption that the electronic coupling between the paramagnetic ions is mediated via excited electronic states in which, roughly speaking, a p electron from the oxygen atom is transferred into an s or a d orbital on Mn^{2+}. The oxygen ion then becomes paramagnetic and mediates the coupling between the metal ions by electronic overlap of the appropriate high-energy oxygen wave functions and the electronic wave functions in the metal ions. This effect is named superexchange and the electronic coupling is apparently sufficiently strong to ensure antiferromagnetic spin alignment in metal ions separated by an oxygen atom. Interesting derivations from this pattern may occur in cases where the cation-anion-cation angle differs from 180°. In such cases superexchange may also occur, but the direct coupling between the metal ions may now be strong enough to compete and line neighbouring spins in a parallel fashion (ferromagnetism). The resulting magnetic behaviour is then expected to be a sensitive function of the lattice parameters and the parameters of the electronic wave functions applied (290,291).

The superexchange mechanism is related to the double exchange mechanism suggested by Zener (23) to account for the ferromagnetic and conduction properties of certain mixed oxides of manganese. In this mechanism an electron is transferred from metal to oxygen 'in cooperation' with an ET from oxygen to another metal ion. This ensures a parallel alignment of the spins of the two metal ions on each side of the oxide ion. The difference from the superexchange mechanism is that the interaction between the metal ions is expressed in terms of two degenerate electronic wave functions corresponding to the localization of the electron on each of the metal ions which are separated by the oxygen atom, whereas the transition probability in the superexchange mechanism also involves excited zero-order states. While the superexchange mechanism thus refers to a higher- order process, the double exchange mechanism refers to a first order process.

(2) Cooperative two-electron transitions are important in certain solid- and liquid-state single-photon radiative processes and in radiation-induced electronic energy transfer between donor and acceptor molecules energy levels of different energy ('off-resonance' transitions) (292,293). For example, two closely spaced absorption bands in molecular crystals HCl at 5313 cm^{-1} and 5465 cm^{-1} were interpreted as an overtone transition in a single HCl molecule and a double excitation of two molecules, respectively (292). The latter process proceeds through an intermediate state corresponding to the absorption of the photon, and the final state is subsequently reached by an intermolecular electronic transition between neighbouring molecules in the crystal.

(3) Two-photon radiative processes of which the raman effect is of primary importance. The transition probability in the raman effect is expressed in terms of matrix elements such as eq.(7.1), in which the perturbations are identified with the electronic transition(dipole) moment. If the energy of the

high-energy virtual intermediate state differs from the peak energy of the incident visible radiation, $h\nu_0$, the normal (off-resonance) raman effect arises. On the other hand the approach of $h\nu_0$ to the energy of the intermediate state corresponds to the resonance raman effect in which a single term in the sum over all the intermediate states dominates (294,295),

7.1 Higher Order Processes in Chemical ET Reactions

Second order and double exchange processes have been considered by several workers as a possible mechanism for AT and inner sphere ET reactions (24,105-107), electrode reactions via active sites on the electrode surface, (106a,300-302), and for long-range ET in biological systems (108-112). Taube and Myers (24) were the first to point out that Zener's double exchange mechanism might be operative in inner sphere reactions in a reaction scheme

$$M_1^{n+} - L - M_2^{m+} \rightarrow M_1^{(n+1)+} - L - M_2^{(m-1)+} \quad (7.3)$$

where M_1 and M_2 are the two metal centres, and L the bridge ligand. George and Griffith (296) suggested the superexchange mechanism as an alternative possibility, which would give the following reaction schemes in the binuclear metal complex

$$M_1^{n+}\text{-L-}M_2^{m+} \nearrow M_1^{(n+1)+}\text{-L}^{-}\text{-}M_2^{m+} \searrow M_1^{(n+1)+}\text{-L-}M_2^{(m-1)+}$$
$$M_1^{n+}\text{-L-}M_2^{m+} \searrow M_1^{n+}\text{-L}^{+}\text{-}M_2^{(m-1)} \nearrow M_1^{(n+1)+}\text{-L-}M_2^{(m-1)+} \quad (7.4)$$

This mechanism obviously corresponds to the participation of intermediate higher energy states in which an electron is transferred to or from the bridge ligand.

These ideas were put into a molecular orbital framework by Halpern and Orgel in an approximate calculation of the electronic coupling matrix elements for inner sphere ET reactions in which the bridge ligand mediates the ET between the metallic centres (297). They assumed that in certain cases the overall ET probability is determined by ET within the binuclear complex and calculated the transition probability between the two discrete electronic levels on the donor and acceptor centres. Furthermore, direct overlap between metallic orbitals was assumed to contribute neglibly as compared with overlap with suitable bridge orbitals. The electronic wave functions of the initial (i), final (f), and intermediate (d) states were represented in three-electron determinantal form

$$\phi_i = (3!)^{-\frac{1}{2}}|\phi_D\phi_L\bar{\phi}_L| \;;\; \phi_f = (3!)^{-\frac{1}{2}}|\phi_A\phi_L\bar{\phi}_L| \tag{7.5}$$

$$\phi_d = (3!)^{-\frac{1}{2}}|\phi'_L\phi_L\bar{\phi}_L| \tag{7.6}$$

where ϕ_D, ϕ_A, and ϕ_L represent one-electron wave functions for the transferring electron being located on the donor, the acceptor, and the bridge ligand, respectively. ϕ_L and $\bar{\phi}_L$ refer to electrons with different spins, and ϕ'_L to an excited ligand orbital. First- and second-order coupling terms then take the form

$$\langle\phi_i|V_i|\phi_i\rangle = \langle\phi_D(1)\phi_L(2)\bar{\phi}_L(3)|V_i|\phi_A(1)\phi_L(2)\bar{\phi}_L(3)\rangle -$$

$$- \langle\phi_D(1)\phi_L(2)\bar{\phi}_L(3)|V_i|\phi_L(1)\phi_A(2)\bar{\phi}_L(3)\rangle \tag{7.7}$$

and

$$(E_i - E_d)^{-1} \langle \phi_D(1)\phi_L(2)\bar{\phi}_L(3) | V_i | \phi'_L(1)\bar{\phi}_L(2)\phi_L(3) \rangle$$

$$\langle \phi'_L(1)\phi_L(2)\bar{\phi}_L(3) | V_d | \phi_A(1)\phi_L(2)\phi_L(3) \rangle \qquad (7.8)$$

respectively, where V_i and V_d are the appropriate perturbations. When several bridge group orbitals in a conjugated double bond system participate, the wave functions are approximated by the form $\phi_t = \sum_k c_{tk}\psi_k$ (t = D, L, A) where ψ_k is the atomic orbital of the k'th atom. The electronic coupling term is then proportional to the mobile bond order (MBO) of the conjugated system, i.e. to the quantity

$$\mathrm{MBO} = \sum_{r,s} c_{tr} c_{ts} \qquad (7.9)$$

The MBO is a measure of the conjugation, and in the present context, also of the electronic 'conductivity' along the inner sphere ET parhway.

This model was the first attempt towards a quantitative theoretical description of inner sphere ET reactions. The model was later extended to include electrostatic interactions between the reactants, and a good correlation between calculated MBO's and experimental rate constants was apparently observed (298). However, it is obvious that the formalism only deals with a single facet of the ET probability, i.e. that of the electronic matrix elements. The overall rate expression (eq.(7.2)) must include the nuclear wave functions as well, and appropriate averaging over the continuous nuclear vibrational spectrum must be performed. The relation between the bridge ligand properties and the ET rate is therefore reflected in both the activation energy and in the pre-exponential factor of the rate expression.

McConnell analyzed the related problem of intramolecular ET between the two phenyl groups, ϕ, in the radical anions of the symmetric α,ω-diphenylalkanes of the general form $[\phi\text{-}(CH_2)_n\text{-}\phi]^-$

(303). In the initial and final states the electron is trapped on one phenyl group due to solvent polarization and C-C bond deformations. Fluctuations in the solvent modes induce degeneracy of the donor and acceptor levels, and the electron is transferred via a set of virtual intermediate states localized on the methylene groups. Since this kind of ET has been the subject of recent extensive investigations, we shall return to a closer examination of McConnell's and later related work.

Within the last few years considerable progress towards the development of a detailed theory for the role of intermediate states in chemical processes has been achieved (105-107,110,112,287,288,300-302,304,305). The development has proceeded within the general framework of the theory of multiphonon processes as outlined above. Thus, such effects as the role of low- and high-frequency nuclear modes (105,106a,287,304), the presence of a continuous electronic spectrum (106a) (as in electrochemical processes), and adiabaticity (305) can be incorporated in the theory. In a number of cases it has furthermore been possible to analyze experimental data reasonably successfully in terms of the theory.

Although no major differences are manifested in the calculations, it is important to distinguish between two different physical situations. The intermediate states may have sufficiently low energies that the activation energy for transition to or from these states is lower than the activation energy for direct transition between the initial and final states. The intermediate state is then a real state which in principle can be detected experimentally, but the nuclear modes do not relax into the equilibrium coordinate values corresponding to the intermediate electronic distribution. Several inner sphere ET and biological ET processes, as well as concerted proton transfer reactions can be analyzed within this conceptual framework. On the other hand, the intermediate states may have such high energies that the activation energy for the direct transition is lower than for transition to or from the intermediate state. In such cases the ET path via the intermediate state may still be operative by providing better electronic or nuclear Franck Condon overlap than the direct transition. This is the role played by the intermediate state in the superexchange mechanism and the earlier attempts to formulate a theory for inner sphere reactions on the basis of this concept. It is clear that in this case transition via the intermediate state is not a real process. The system cannot be detected in this state, but when it undergoes an electronic transition at the reaction hypersurface between the initial and final state potential energy surfaces it does so by a probability which is partly determined by the wave functions of the intermediate state.

7.2 Theoretical Formulation of Higher Order Rate Probability

7.2.1 Semiclassical Methods.

Most of the calculations on higher order chemical processes have been performed within a semiclassical formalism. We shall illustrate these results by considering at first the one-dimensional classical motion along a single reaction coordinate from an initial to a final electronic state via a single intermediate state. The reaction coordinate mimics for example the characteristic solvent motion, and it is also assumed that subsequent to the process the excess energy is dissipated. We consider at first the nonadiabatic limit (105,304).

For each of the electronic states a one-dimensional potential energy surface can be defined

$$U(q) = U_{no} + f_n(q - q_{no}) \qquad (7.10)$$

where the index n refers to the initial (i), intermediate (d), and final (f) state. q is the reaction coordinate, and q_{no} the equilibrium values of this coordinate in the various electronic states. U_{no} thus incorporates both the electronic and the equilibrium solvation energies. The transition probability is calculated by means of the second order perturbation theory expression, eq.(7.2) which can be written more explicitly in the form

$$W_{fi} = \frac{2\pi}{\hbar} \Big(\sum_i \exp(-E_i^o/k_B T)\Big)^{-1} \sum_{i,f} \exp(-E_i^o/k_B T) \qquad (7.11)$$

$$\left|\sum_d \frac{\langle \Psi_f(\vec{r},q)|V_d|\Psi_d(\vec{r},q)\rangle\langle\Psi_d(\vec{r},q)|V_i|\Psi_i(\vec{r},q)\rangle}{E_i^o - E_d^o + i\gamma}\right|^2 \delta(E_i^o - E_f^o)$$

where $\vec{r}$ refers to the electronic coordinates, and $\Psi_n(\vec{r},q)$ (n = i,d,f) is the total wave function of the system in the state n. In view of the continuous nature of the solvent vibrational spectrum it is, however, legitimate to replace all summations in this equation by integrations, i.e. to perform the operation

$$\sum_n \rightarrow \int \rho_n(E_n^o)\, dE_n^o \qquad (7.12)$$

where $\rho_n(E_n^o)$ is the vibrational level density. For harmonic potentials of frequency ω_n, $\rho_n(E_n^o) = (\hbar\omega_n)^{-1}$. In the Born-Oppenheimer and Condon approximations eq.(7.10) then takes the following form

$$W_{fi} = \frac{2\pi}{\hbar}\Big[\int \rho_i \exp(-E_i^o/k_B T)\Big]^{-1} |V_{fd}|^2 |V_{di}|^2 \int \rho_i\, dE_i^o \qquad (7.13)$$

$$\exp(-E_i^o/k_B T)\, \rho_f\, dE_f^o\, \delta(E_i^o - E_f^o)\, |M(E_i^o, E_f^o)|^2$$

V_{di} and V_{fd} are here the electronic factors which couple the zero order electronic intermediate state with the initial and final states, respectively. $M(E_i^o, E_f^o)$ contains the overlap integrals of the nuclear wave functions, $\chi_n(q)$,

$$M(E_i^o, E_f^o) = \int_{U_{do}}^{\infty} \rho_d dE_d^o \frac{(\chi_f(E_f^o), \chi_d(E_d^o))(\chi_d(E_d^o), \chi_i(E_i^o))}{E_i^o - E_d^o + i\gamma} \qquad (7.14)$$

The presence of the delta function furthermore simplifies eq.(7.14) to

$$W_{fi} = \frac{2\pi}{\hbar}\left[\int \rho_i \exp(-E_i^o/k_BT)\right]^{-1} |V_{fd}|^2 |V_{di}|^2 \int_{\max\{U_{io}, U_{fo}\}}^{\infty} \rho_i \rho_f dE_i^o$$

$$\exp(-E_i^o/k_BT)|M(E_i^o, E_i^o)|^2 \qquad (7.15)$$

where the lower integration limit with respect to E_i^o is determined by the higher of U_{io} and U_{fo} .

When proceeding further the nuclear overlap functions (χ_d, χ_i) and (χ_f, χ_d) are calculated by representing the nuclear wave functions, $\chi_n(q)$, in the quasiclassical approximation (142). In the classical region, i.e. 'inside' the potential well these wave functions have the general form

$$\chi_n(q) = C_n(E_n^o)[p_n(q)]^{-\frac{1}{2}}\cos\left\{\int_{a(E_n^o)}^{q} p_n(q)dq - \frac{\pi}{4}\right\} = \qquad (7.16)$$

$$C_n(E_n^o)\,[p_n(q)]^{-\frac{1}{2}}\cos\Phi_n$$

We have introduced the dimensionless momentum, $p_n(q)$, related to the real momentum, $P_n(q)$, by the transformation $p_n(q) = (\mu\hbar\omega_n)^{-\frac{1}{2}}$ $P_n(q)$, where

$$P_n(q) = [2\mu(E_n - U_n(q)]^{\frac{1}{2}} \qquad (7.17)$$

$a(E_n^o)$ is the left hand turning point for motion in the classical region, and $C_n(E_n^o)$ is a normalization constant to be determined from the condition

$$\int_{-\infty}^{\infty} |\chi_n(q)|^2 dq \quad \int_{a(E_n^o)}^{b(E_n^o)} [C_n(E_n^o)]^2 [p_n(E_n^o)]^{-1} \cos^2 \Phi_n \, dq = 1 \qquad (7.18)$$

where $b(E_n^o)$ is the right hand turning point. Since, in the classical region, $\cos^2 \Phi_n$ is a rapidly oscillating function, it can be replaced by its average value, 1/2, in the integration region.

The integration of eq.(7.18) can then be performed to give (142)

$$\frac{1}{2} C^2 (\mu\omega_n) \int_a^b P^{-1} dx = \pi/2 \; ; \; C = (2/\pi)^{1/2} \qquad (7.19)$$

where x ($= (\hbar/\mu\omega_n)^{1/2} q$) is now the reaction coordinate in units of length.

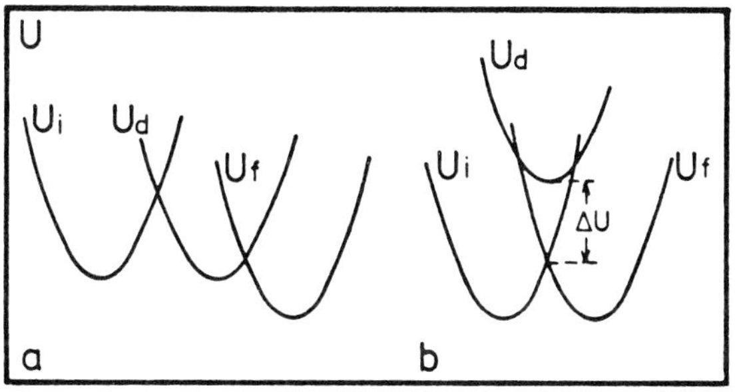

Fig (7.1)

We now leave out computational details concerning the integrations with respect to q to provide the overlap functions of the quasiclassical nuclear wave functions and the subsequent integrations with respect to E_d^0 and E_i^0 to provide the final rate expressions (105). We can then summarize the results as follows:

(A) If the potential energy surfaces are located as in fig.(7.1a), i.e. when the energy of the intersection point between the intermediate and either the initial or the final state surface is higher than the energy of the intersection point between the initial and final state surfaces, then

$$W_{fi} = 2\pi^{\frac{3}{2}} |V_{fd}|^2 |V_{di}|^2 \left\{\hbar^3 k_B T [|U_{fd} - U_{di}|/k_B T]^{\frac{1}{2}}\right\}^{-1} \tag{7.20}$$

$$\exp[-(U^* - U_{io})/k_B T]/[U_i'(q) - U_d'(q)]_{q=q^*_{di}} [U_d'(q) - U_f'(q)]_{q=q^*_{fd}}$$

$U_n'(q)$ is the derivative of $U_n(q)$ with respect to q, q^*_{di} and q^*_{fd} the coordinates of the intersection point between the intermediate state potential energy surface and the initial and final state potential energy surfaces, respectively, and U^* (= U_{fd} or U_{di}) the energy value in the intersection point corresponding to the higher energy. Finally, $Z_i = \int \rho_i \exp(-E_i^0/k_B T) dE_i^0$ is the classical partition function in the initial state. For the particular case where the zero-order states can be represented by displaced harmonic potential energy surfaces of uniform frequency, ω , eq.(7.20) becomes

$$W_{fi} = \pi^{\frac{3}{2}} |V_{fd}|^2 |V_{di}|^2 / \left\{\hbar(\hbar\omega)(E_r^{di} E_r^{fd})^{\frac{1}{2}} [|U_{fd} - U_{di}|/k_B T]^{\frac{1}{2}}\right\}$$

$$\exp[-(U^* - U_{io})/k_B T] \tag{7.21}$$

where E_r refers to the reorganization energies defined previously and for the transitions indicated by the superscripts.

(B) If the potential energy surfaces are located as in fig.(7.1b), i.e. if the intermediate state is now of high energy, the rate expression corresponding to eq.(7.21) becomes

$$W_{fi} = \pi^{\frac{1}{2}} |V_{fd}|^2 |V_{di}|^2 [\hbar(\Delta U)^2 (E_r^{fi} k_B T)^{\frac{1}{2}}]^{-1} \exp[-U_{fi} - U_{io})/k_B T] \tag{7.22}$$

(cf.eq.(7.44)) where ΔU is the energy distance between the energy of the intersection point between the initial and final state potential energy surfaces and the minimum energy of the intermediate state potential energy surface.

From these rate expressions we notice the following about the role of intermediate states in elementary rate processes:

(1) If the energy of the intermediate state is high ($U_{do} > U_{fi}$) the activation energy of the process is determined by the initial and final state potential energy surfaces only, as for the two-level systems. However, the electronic coupling term in the pre-exponential factor is now determined by the coupling between the intermediate state and both the initial and final states rather than by the direct coupling between the initial and final states.

(2) When the energy of the intermediate state is low ($U_{do} < U_{fi}$) the transition via the intermediate state occurs not only because of favorable electronic coupling compared with the direct transition, but also because of a lower activation energy. In this case the system is actually located in the intermediate state, and transitions to or from this state are now energy-conserving real physical processes. The system could thus in principle be detected experimentally when existing in this state.

(3) The nature of second order transitions via a low-lying intermediate state is different from that of two consecutive two-level transitions. In the latter case each process proceeds independently of all previous steps. On the other hand, in the

former case the transition proceeds in a single step, and the energy released after each electronic transition at an intersection point between two potential energy surfaces is not dissipated before the system relaxes into its final state equilibrium nuclear configuration. The rate expression here contains no averaging over the intermediate states, and its final appearance is therefore different from that of consecutive reactions.

The results obtained for a one-dimensional motion can be generalized to motion in any number of dimensions (304). However, the trajectory of the system is then obviously more complicated, and the activation energy is no longer solely determined by the saddle points of the reaction hypersurfaces but also by the mutual position of the minima of the potential surfaces.

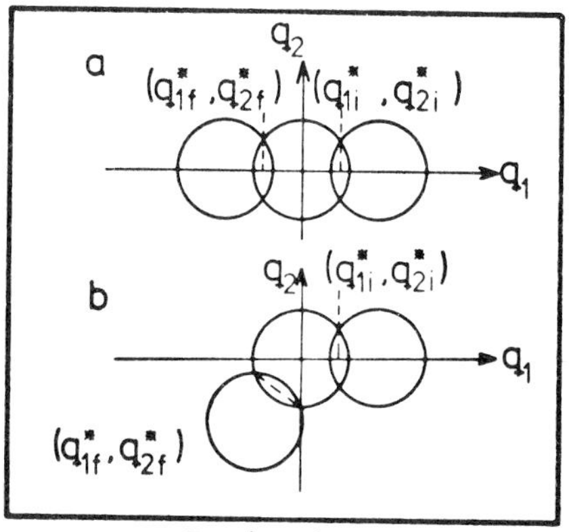

Fig (7.2)

This can be illustrated with reference to fig.7.2 which shows cross sections of two-dimensional potential energy surfaces for different positions of the minima. In fig.7.2a only one equilibrium coordinate value is shifted. The trajectory then follows the q_1 axis, and the activation energy is determined by the saddle point of highest energy. However, when both equilibrium coordinates are shifted (fig. 7.2b), the reaction path is curved which may require an additional contribution to the activation energy which is thus no longer solely defined by the energies of the appropriate saddle points (304).

7.2.2 The Effect of High-Frequency Modes.

The effect of high-frequency intramolecular quantum modes has also been investigated within the semiclassical formalism (106b,287). Reorganization of such modes are of course no less important for second order processes than for first order processes. For example, if concerted proton transfer reactions are interpreted as second order processes, strong equilibrium coordinate shifts occur in at least one high-frequency mode, the proton stretching mode. When reorganization of quantum modes are incorporated in the second order ET formalism we can define potential energy surfaces for each of the three electronic states in complete analogy to first order processes. Thus, for each electronic state n

$$U_n(Q,q) = U_{no} + f_n^S(q) + g_n^P(Q) \tag{7.23}$$

where $f_n^s(q)$ and $g_n^P(Q)$ refer to separate contributions from the low- (medium) and high-frequency (intramolecular) modes, respectively (cf. eqs(4.1) and (4.2)). Similarly, by a procedure analogous to the one outlined above, the thermally averaged second order transition probability can be written in the form

$$W = \frac{2\pi}{\hbar}|V_{fd}|^2|V_{di}|^2(Z_i^P)^{-1}(Z_i^S)^{-1}\sum_{\substack{\varepsilon_i^P,\varepsilon_i^s\\ \varepsilon_f^P,\varepsilon_f^s}}\exp[-(\varepsilon_i^P+\varepsilon_i^S)/k_BT]$$

$$|M(E_i^o,E_f^o)|^2\,\delta(E_i^o - E_f^o) \tag{7.24}$$

where Z and $\mathcal{E}$ refer to the partition function and the nuclear vibrational energy levels of the state indicated by the subscript and the particular nuclear subsystem indicated by the superscript (P the fast and S the slow subsystem). Furthermore, $M(E_i^o, E_f^o)$ now contains nuclear energy contributions and wave functions referring to both the fast and the slow nuclear subsystems (cf.eq.(7.13)).

Further evaluation of eq(7.24) now depends on the concrete properties of the system to be investigated. We shall therefore specifically consider the two coupled proton transfer steps (concerted proton transfer) shown schematically in fig.7.3. In the initial and final states the proton is localized on the donor (D) and acceptor (A), respectively. A third molecular species (catalyst) assists the overall proton transfer and is able to mediate the proton transfer in two ways as indicated in the figure. We shall also invoke the following two plausible approximations which strongly simplifies the further calculations:

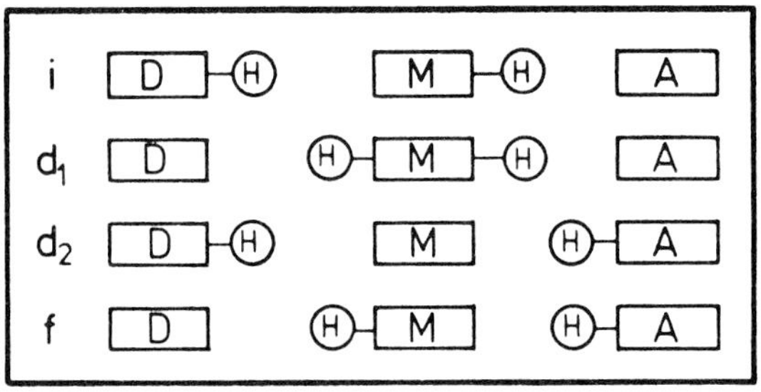

Fig (7.3)

Models for concerted proton transfer from donor (D) to acceptor (A) via a catalyst molecule (M). Two possible intermediate states, d_1 and d_2.

(a) We shall assume that the process is totally nonadiabatic and that the double adiabatic approximation is adequate. The total electronic-nuclear wave function of the system in the various electronic states can then be written in the form

$$\Psi_n(\vec{r},\vec{Q},\vec{q}) = \psi_n(\vec{r};\vec{Q};\vec{q})\varphi_n^{v_j}(\vec{Q};\vec{q})\chi_{N_n}(\vec{q}) \tag{7.25}$$

where ψ, φ, and χ are the wave functions of the electrons, the protons, and the solvent, respectively, v_j refers to the quantum number of the j'th proton ($j = 1,2$, and $v = k,l,m$ for $n = i, f, f$), and N_n to the solvent nuclear quantum number in the electronic state n.

(b) When one proton is transferred, its nuclear vibrational quantum number is generally subject to a change. However, we shall assume that this does not affect the nuclear quantum numbers of the other proton which is not transferred during the particular single-proton transfer step.

With these assumptions $M(E_i^\circ, E_f^\circ)$ can be written in the form

$$M(E_i^o, E_f^o) = \sum_{N_d, l_1, l_2} \langle \varphi_f^{m_2} \varphi_f^{m_1} | \varphi_d^{l_2} \varphi_d^{l_1} \rangle \langle \varphi_d^{l_2} \varphi_d^{l_1} | \varphi_i^{k_2} \varphi_i^{k_1} \rangle \tag{7.26}$$

$$(\chi_{N_f}, \chi_{N_d})(\chi_{N_d}, \chi_{N_i}) / \left\{ U_{io} - U_{do} + \varepsilon_i^s(N_i) - \varepsilon_d^s(N_d) + \right.$$

$$\left. + \varepsilon_i^p(k_1, k_2) - \varepsilon_d^p(l_1, l_2) + i\gamma \right\}$$

where k_j, l_j, and m_j are the nuclear vibrational quantum numbers of the j'th proton in the initial, intermediate, and final electronic state, respectively, and $\langle\ \rangle$ and (,) refer to integrations with respect to the proton and solvent coordinates, respectively.

Assumption (b) now implies that only terms for which $k_2 = l_2$ and $l_1 = m_1$ survives in the sum over l_1 and l_2. Eq.(7.22) can then be recast in the following form which bears a close formal resemblance to the corresponding expression for first order processes (eqs.(4.9) and (4.12))

$$W = (Z_i^p)^{-1} \sum_{k_1,k_2} \sum_{m_1,m_2} \exp[-\varepsilon_i^p(k_1,k_2)/k_B T] S^{fd}_{k_2,m_2} S^{di}_{k_1,m_1}$$

$$W_m(k_1,k_2,m_1,m_2) \qquad (7.27)$$

where S refers to the Franck Condon nuclear overlap integrals for each proton corresponding to the electronic states and nuclear vibrational quantum numbers given by the superscripts and subscripts, respectively, and W_m is the second order transition probability determined for the classical modes only, but referring to particular values of the vibrational quantum numbers in the initial and final states. $W_m(k_1,k_2,m_1,m_2)$ is thus given by eqs. (7.18)-(7.20), but with the energy gaps and intersection point energies modified by the vibrational energies of the proton modes.

Eq.(7.27) constitutes the theoretical expression for nonadiabatic second order rate processes when reorganization in both low-frequency medium modes and high-frequency intramolecular modes is important. In subsequent applications of this expression it is necessary to introduce explicitly given potential energy surfaces in order to calculate $W_m(k_1,k_2,m_1,m_2)$ and the Franck Condon factors of the high-frequency modes. We notice, however, a general implication of eq(7.27). This equation refers to a triple manifold of classical potential energy surfaces in each electronic state corresponding to all values of the high-frequency vibrational quantum numbers. The reaction probability is a thermally averaged sum of all independent transitions from a given initial to a given final state via a given intermediate state. While the Gibbs factor decreases with increasing values of the quantum numbers, the Franck Condon factors increase, but for sufficiently high vibrational frequencies this increase is weaker than the decrease of the Gibbs factor. When $U_{di}^{oo} > U_{fd}^{oo}$ (where the superscripts refer to the quantum numbers of the two protons) (fig.7.5) this means that the major contribution to the sum is provided by $k_1 = k_2 = m_1 = 0$. On the other hand, m_2 only enters the Franck Condon factors which, as noted, increase with increasing quantum number. The sum of eq.(7.27) therefore effectively only contains terms from different m_2, of which furthermore a single term, m_2^*, dominates, i.e. the one for which the energy of the corresponding intersection point, $U_{fd}^{om_2^*}$, of the potential energy surfaces is approximately equal to U_{di}^{oo} (fig.7.4)). In this case eq.(7.27) takes the simpler form

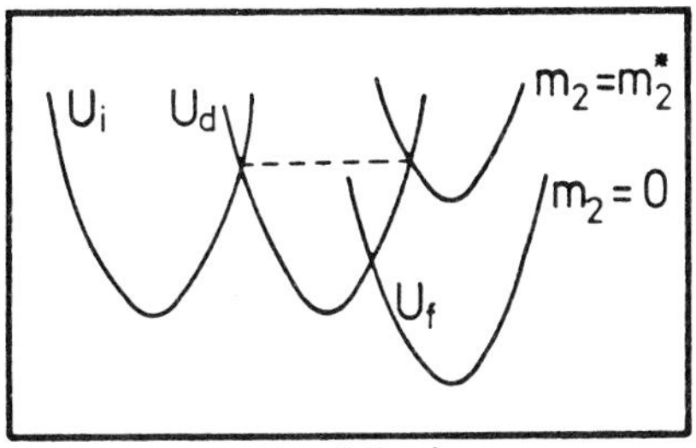

Fig(7.4)

$$W_{fi} \approx Z_p^{-1} \sum_{m_2} S_{o,m_2}^{fd} S_{o,o}^{di} W_m(o,o,o,m_2) \approx \qquad (7.28)$$

$$Z_p^{-1} S_{o,m_2^*}^{fd} S_{o,o}^{di} W_m(o,o,o,m_2^*)$$

where $\varepsilon_f^p(m_2^*) \approx U_{di}^{oo} - U_{fd}^{oo}$. Similarly, for $U_{di}^{oo} < U_{fd}^{oo}$ the dominating term in the sum of eq.(7.27) is

$$W_{fi} \approx Z_p^{-1} \exp[-\varepsilon_i^p(k_1^*)/k_B T] S_{o,o}^{fd} S_{k_1^*,o}^{di} W_m(k_1^*,o,o,o) \qquad (7.29)$$

where $\varepsilon_i^p(k_1^*) \approx U_{fd}^{oo} - U_{di}^{oo}$. In contrast to first order processes, second order processes are thus commonly expected to proceed by participation of excited high-frequency nuclear vibrational states.

7.2.3 Adiabatic Second Order Processes

We have so far assumed that all transitions in the crossing region between the zero order potential energy surfaces were totally nonadiabatic. However, within the semiclassical approximation, rate expressions for higher order adiabatic processes can also be provided (305). We shall illustrate this for a second order process in which the potential energy surfaces are again characterized by a single (classical) nuclear coordinate q. This reaction scheme is in fact rather similar to that discussed by Laidler many years ago (306) for the nonadiabatic quenching of excited sodium, mercury, and cadmium atoms by different atoms and molecules in the gas phase.

For a single classical passage across the intersection region the reaction probability is given by the Landau-Zener expression (cf. chapter 5)

$$1 - \exp(-2\pi\gamma) \; ; \; \gamma = (\Delta U_{\pm}/2)^2/\hbar v|U'_i - U'_f| \qquad (7.30)$$

$\Delta U_{\pm}$ is here the splitting of the zero order surfaces at the intersection point, and the other symbols were all defined in chapter 5.

For sufficiently low values of $\Delta U_{\pm}$ the system must pass the crossing point several times before reaching the final state.

The total reaction probability is then calculated as a sum of all the probabilities corresponding to each passage. If the slopes of the terms in the crossing region have the same sign (fig.7.5) the overall passage can only proceed in one of the two ways shown in the figure. For such transitions the overall transmission coefficient for passing a single crossing point is

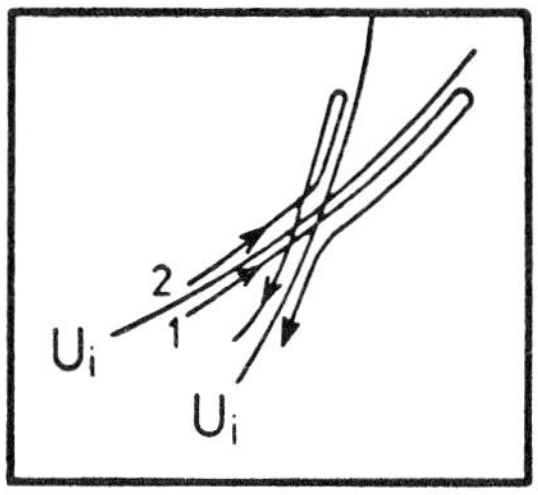

Fig(7.5)

$$2\exp(-2\pi\gamma)[1-\exp(-2\pi\gamma)] \tag{7.31}$$

On the other hand, if the sign of the slopes are different infinitely many trajectories are possible. Thus, if the system passes the crossing point without reacting it may perform any number of oscillations on the upper adiabatic surface. The transmission coefficient here takes the form of an infinite geometric converging sum

$$f(\gamma) = [1-\exp(-2\pi\gamma)]/[1-\frac{1}{2}\exp(-2\pi\gamma)] \tag{7.32}$$

We shall apply this result to the arrangement of the potential energy surfaces considered above. For a low-lying intermediate state (fig.7.1) the overall transmission coefficient is then $f(\gamma_{di})f(\gamma_{fd})$. For the two limiting cases of totally nonadiabatic ($\gamma \ll 1$) and adiabatic ($\gamma > 1$) processes this product becomes $16\pi^2\gamma_{di}\gamma_{fd}$ and unity, respectively. When the transition probability is averaged over all initial vibrational energies and velocities as in chapter 5 the resulting expression for the totally nonadiabatic reaction becomes

$$W_{fi}^{na} = (2\pi\hbar)^{-1} Z_i^{-1} 16\pi^2 \exp(-E_A^{na}/k_B T)\, \mu \tag{7.33}$$

$$\int_0^{\infty} \gamma_{di}\, \gamma_{fd}\; v(E) \exp(-\frac{1}{2}\mu v^2/k_B T)\, dv$$

where E_A^{na} is the activation energy determined previously, Z_i the classical partition function of the initial state and u the mass associated with the nuclear motion. Inserting for the lower intersection point $v \approx (2|U_{fd}-U_{di}|/\mu)^{\frac{1}{2}}$, and putting $v(E) = v_T$, i.e. the thermal velocity, subsequently gives an expression identical to eq.(7.20) except that it is higher by a factor of 2. This is associated with the fact that we have here incorporated the sum of all trajectories in the passage of the two intersection points.

For purely adiabatic processes we obtain similarly

$$W_{fi}^{ad} = (2\pi\hbar\, Z_i)^{-1} k_B T \exp(-E_A^{ad}/k_B T) \tag{7.34}$$

where E_A^{ad} is now the highest point on the lower adiabatic surface (cf. chapter 5). Finally, if one of the parameters is small (e.g. $\gamma_{fd} \ll 1$) and the other one large ($\gamma_{di} > 1$) the rate expression takes the form

$$W = (2\mu)^{\frac{1}{2}} k_B T(\Delta U_{\pm}^{di}/2)^2 \left\{ Z_i \hbar |U_i' - U_f'| |U_{fd} - U_{di}|^{\frac{1}{2}} \right\}^{-1}$$

$$\exp(-E_A/k_B T) \tag{7.35}$$

where E_A is the energy of the higher of the intersection points corrected for the amount of 'splitting' if the adiabatic transition occurs at this point.

We consider finally the second order processes through high-energy intermediate states (fig.7.1). At the intersection point between the initial and final state potential energy surface W = f(γ_{fi}) where

$$\gamma_{fi} = (\Delta U_{\pm}^{fi}/2)^2/\hbar v|U'_i - U'| \tag{7.36}$$

and $\Delta U_{\pm}^{fi}$ is now the splitting between these two surfaces in the intersection region. We shall assume that this splitting is negligible in the absence of the third molecular species which provides the intermediate electronic state. However, in the presence of this species the splitting assumes a finite value which can be determined from second order perturbation theory to be

$$\Delta U_{\pm}^{fi} = 2\, V_{fd}\, V_{di}/\Delta U \tag{7.37}$$

where all the symbols have been defined previously (cf. eq.(7.22)). If we furthermore only consider transitions at the intersection point between the initial and final state potential energy surfaces, i.e. we assume that the system is not thermally excited so that it could actually exist in the high-energy intermediate state, then an averaging procedure similar to the one represented by eq.(7.33) gives for W_{fi}:

$$W_{fi} = (2\pi\mu k_B T)^{\frac{1}{2}}(\Delta U_{\pm}^{fi}/2)^2[\hbar^2 Z_i|U'_i - U'_f|]^{-1}\exp(-E_A^{fi}/k_B T) \tag{7.38}$$

where E_A^{fi} is now the energy of the intersection point between the initial and final states, possibly corrected for the amount of splitting. If we here represent the potential surfaces by harmonic oscillators of frequency ω, and invoke the condition $\hbar\omega \ll k_B T$, this equation can be transformed to eq.(7.22).

7.2.4 Quantum Mechanical Formulation.

The semiclassical formalism represents a rather general basis for a theoretical description of second order processes. However, this theory is subject to the general reservations about semiclassical methods which we discussed briefly in chapter 5. In particular it requires that a classical nuclear subsystem is present and that the quantum and classical subsystems remain as such during the process. While this is a valid assumption for most chemical processes at room temperature it nevertheless represents a fundamental limitation of the theory of second order processes, and it would therefore be desirable to formulate a proper quantum mechanical theory for these processes.

Heavy mathematical difficulties have, however, so far prevented the development of such a theory to a level comparable to that for first order processes. Calculations have thus been successfully performed for a model system consisting of a single displaced low-frequency harmonic nuclear mode (107) to provide eq.(7.21), and this same approach was recently extended to incorporate high-frequency modes as well (307).

7.3 Relation to Experimental Data

Several electron transfer reactions can be interpreted in terms of the theoretical framework of higher order processes, even though a proof that this is the actual mechanism is hardly possible to obtain. Some of the most convincing evidence for the occurrence of higher order processes relates to electrochemical and biological processes. We shall deal with these systems in separate chapters, however, and at present in turn consider the following homogeneous ET systems:

(A) Inner sphere electron transfer and related processes.

(B) The 'cation effect' in outer sphere electron transfer processes.

(C) Intramolecular electron transfer in organic radicals.

(A) Inner Sphere Electron Transfer. It was suggested early (24,296,297) that second order effects may constitute the basic ET mechanism in inner sphere electron transfer processes. The intermediate state must then, however, correspond to a high-energy electronic level in all cases where the bridge ligand is a small molecule such as water or halide ions. Thus, the intermediate state would correspond to the localization of the electron in an antibonding bridge ligand orbital or an electron being transferred from a bonding orbital on the bridge to the acceptor centre. For small ligand molecules the energies involved in either of these steps would, however, be far higher than any measured activation energies for these processes (307). On the other hand, a second order transition via a high-energy intermediate state might be conceivable since the electronic coupling term in this case would involve the overlap between orbitals located on neighbouring atoms (metal and bridge), whereas a first order transition would involve a weaker overlap between a donor and an acceptor orbital separated by a bridge ligand. However, except for certain inner sphere ET processes which may involve spin-orbit coupling (in particular those involving Co(III)/Co(II) redox couples), we have seen that the majority of inner sphere ET processes are generally more likely to correspond to the adiabatic limit.

Electron transfer through more extensive bond systems of coordinated groups (a 'remote attack' mechanism) was suggested in a number of cases on the basis of comparisons between reaction rates of complexes containing ligands with conjugated π-bond systems and related complexes containing σ-bond ligands (308). However, a definite proof for such a mechanism was only provided relatively recently, for the reaction between $[Cr(H_2O)_6]^{2+}$ and the pentaammine isonicotinamide complex of Co(III), i.e. (309)

$$[(NH_3)_5 \; Co \; N\langle\bigcirc\rangle -\underset{\underset{NH_2}{|}}{C} = O \;]^{3+} + [Cr(H_2O)_6]^{2+} \; \overset{H^+}{\rightarrow}$$

$$[Co(H_2O)_6]^{2+} + [HN\langle\bigcirc\rangle -\underset{\underset{NH_2}{|}}{C} = OCr(H_2O)_5]^{4+} + 5NH_4^+$$

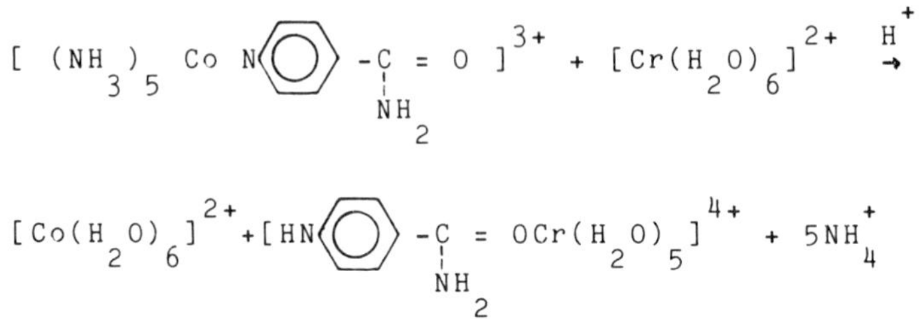

(7.39)

In this and other systems (310) there is good evidence for a 'chemical' mechanism with ET through a low-lying intermediate state in which the electron is localized on the isonicotinamide ligand. Thus, the rate constants for ET from $[Cr(H_2O)_6]^{2+}$ to the isonicotinamide complexes of Co and Cr are 17.4 and 1.8 $(dm^3 M^{-1} s^{-1})$, respectively (309). This small difference was observed in spite of the fact that the free energy of reaction in the former case is about 0.6 ev more negative than in the latter in striking contrast to the fact that the corresponding rate constant ratio for complexes with small irreducible bridge ligands amounts to several millions (17,307,309). The difference is understandable on the basis of a second order mechanism with ET through a low-lying intermediate state. The highest crossing point between the zero order surfaces is that between the initial and intermediate states corresponding to ET from Cr(II) to the bridge (fig.7.1). The rate constant for reaction with the corresponding Ru complex is $3.5 \cdot 10^5$ $dm^3 M^{-1} s^{-1}$, although the free energy of this reaction is approximately the same as for the Co complex (311). This is usually explained by the different symmetries of the acceptor orbitals. The acceptor orbital for Ru(III) is of t_{2g} symmetry which overlaps well with the intermediate state ligand Υ-orbitals, whereas in the Co(III) complex the electron goes to an e_g orbital of low overlap with the bridge orbital (307). This effect is however, only important if the reaction of the Co(III) complexes is nonadiabatic. In addition, the reaction of the Ru complex has a substantially smaller intramolecular reorganization energy which can account for a factor up to about 10^2 in the difference between the rate constants (cf. section 5).

Intramolecular ET induced by certain oxidation processes on the 'remote' end of the bridge ligand is related to the inner sphere ET reactions of the kind discussed. In such complexes as

$[(NH_3)_5$ Co N  $-CH_2OH]^{3+}$ (7.40)

the intramolecular ET from the reductive end $-CH_2OH$ to the oxidative Co(III) is very slow. However, if a strong external oxidant such as MnO_4^- or Ce(IV) is added, the $-CH_2OH$ group is oxidized to a radical, which can be oxidized further to $-CH{=}O$ either by an additional external redox equivalent or by intramolecular ET to Co(III). The oxidation product in the two cases is different, since the resulting Co(III) in the intramolecular ET rapidly hydrolyzes to $[Co(H_2O)_6]^{2+}$.

The general reaction pattern of these induced ET processes involves in fact two intermediate radicals formed by parallel elementary steps (312). One radical has a fairly long life-time (10^{-7} - 10^{-3} s), and ET to external oxidants therefore effectively competes with the intramolecular process. On the other hand, the second radical undergoes intramolecular ET so rapidly that the external oxidant cannot follow this step. Since the ET between the external oxidant and the intermediate radical would be practically diffusion controlled, the decay of this 'nontrappable' intermediate state may well correspond to a second order process.

The discussion of the isonicotinamide complexes suggests an analysis of bridge-assisted ET reactions on the basis of free energy relationships (313). If we consider a series of 'closely related' ET reactions, in which the acceptor is varied, then, provided that E_r stays constant, only the final state potential energy surface is shifted vertically relative to the initial and intermediate potential energy surfaces, whereas the relative position of the latter two is maintained. If the left-hand crossing point is the higher (fig.7.1) the activation energy is determined by this point, corresponding to ET from donor to bridge, and therefore shows no dependence on the overall heat of reaction. If at some acceptor strength the right-hand crossing point becomes the higher, the activation energy is now determined by this point (corresponding to ET from bridge to acceptor) and now increases with increasing heat of the overall reaction. The resulting qualitative form of the free energy plot then consists of two rather sharply separated (within a few multiples of $k_B T$) regions of different slopes. An analogous conclusion can be drawn if the order of the ET steps is inverted (the pull push mechanism).

These principles can only be applied in a qualitative way to proper inner sphere ET reactions both because it is rarely, if ever, possible in practice to vary the donor or acceptor metal centres over extensive ranges, and because of the general lack of information about the thermodynamic parameters for the binuclear complexes. However, if the electron is transferred from the donor to a ligand and subsequently transmitted to the acceptor in an outer sphere step, then the change of the heat of reaction can be obtained from the standard redox potentials of the acceptor couples. This was the basis of a recent investigation of the reactions of the complex (313)

$$[(2,2'\text{-bipy})_2(H_2O)Cr^{II}(4,4'\text{-bipy-}H^+)]^{3+} \qquad (7.41)$$

with a series of luteo Co(III) complexes (bipy = bipyridyl). In comparison with the ET reaction between $[Cr(2,2'\text{-bipy})_3]^{2+}$ and the same Co(III) complexe the complex (7.41) has a much lower activation energy (2-4 kcal against about 10 kcal) and a much more negative activation entropy -(25-30) cal K^{-1} against about -(10-15) cal K^{-1}. This was interpreted as a loss of translational entropy corresponding to attack by Co(III) at a particular site of the Cr(II) complex, followed by an energetically facile ET. The free energy plot is shown in fig.7.6 and looks qualitatively as predicted if the total ET path is viewed as a second order process in which the first step is ET from Cr(II) to 4,4'-bipy and the second, ET from 4,4'-bipy to Co(III) (a push pull mechanism).

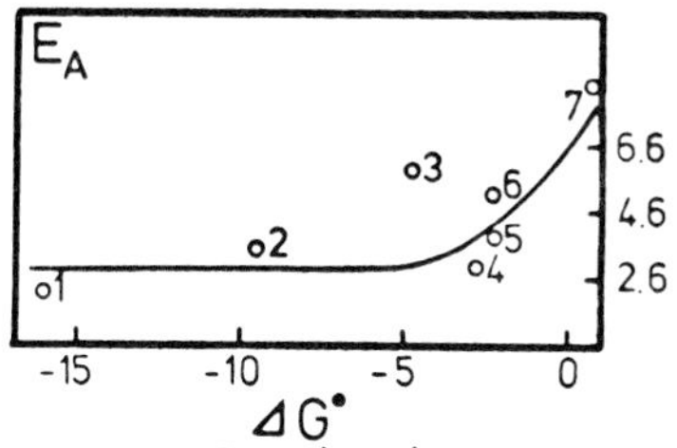

Fig (7.6)

E_A (kcal) plotted against $\Delta G°$ for the reactions of the compound (7.41) with Co(III) complexes. 1: acetatopentaammine. 2: hexaammine. 3: tris-meso-butanediammine. 4: bis-ethylenetriammine. 5: tris-ethylenediammine. 6: tris-(racemic)-propanediammine. 7: tris-(—)-butanediammine.

(B) The Cation Effect. Both homogeneous ET processes involving electron exchange in the $[Fe(CN)_6]^{3-/4-}$ (316) and $MnO_4^{-/2-}$ (317) couples and in several heteronuclear systems (318), and the electrochemical reduction of $[Fe(CN)_6]^{3-}$ at metal (319) and insulator electrodes (65) are subject to strong increases of the rate constants with increasing addition of alkali, earth alkali, and quaternary ammonium ions. For the alkali metal ions the order of 'efficieny' is $Li^+ < Na^+ < K^+ < Rb^+ < Cs^+$, and the rate constants increase by factors of 20-100 when the 'inert' cation concentrations increase from very small values up to about 3 M dm^{-3}. These effects are larger than the effects due to electrostatic screening as estimated from simple models, and a trinuclear collision complex consisting of an ion pair of one of the reactants and a cation, and the second reactant has therefore been suggested. In this complex the electron passes through an intermediate state localized on the cationic bridge. For the $[Fe(CN)_6]^{3-/4-}$ and $[MnO_4^{-/2-}]$ systems such a state which corresponds to an alkali metal atom in the solution, must have a very high energy. (This is not necessarily so, however, for the alkali metal ion catalyzed ET between naphtalene and its anion (320) where the states corresponding to the localization of the excess electron on either the alkali ion or the naphtalene molecule have approximately the same energy). It is therefore clear that if such states participate, they do so by providing a better electronic overlap than for the direct ET.

It is also clear that under the conditions where appreciable cation effects are observed, a possible bridge mechanism is only one of several effects which could give rise to a monotonous rate increase. Other effects are : (a) electrostatic screening; (b) change of the standard redox potential due to preferential ion pairing with one redox component; (c) change of bond distances and frequencies in the intramolecular modes by the electric field of the cations, and (d) freezing of part of the solvent space due to hydration of the cations(space correlation effects, cf. chapter 2). In order to estimate the importance of a second order effect, the two electronic matrix elements $V_{fi}^{(1)}$ and $V_{fd}^{(1)} V_{di}^{(1)}/(E_i - E_d)$ should be compared for the average geometry of the termolecular collision complex. However, this can only be done in an extremely approximate way. If we use the estimate of $V_{cc'}^{(1)} \simeq 10^{-3}$ ev for an ET between ions at 'contact distance' (91), $(E_i - E_d) \simeq 1$ ev, and a Slater orbital exponential dependence on the ET distance, it is seen that $V_{fi}^{(2)} \simeq 10^{-3} V_{fi}^{(1)}$ for the second order mechanism to become effective. Using furthermore the two limiting values estimated for the orbital exponent (91), 4.7 A^{-1} and 1 A^{-1}, the ET distance for the direct transition must be larger than the ET distance to the bridge by 1.5 A and by 7 A, respectively, or about an ionic diameter for the latter to contribute significantly.

(C) Intramolecular ET in Organic Radicals. Intramolecular electron transfer in organic anionic radicals $\varphi\text{-}(CH_2)_n\text{-}\varphi^-$ where φ is a phenyl, naphtyl or related radical, and n varies from zero to 20, has been the subject of extensive experimental investigation (321-323). In polar solvents the excess electron is trapped at one of the terminal aromatic groups. However, fluctuations in the solvent and intramolecular modes induce degeneracy between the electronic levels of the two aromatic groups and a subsequent electron transfer between them. The velocity of this process is commonly of such a value (10^6-10^7 s^{-1}) that it can be estimated by the broadening of the electron spin resonance signal of the odd electron.

The intramolecular electron transfer can proceed by two mechanisms. If the chain which connects the donor and acceptor groups is sufficiently flexible (large n), the process is most likely a direct electron transfer involving diffusion of the end groups up to a favorable relative distance at which the electron transfer occurs. The activation energy is here expected to be close to the activation energy for the bimolecular process involving freely mobile donor and acceptor molecules similar to the bound groups. However, the presence of the 'molecular' chain means that the values of the rates correspond approximately to a concentration of one molecule per sphere having a radius of the extended chain. Therefore, the smaller the chain length the larger the 'effective' concentration of donor and acceptor groups, and the larger the intramolecular rate constants. These expectations are borne out by the experimental rate constants for the (α-naphtyl)-$(CH_2)_n$-(α-naphtalenide) anions in hexamethylphosphorictriamide (HMPA). For n = 12 the effective concentration of donor and acceptor groups is here 0.07 M dm^{-3}, while the rate constant (15 °C) and activation energy is $1.75 \cdot 10^7$ s^{-1} and 3.6 kcal, respectively (321). The corresponding values for the intermolecular ET between n-butyl-α-naphtalene and its anion are $3.5 \cdot 10^7$ s^{-1} - or a rate constant of $6 \cdot 10^8$ $dm^3 mol^{-1}$ s^{-1} - and 4.8 kcal, both of which are close to the diffusion controlled limit. The esr spectrum for n = 4 at -15°C is practically identical to that for n = 12 at 45°C showing the effect of the decreased chain length but otherwise a similar direct ET mechanism (321).

When n < 2, or for rigid molecular 'chains', unfavorable electronic overlap between the donor and acceptor orbitals makes a direct ET less efficient. McConnell firstly suggested that the electron transfer in these systems proceeds through a set of high-energy virtual intermediate (3d) orbitals localized at the methylene links of the chain (303). Initially the electron is trapped at the donor group. Fluctuations in all the nuclear coordinates subsequently induces approximate degeneracy of the donor and acceptor levels. The activation energy of the process is thus determined by the energy required to create this nuclear state from the initial equilibrium configuration. Provided that direct interaction between the donor and acceptor orbitals, ψ_d and ψ_a, can be neglected, the symmetric and antisymmetric combinations, $(\psi_d + \psi_a)/2^{\frac{1}{2}}$ and $(\psi_d - \psi_a)/2^{\frac{1}{2}}$, represent degenerate zero order electronic wave functions of the system. However, coupling to the intermediate state orbitals lifts the degeneracy and provides an energy gap, g, between the two lowest electronic states, for which calculation by perturbation theory gave the following plausible form

$$g = -(2T^2/D)(-t/D)^{N-1} \tag{7.42}$$

T is here the electron exchange energy between the donor or acceptor orbital and its nearest 3d neighbour in the methylene chain, t the exchange energy between neighbouring 3d orbitals, D the energy gap between a noninteracting 3d orbital and ψ_d or ψ_a, and N-2 is the number of links (methylene groups) in the chain. The ET rate is subsequently determined as $h|g|$ and seen to decrease by a factor of $t/D \approx 0.1$ for each additional link added to the chain.

This mechanism can be reformulated in terms of the theory outlined in the present chapter. Thus, in view of the small size of the methylene group we notice at first that the coupling between the electronic and nuclear motion is expected to be strong not only in the initial and final states such as noted by McConnell, but also in all the intermediate states. The electron transfer can then proceed in the following two ways:

(1) For a low-lying intermediate state (in the sense discussed above) the results of McConnell need modification. Thus, the ET rate between two identical groups separated by a single methylene group is given by eq.(7.20) where the activation energy is determined by either intersection point ($U_{di} = U_{fd}$), i.e.

$$U_{di} = U_{fd} = (E_r^{di} + U_{di})^2 / 4E_r^{di} \qquad (7.43)$$

For two identical intermediate states we can perform a similar semiclassical analysis for which, however, we give here only the expression for the activation energy. If the intersection point between the initial and first intermediate states is the higher, the activation energy is again given by eq.(7.43). On the other hand, if the intersection point corresponding to ET from one methylene group to another is the higher, then the activation energy takes the form.

$$E_A = E_r^{d_2 d_1} / 4 + U_{d_1 i} \qquad (7.44)$$

This analysis can be generalized to any number of intermediate states, to asymmetric processes, and to many-dimensional potential energy surfaces.

(2) If the intermediate state is of such high energy that the activation energy to or from this state is higher than the acti-

vation energy for the direct transition, then the intermediate state assists the electron transfer by providing a better electronic overlap (i.e. via the methylene groups) than for the direct transition. In this case the rate constant takes the form

$$W = \frac{(\pi\mu k_B T)^{1/2} (\Delta U_{fi}/2)^2}{Z(U'_i - U'_f)\hbar^2 2^{3/2}} \exp(-E_A/k_B T) \qquad (7.45)$$

where all the symbols have been defined before. $E_A = U_{fi} = E_r^{fi}/4$, while the form of the electronic coupling factor depends on the number of intermediate states (methylene groups). Thus, by higher order perturbation theory (142), for a single intermediate state

$$\Delta U_{fi} = 2V_{di} V_{fd} / (U_d^* - U_{fi}) \qquad (7.46)$$

(cf. eq.(7.7a)). For two identical intermediate states located as in fig.7.7b

$$\Delta U_{fi} = 2 V_{d_2 d_1} V_{d_1 i}^2 / [(U_{d_2 d_1} - U_{fi})^2 - V_{d_2 d_1}^2] \qquad (7.47)$$

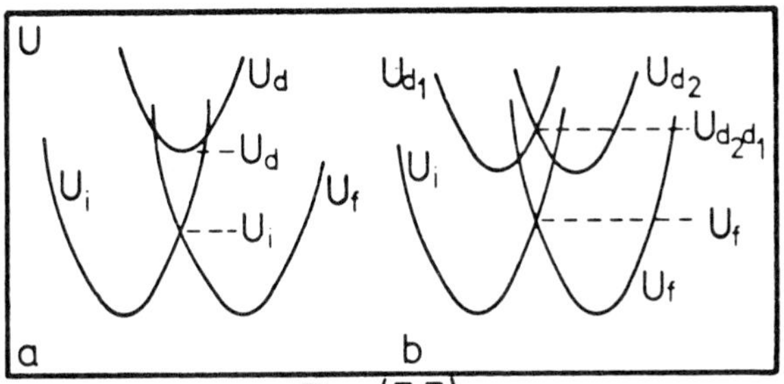

Fig (7.7)

ET via one (a) and two (b) high-energy intermediate states. $V_{d_2 d_1}$ (fig.(7.7b) is half the splitting between the intermediate potential surfaces d_1 and d_2.

It is possible in principle to determine experimentally which of the two mechanisms prevails in a given system. For the low-lying intermediate states the activation energy depends directly on the energy gap between the initial and intermediate states as reflected for example in the difference between the electron affinities of the donor group and the methylene group. For high-energy intermediate states this dependence is weaker and furthemore temperature independent, being now solely reflected in the denominator of the pre-exponential factors(eqs.(7.45)-(7.47)).

The experimental data available at present are generally insufficient to permit such a distinction with certainty. However, it is notable that the intramolecular ET rate between two naphtyl moieties linked together in the para-positions of the rigid cyclohexane, i.e. trans-1,4-bis-(α-naphtylmethyl)-cyclohexane is $9 \cdot 10^6$ s^{-1} at 15 C in HMPA. For the corresponding ET between two phthalimide groups only the rate at 45 ° C is known being $1 \cdot 10^6$ s^{-1} (323). The ET rate between phthalimide groups in the 1,3 positions is $4 \cdot 10^6$ s^{-1}, and if the activation energy is approximately the same for the 1,4- and 1,3-compounds the rate at 15 ° C for the 1,4-compound would be $3 \cdot 10^5$ s^{-1}, i.e lower by a factor of approximately 30 than for the 1,4-dinaphtyl compound.

However, the electron affinities of naphtalene and phthalimide are 0.15 ev and 1.34 ev, respectively (324). This would give both a far higher rate difference and a far higher activation energy than commonly observed for these processes ($E_A \approx 6$ kcal for the phthalimide system) if the intermediate states had sufficiently low energy that the system actually proceeds through these states. A consideration of the experimental data in terms of the theory of higher order processes therefore suggests that the intermediate states are of high-energy virtual nature for which a rate difference of a factor of 30 is furthermore compatible with the form of the pre-exponential factors. This mechanism could thus transmit the electron through the rigid nuclear cyclohexane framework, i.e. over a distance of about 9 Å.

8 ELECTROCHEMICAL PROCESSES

8.1 Fundamental Properties of Electrochemical Reactions

The fundamental analogies between electron and proton transfer reactions in homogeneous solution and electrochemical processes were recognized early (48) and also incorporated in the multi-phonon theory of elementary rate processes at an early stage of the development of this formalism (12,42,83,84).

This theory applied to electrochemical processes has at present reached a level at which certain general features, such as the dependence of the current density on overpotential and temperature, and the role of low- and high-frequency nuclear modes are reasonably well understood. However, studies of electrochemical processes are presently entering a prospective new stage. This new development is associated firstly with the introduction of new techniques for the experimental studies of the interphase, such as electronic and raman spectroscopy of surface layers and adsorbed species, photoemission of the electrode-electrolyte system, and surface spectroscopic techniques developed for solid-vacuum systems. Secondly, it is associated with the adoption of new theories which incorporate concepts from both solid-state and surface physics, from the theory of 'nonlocal' (i.e. nonuniform and anisotropic) dielectrics, and from the theory of chemical processes.

We should, however, recognize that a metallic or semiconductor electrode in contact with a dielectric medium represents a system which is substantially more complicated than systems commonly studied in the field of surface physics. Consequently, the assumptions which at present have to be invoked in the theoretical description of electrochemical phenomena are far more drastic than those involved in both solid-vacuum systems and homogeneous systems. These differences are associated in particular with the following features of the heterogeneous metal-electrolyte system:

8.1.1 The nonuniform dielectric medium.

Close to the electrode surface the structure of the solvent is different from the structure of the bulk solvent. This effect is in a sense analogous to the nonuniformity induced by the presence of ions of a finite geometric extension in the medium, and the extension of this region is expected to be approximately similar to the extension of the microstructure of the medium. Incorporation of the medium nonuniformity effects would require that the description of the homogeneous isotropic medium outlined in chapter 2 is modified essentially in two ways (325,326). Firstly, the fact that the dielectric permittivity and response functions, $\varepsilon(\vec{r},\vec{r};\ t-t')$ and $G(\bar{r},\bar{r}';\ t-t')$ are no longer solely dependent on $\vec{r}-\vec{r}'$ means that no simple relationship between the polarization vector and the electric field is now available. Secondly, even in the absence of structural effects the presence of a sharp boundary affects the polarization response of the medium by causing reflection of travelling polarization waves from the bulk. In this way the sharp boundary itself represents an 'impurity'.

Attempts towards a theoretical formulation of the 'nonlocal' dielectric medium near a surface have at present appeared (325,326). We shall, however, make no attempt to review the situation of this difficult and so far unsettled topic, but take the more practical approach most commonly applied for the electrochemical interphase region. We shall thus view the medium as a bulk medium up to the surface. However, the structural effects in the interphase region can be incorporated by means of suitable molecular and continuum models for the double layer (327-329) which can provide the reorganization and work terms required for the general theory. For example, one of the most commonly accepted views of the double layer is that the electrode is covered by a layer of solvent molecules oriented according to the charge of the electrode (the inner Helmholtz layer). Unless the depolarizer ions are specifically adsorbed, the discharge is then assumed to involve ions located outside this layer (the outer Helmholtz layer). This compact region is followed by the 'diffuse' part of the double layer in which the ionic atmosphere differs from that in the bulk solution (327). Other models (328,329) allow for the presence and orientation of different kinds of solvent molecules on the surface, i.e. both individual and small clusters of molecules.

8.1.2 The continuous electronic spectrum

In contrast to homogeneous processes, electrochemical processes involve a continuous electronic spectrum in one of the reactants, i.e. the metal or semiconductor electrode. Even if homogeneous and heterogeneous ET processes are both viewed as multiphonon radiationless processes, this will cause some important qualitative differences between the two classes. The continuous electronic spectrum is most frequently handled within the 'single-electron approximation', i.e. by viewing the interaction of a given electron with all the other electrons as equivalent to the formation of a set of 'quasiparticles' characterized by a given distribution function $n(\varepsilon)$ and level density $\rho(\varepsilon)$ (where ε is the energy of a given quasiparticle level). The distribution function is moreover commonly identified with the Fermi distribution function, i.e.

$$n(\varepsilon) = \left[1 + \exp[(\varepsilon - \varepsilon_F)/k_B T]\right]^{-1} \qquad (8.1)$$

where ε_F is the Fermi energy. For $\varepsilon > \varepsilon_F$ $n(\varepsilon)$ thus decays exponentially with increasing energy, whereas $n(\varepsilon)$ is approximately constant and equal to unity for $\varepsilon < \varepsilon_F$. For metals $n(\varepsilon)$ is thus a rapidly varying function near the Fermi level, whereas it is zero for semiconductors for which ε_F is located in the band gap. In contrast, for metals $\rho(\varepsilon)$ is a slowly varying function near ε_F. For freely mobile electrons $\rho(\varepsilon)$ is for metals

$$\rho(\varepsilon) = (m_e V_e / 2\pi^2 \hbar^3)\,[2m_e(\varepsilon - U_c)]^{\frac{1}{2}} \qquad (8.2)$$

where m_e is the mass of the electron, V_e the volume of the electrode, and U_c the energy at the bottom of the conduction band. For semiconductors which display electronic conduction $\rho(\varepsilon)$ has a similar form, whereas for hole conduction (or electron conduction through the valence band) m_e is replaced by the effective 'hole mass', m_h and $\varepsilon - U_c$ by $U_v - \varepsilon$, where U_v is the energy at the top of the valence band. The single-particle approximation is known to be adequate for metals when the particle energy is close to the Fermi energy and for semiconductors when the concentration of electrons (holes) in the conduction (valence) band is small (330). We shall adopt this approximation and the Fermi distribution function in the following, but notice that several important conclusions about current-voltage relationships, some of which we shall derive below are also valid for less restrictive assumptions about the distribution function, including in fact the single-particle approximation itself (331).

We shall now proceed to derive some general features of electrochemical processes. For this aim we invoke the following three general and plausible assumptions (33,34,331):

(1) The total current is viewed as a sum of independent ET's to or from individual electrode levels.

(2) The rate determining step of the overall electrochemical process is the electron transfer, i.e. we shall ignore diffusion of the molecular reactants from the bulk up to the electrode surface.

(3) We shall assume that the probability per unit time of ET between a discrete molecular (depolarizer) level and an individual electrode level, ε , and for given values of the high-frequency nuclear quantum numbers of the depolarizer molecule in the initial and final states (v and w) is given by an expression formally identical to the rate expression of absolute rate theory (45)

$$W_{wv}(\varepsilon) = \frac{k_B T}{h} \varkappa_{wv} \exp[-G_A^{wv}(\varepsilon)/k_B T] \qquad (8.3)$$

$\varkappa_{vw}$ is the transmission coefficient, including the appropriate Franck Condon factor for the high-frequency modes, and G_A^{vw} the free energy of activation for a given microscopic process. For the sake of simplicity we shall furthermore only consider cathodic processes. The overall current density, $i(\eta)$, is then

$$i(\eta) = eC\, Z^{-1} \frac{k_B T}{h} \sum_{v,w} \exp(-\varepsilon^v/k_B T)\varkappa_{wv} \qquad (8.4)$$

$$\int_{-\infty}^{\infty} (\varepsilon) n(\varepsilon) \exp[-G_A^{wv}(\varepsilon,\eta)/k_B T] d\varepsilon$$

where ε^v and Z are the energy levels and statistical sum of the high-frequency modes, η the overvoltage of the electrode ($= \varphi_m - \varphi_m^o$, where φ_m is the actual potential of the electrode and φ_m^o the equilibrium potential), e the (numerical) electronic change, and C the concentration of the depolarizer at the site of electron transfer. By means of eq.(8.4) we can find the 'microscopic' transfer coefficient, $\alpha_{wv}(\varepsilon,\eta)$

$$\alpha_{wv}(\varepsilon,\eta) = \partial G_A^{wv}(\varepsilon,\eta)/\partial[\Delta G_{fi}(\varepsilon,\eta)] \tag{8.5}$$

where $\Delta G_{fi}(\varepsilon,\eta)$ is the free energy of reaction.

The electron is initially located at the level ε. At the electrode potential ϕ_m its energy is shifted by an amount $e\phi_m$ which implies that the free energy of the initial state, G_i, is

$$G_i = \varepsilon - e\phi_m + I_{io} + g_i \tag{8.6}$$

where I_{io} contains the electronic energy and equilibrium solvation energy of the depolarizer ion, and g_i (= $-k_B T \ln \sum_j \exp(-\varepsilon_j^i/k_B T)$ where j refers to the states of both the classical and the high-frequency nuclear modes) the free energy contribution of the total nuclear subsystem reckoned from the ground level I_{io}. Similarly, in the final state the electron is located at the depolarizer, and the free energy then takes the form

$$G_f = I_{fo} + g_f \tag{8.7}$$

where the meaning of the symbols is analogous to that of eq.(8.6).

We have implicitly assumed that the depolarizer ion is located at the outer Hemholtz plane and that the whole potential drop from the metal to the solution occurs inside this part of the double layer. For this reason we have omitted work terms of the form $ze\psi_1$ (where z is the ionic charge, and ψ_1 the potential at the site of the depolarizer ion. In addition, the vibrational energy spectrum of the depolarizer ion might be affected by the electrode potential. A change of the electrode potential might thus affect the force constants of the appropriate potential energy surfaces giving rise to a potential dependence of g_i and g_f. These effects could be incorporated by a suitable model for the electrical double layer, but in line with our omission of the work terms we shall refrain from doing so. We can then write the 'microscopic' free energy of reaction as

$$\Delta G_{fi}(\varepsilon,\eta) = \Delta I^{o}_{fi} - (\varepsilon - e\phi_m) + \Delta g_{fi} = e\eta - (\varepsilon - \varepsilon_F) + \Delta G^{(fi)}_{oF} \tag{8.8}$$

where

$$\Delta I^{o}_{fi} = I_{fo} - I_{io} \ ; \Delta g_{fi} = g_f - g_i$$

and ΔG $(= \Delta I + e\phi^{o}_{m} - \varepsilon_F + \Delta g_{fi})$ is the free energy change of the process at the equilibrium potential (η = 0) and when the electron is initially located at the Fermi level. This quantity is of macroscopic (thermodynamic) nature. Thus, introducing formally the equilibrium between the oxidized and reduced molecular species, and the electron and hole concentrations, i.e.

$$Ox + e \rightleftarrows Red \tag{8.9}$$

$$K = C_R\,[1-n(\varepsilon)]/C_O\,n(\varepsilon) \tag{8.10}$$

the equilibrium potential is determined by the equation $\Delta G_{fi}(\varepsilon, 0) = -k_B T\ln K$. Inserting eq.(8.1) then gives

$$\Delta G_{fi}(\varepsilon,0) = -(\varepsilon - \varepsilon_F) - k_B T\ln(C_R/C_O) \tag{8.11}$$

or, in view of eq.(8.8)

$$\Delta G_{fi}(\varepsilon_F,0) = \Delta G^{fi}_{oF} = -k_B T\ln(C_R/C_o) \tag{8.12}$$

We can now proceed to a calculation of the important 'macroscopic' phenomenological current-voltage relationship. From eq.(8.8), we can rewrite eq.(8.5) in either of the forms

$$\alpha_{wv}(\varepsilon,\eta) = \partial G^{wv}_{A}(\varepsilon,\eta)/\partial(e\eta) = -\partial G^{wv}_{A}(\varepsilon,\eta)/\partial\varepsilon \tag{8.13}$$

With reference to our discussion in chapters 3 and 4 we realize that this microscopic transfer coefficient is not restricted to values between zero and unity. On the other hand, the macroscopic, experimentally determined transfer coefficient, α_{ex}, is given by the equation

$$\alpha_{ex} = -k_B T d\ln i/d(e\eta) \qquad (8.14)$$

and contains (cf. eq.(8.4)) an averaging over all nuclear electronic states. From eqs.(8.4) and (8.12) we can then rewrite eq.(8.13) in the form

$$\alpha_{ex} = [i(\eta)]^{-1} Z^{-1} \sum_{v,w} \exp(-\varepsilon^{v}/k_B T) \int_{-\infty}^{\infty} \alpha_{wv}(\varepsilon,\eta) \cdot \qquad (8.15)$$

$$i_{wv}(\varepsilon,\eta) d\varepsilon = \langle \alpha_{wv}(\varepsilon,\eta) \rangle$$

where the microscopic' current density, $i(\varepsilon,\eta)$ (involving a single level in the electrode), is

$$i_{wv}(\varepsilon,\eta) = eC\rho(\varepsilon)n(\varepsilon)\frac{k_B T}{h} \kappa_{wv} \exp[-G_A^{wv}(\varepsilon,\eta)/k_B T] \qquad (8.16)$$

This equation therefore shows that α_{ex} can be viewed as the microscopic transfer coefficient averaged with respect to all microscopic current densities. The equation furthermore leads to a very important relation for metal electrodes. We can thus insert the expression for $i_{wv}(\varepsilon,\eta)$ (eq.(8.16)), $\alpha_{wv}(\varepsilon,\eta)$ as the derivative of $G_A^{wv}(\varepsilon,\eta)$ with respect to ε (eq.(8.13)) in eq.(8.16) and view $\rho(\varepsilon)$ essentially as a constant. A subsequent integration by parts with respect to ε gives the following relationship between $\alpha_{ex}(\eta)$ and the distribution function $n(\varepsilon)$

$$\alpha_{ex} = [i(\eta)]^{-1} Z^{-1} \sum_{v,w} \exp(-\varepsilon^{v}/k_B T) \int_{-\infty}^{\infty} i_{wv}(\varepsilon,\eta)[1-n(\varepsilon)]d\varepsilon =$$

$$Z_{int}^{-1} \sum_{v,w} \exp(-\varepsilon^{v}/k_B T)[1 - \langle n(\varepsilon)\rangle] \qquad (8.17)$$

where $\langle n(\varepsilon)\rangle$ is $n(\varepsilon)$ averaged with respect to all the microscopic current densities in the same way as eq.(8.15).

From eq(8.17) we can draw the important conclusion that since $0 < \langle n(\varepsilon)\rangle < 1$, the experimental transfer coefficient for metal electrodes must also be located between zero and unity. This result is different from homogeneous processes where the transfer coefficient can be negative in the strongly exothermic region, and it is physically associated with the presence of the continuous electronic spectrum in the metal electrode. Fig.8.1 thus shows three relative positions of the initial and final state potential energy surfaces when the electrode is initially located at the Fermi level of the electrode. For small overvoltages (a in fig.8.1) $\langle\alpha\rangle \approx 0.5$, and electrons near the Fermi level contribute most to the overall current density. For large negative overvoltages, however, the activation energy for transitions of electrons at energies below the Fermi level is smaller than for electrons at the Fermi level. In contrast to processes in homogeneous solution the strongly exothermic region therefore corresponds to a vanishing activation energy, and $\langle\alpha\rangle \approx 0$. (the activationless overpotential region). Such a region is always found for metal electrodes since the continuous electronic spectrum corresponds to a continuous manifold of potential energy surfaces shifted vertically relative to eachother. For large positive overvoltages electronic transitions from levels above the Fermi level have a smaller activation energy than transitions from the Fermi level. In this region the free energy of activation coincides with the free energy of reaction, and $\langle\alpha\rangle \approx 1$. This region is called the barrierless overpotential region, since the electrochemical process, although having a large activation energy does not exhibit an activation barrier.

We shall postpone more specific calculations of the current-overpotential relationship to a subsequent section and complete the present section by some general observations on ET processes at semiconductor electrodes. Metal and semiconductor electrodes display two important differences. Firstly, the concentration of mobile charge carriers, i.e. electrons in the conduction band and holes in the valence band, is much lower for semiconductors than for metals. Moreover, the charge carrier concentration (10^{13} -10^{17} cm^{-3}) is typically much smaller than the electrolyte concentration (10^{19} cm^{-3} for a 0.01 mol dm^{-3} solution).

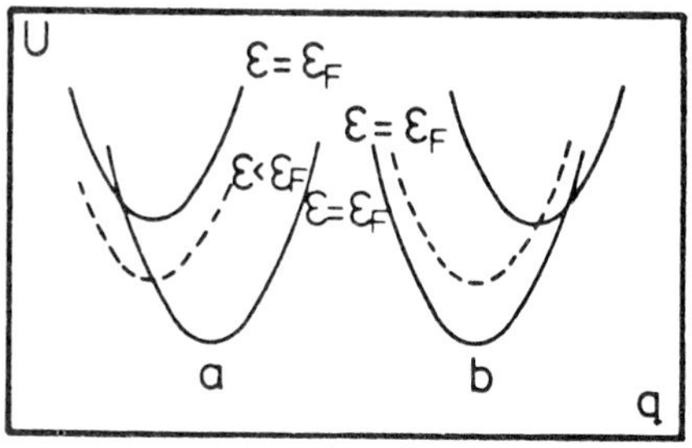

Fig (8.1)

Since the extension of the diffuse double layer is inversely proportional to the aquare root of the carrier concentration, this implies that the potential drop in the electrode-electrolyte region occurs primarily on the electrode side, in contrast to metal electrodes for which it occurs on the electrolyte side. When the overpotential of the bulk semiconductor varies, the surface (contact) potentials of the electrode corresponding to the bottom of the conduction band and the top of the valence band, are therefore only slightly affected by the potential variation. The current-voltage characteristics are instead determined by the carrier concentration which is strongly affected by the overpotential (332).

Secondly, for semiconductors the Fermi level is located in the energy gap between the valence and conduction bands. This means that the energy levels which provide the dominating contributions to the current density are now located either in the conduction or the valence band. The overall current density in either cathodic or anodic direction then consists of additive contributions from the conduction and valence bands, commonly named electron (i_e) and hole (i_h) currents, respectively. In the following we shall consider these contributions separately, and for the sake of simplicity we furthermore consider at first cathodic currents only. In line with our derivation of the current density expressions for metal electrodes the analogous expressions for the cathodic current densities are (29,34)

$$i_e(\eta) = eC\frac{k_BT}{h}\int_{\varepsilon_F+\Delta_c+e(\eta-\eta_c)}^{\infty} \rho_e(\varepsilon + e\eta - e\eta_c)n(\varepsilon)\exp[-G_A(\varepsilon,\eta)/k_BT]d\varepsilon \tag{8.18}$$

$$i_h(\varepsilon) = eC\frac{k_BT}{h}\int_{-\infty}^{\varepsilon_F-\Delta_v+e(\eta-\eta_v)} \rho_h(\varepsilon - e\eta + e\eta_v)n(\varepsilon)\exp[-G_A(\varepsilon,\eta)/k_BT]d\varepsilon \tag{8.19}$$

(cf. eq.(8.4)) where, for the sake of simplicity, we have ignored the high- frequency modes. The integration limits are here the lower edge of the conduction band and the upper edge of the valence band for the electron and hole current, respectively, at the electrode-electrolyte contact. η and η_c or η_v are the overvoltage in the bulk of the semiconductor and at the contact for the lower edge of the conduction band and the upper edge of the valence band, respectively, and the remaining symbols have all been defined previously or in fig. 8.2.

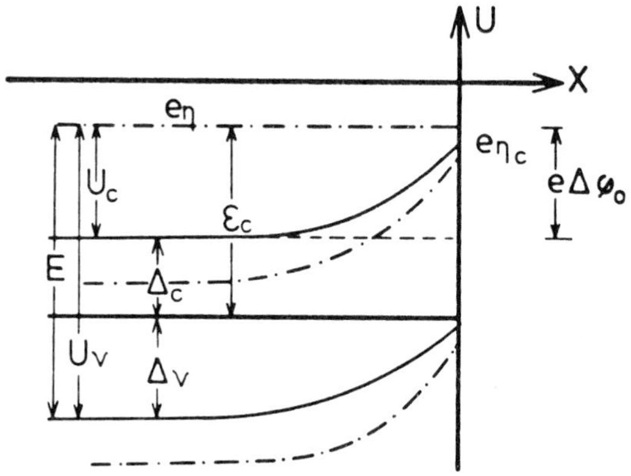

Fig (8.2)

Energy band diagram for the semiconductor-electrolyte interphase region.

Since $\rho(\varepsilon) = 0$ in the band gap we can not straightaway ignore the dependence of $\rho(\varepsilon)$ on ε. If we use the explicit forms given use eq.(8.2) we can, however, use a procedure almost exactly similar to the one leading to eqs.(8.15) and (8.17). For semiconductor electrodes these equations are

$$\alpha^e_{ex} = \langle\alpha^e(\varepsilon)\rangle + \langle k_B T/2(\varepsilon - \varepsilon_c)\rangle(1-\partial\eta_c/\partial\eta) = \tag{8.20}$$

$$1 - \langle n(\varepsilon)\rangle - (k_B T/2(\varepsilon - \varepsilon_c)\rangle\ (\partial\eta_c/\partial\eta)$$

The corresponding equations for α^h_{ex} are obtained by replacing $(\varepsilon-\varepsilon_c)$ by $(\varepsilon_v-\varepsilon)$ and η_c by η_v. From these equations we can draw the following conclusions about the phenomenological behaviour of semiconductor electrodes:

(1) If the carrier density is low, then $\partial\eta_{c,v}/\partial\eta = 0$. Since furthermore $\langle n(\varepsilon)\rangle$ is practically zero or unity in the conduction and valence band, respectively, α_{ex} is unity and zero for cathodic electron and hole currents, respectively. For anodic currents the reverse effects would be predicted. These effects which contrast with the behaviour of metal electrodes are associated with the fact that since the contact potential only depends weakly on the bulk overpotential, the density for cathodic (anodic) electron (hole) currents is primarily determined by the carrier concentration which depends exponentially on the overpotential (see further below).

(2) If the charge carrier concentration is large compared with the electrolyte concentration, then $\partial\eta_c/\partial\eta \approx 1$. Since the dominating contributions to the current are provided by values of ε in the interval $\varepsilon - \varepsilon_c \approx k_B T$, this implies that $\alpha \approx 0.5$, or that the behaviour of the semiconductor electrode is now similar to that of a metal electrode.

8.1.3 Adiabaticity effects in many-potential surface systems.

We have so far viewed the elementary chemical processes as independent electronic transitions between pairs of potential energy surfaces. However, for electrochemical systems both the initial and final states are represented by practically continuous manifolds of potential energy surfaces shifted vertically relative to eachother. Since the spacing between the surfaces in each state is much smaller than the thermal energy $k_B T$, many electronic levels commonly participate in the reaction. This provides again new contents to the question of the adiabatic or nonadiabatic nature of electrochemical processes.

We notice at first that the ET processes between the depolarizer molecule and the individual electronic levels of the electrode are necessarily nonadiabatic due to the delocalized character of the electronic wave functions in the electrode (29,333). For example, if the electrons are represented by free electron wave functions of the form $V_e^{-\frac{1}{2}} \exp(i\,\vec{k}\cdot\vec{r})$ where $k = \hbar^{-1}[2m_e(\varepsilon - U_c)]^{\frac{1}{2}}$, then the exchange integral becomes vanishingly small due to the macroscopic quantity V_e. On the other hand, a macroscopic number of levels participate in the reaction due to the large density of levels. Secondly, the electronic coupling is determined by the overlap of the delocalized wave function of the electron in the electrode and a localized wave function of a certain characteristic extension a_0 (a few A). The product of the variations of the two wave functions is determined approximately by $|\Delta k\, a_0|^2 \approx |(\partial k/\partial\varepsilon)\Delta\varepsilon\, a_0|^2 \approx m_e a_0^2 (k_B T)^2 / 2\hbar^2(\varepsilon - U_c) \ll 1$, where $\Delta\varepsilon \simeq k_B T$ is the energy interval of electrode levels which contribute to the process. The variation of the electronic coupling factor is thus expected to be small (29) and will be considered as constant in the following.

Analysis of electronic transitions from a given initial state potential energy surface to a group of final state potential surfaces is in general difficult, but conditions for which the transitions proceed independently have been provided within a generalization of the Landau-Zener formalism (cf. chapter 5) (221,335).

When the initial and final states are represented by two manifolds of electronic levels, ε_i and ε_f, the overall passage of an electron from the electrode to a depolarizer ion takes a complicated course on the potential energy surfaces $U(\varepsilon_i, \varepsilon_f, q)$. Thus, when at first the system reaches the intersection point for the surfaces corresponding to given values of ε_i and ε_f, $U(\varepsilon_i, \varepsilon_f, q^*)$, there is a certain probability of ET from ε_i to ε_f, i.e. from an electrode to a depolarizer level. On further passage the system reaches another intersection point $U(\varepsilon_i', \varepsilon_f)$

where the system may again either remain in the final state or return to an electrode level, which now differs from the level at which the electron was initially located. For a quantitative description all possible reaction paths, i.e. involving all pairs of energy levels ε_i and ε_f , must in principle be included. Incorporation of these continuity effects in the analysis of the adiabatic or nonadiabatic character of the process then leads to the following, intuitively expected result (333). If the inequality

$$k_B T \rho(\varepsilon^*)\varkappa(\varepsilon^*) < 1 \qquad (8.21)$$

is valid, then the overall reaction probability by passage of the manifolds of intersection points is small. This inequality implies that all transitions between individual electronic levels occur independently. If the inverse inequality is valid, the overall probability is unity, provided that the energy of the system exceeds the energy corresponding approximately to the intersection point of potential surfaces for which the electron is initially located around the Fermi level. The reaction probability is then formally given by eq.(5.48), where E_A is now the energy of this intersection point. However, the overall adiabaticity now arises as a consequence of a manifold of otherwise nonadiabatic microscopic electronic transitions.

8.2 Quantum Mechanical Formulation of Electrode Kinetics

8.2.1 Metal electrodes.

We now proceed to a quantum mechanical approach to a description of simple electrochemical processes. We shall concider the nonadiabatic limit only and use the one-electron approximation and Fermi distribution function. We shall furthermore ignore the nonuniformity of the medium and, for the sake of simplicity, the presence of high-frequency modes. We can then view the cathodic process at a metal electrode in the following way:

The position of the molecular acceptor level is determined by the electronic structure of the molecule and by the strong interaction with the molecular and medium modes. On the other hand, the electron in the electrode is delocalized and interaction with the medium modes can be ignored (unless discrete surface levels are important). Furthermore, the acceptor level must initially be located well above the Fermi level, since the ET would otherwise occur with a large probability and in an activationless fashion. At this high energy the concentration of electrons is very small. However, fluctuations in the medium modes induce a temporary degeneracy between the molecular acceptor level and donor levels around the Fermi level where the electronic distribution function assumes appreciable values. At these values of the nuclear coordinates the ET occurs and is followed by a subsequent relaxation to the equilibrium values corresponding to the final state well below the Fermi level.

The process is thus viewed as being microscopically equivalent to homogeneous ET process. We therefore proceed directly to the expression for the microscopic ET probability in the nonadiabatic limit and with our previous assumptions invoked, i.e. for the cathodic process (29,83)

$$W(\varepsilon,\eta) = |V|^2(\pi/k_BT\hbar^2E_r)^{\frac{1}{2}}\exp\left\{-[E_r-k_BT\ln(C_R/C_o)\right.$$

$$\left.-(\varepsilon-\varepsilon_F)+e\eta]^2/4E_rk_BT\right\} \tag{8.22}$$

where all symbols have been defined previously. With reference to eq.(8.4) this gives for the overall current density

$$i = eC|V|^2(\pi/k_BTh^2E_r)^{\frac{1}{2}}\int_{-\infty}^{\infty}\rho(\varepsilon)n(\varepsilon) \tag{8.23}$$

$$\exp\left\{-[E_r-k_BT\ln(C_R/C_o)-(\varepsilon-\varepsilon_F)+e\eta]^2/4E_rk_BT\right\}$$

which represents the expression to be eventually compared with experimental data. For metal electrodes we can now insert $\rho(\varepsilon) \approx \rho(\varepsilon^*)$ and $n(\varepsilon)$ from eq.(8.1) where ε^* is the value of ε which gives the dominating contribution to the integral with respect to ε. We notice already here that for small overvoltages the Gaussian function of eq.(8.23) has a maximum at values of ε much higher than ε_F. For small η the Gaussian function and $n(\varepsilon)$ in the integrand are therefore both appreciable only in a small ε-range around ε_F.

We can derive the current-voltage relationship more explicitly by invoking the Laplace integration (equivalent to the saddle point method). Thus, we write the integrand as $\exp[f(\varepsilon)]$ (i.e. $n(\varepsilon) = \exp[k_BT\ln n(\varepsilon)/k_BT]$) and notice then that the total exponent in eq.(8.23), and therefore the integrand, is maximum at a value $\varepsilon = \varepsilon^*$ determined by differentiating the exponent with respect to ε, i.e. by the equation (29,83,95)

$$n(\varepsilon^*) = \frac{1}{2} + \frac{k_B T \ln(C_R/C_o)}{2E_r} + \frac{\varepsilon - \varepsilon_F}{2E_r} - \frac{e\eta}{2E_r} \tag{8.24}$$

In general $-k_B T\ln(C_R/C_O)$ is much smaller than E_r. When solving eq.(8.24) with respect to ε^* we can then distinguish between three overvoltage regions (fig.8.3):

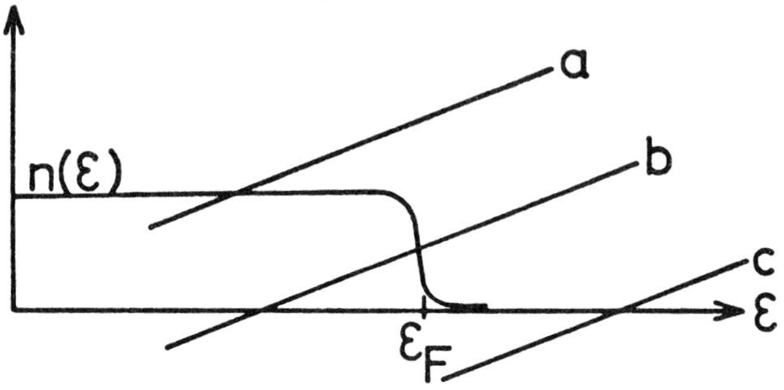

Fig 8.3

(1) For $|e\eta| < E_r + k\,T\ln(C_R/C_O)$, $\varepsilon^* \approx \varepsilon_F$. By noting that

$$\int_{-\infty}^{\infty} \exp[f(\varepsilon)]d\varepsilon \approx [2\pi/f''(\varepsilon^*)]^{\frac{1}{2}} \exp(f(\varepsilon^*)] \tag{8.25}$$

we can rewrite eq.(8.23) in the form

$$i(\eta) \approx eC(\pi/k_B T\hbar^2 E_r)^{\frac{1}{2}} \Delta\varepsilon^* \rho(\varepsilon^*) \cdot \tag{8.26}$$

$$\exp\left\{ -[E_r - k_B T\ln(C_R/C_o) + e\eta]^2/4E_r k_B T \right\}$$

where

$$\Delta\varepsilon^* = k_B T \left\{ 2\pi (k_B T/2E_r + n(\varepsilon_F)[1-n(\varepsilon_F)])^{-1} \right\}^{\frac{1}{2}} \approx (8\pi)^{\frac{1}{2}} k_B T \tag{8.27}$$

Eq.(8.27) represents the current density in the 'normal' overpotential region. The overvoltage dependence of the current density is primarily reflected in the exponential factor giving for the transfer coefficient

$$\alpha_{ex} \approx \frac{1}{2} + e\eta/2E_r \tag{8.28}$$

(cf. chapter 3), i.e. the transfer coefficient is expected to vary from zero to unity when $e\eta$ varies from $-E_r$ to E_r. In particular, in view of the fact that E_r is commonly one or several ev, α_{ex} is expected to be close to 0.5 in a fairly wide potential range on both sides of the equilibrium potential. This range may furthermore be substantially widened if anharmonic classical modes are subject to reorganization, or if high-frequencv modes are present. We notice finally that the number of levels effectively contributing to the current, is approximately $\Delta\varepsilon^* \rho(\varepsilon^*) = (8\pi)^{\frac{1}{2}} k_B T \rho(\varepsilon_F)$.

(2) For $0 > e\eta > E_r + k_B T \ln(C_R/C_O)$ we see (fig.8.3) that $n(\varepsilon^*) \approx \exp[-(\varepsilon^* - \varepsilon_F)/k_B T] \approx \exp[-(e\eta - k_B T \ln(C_R/C_O) - E_r)/k_B T] \approx 0$. Therefore $\alpha_{ex} \approx \alpha(\varepsilon^*) \approx 1$, and

$$i \approx 2eC|V|^2 \hbar^{-1} \rho(\varepsilon^*) \exp\left\{ -[e\eta - k_B T \ln(C_R/C_o)]/k_B T \right\} \tag{8.29}$$

This is the barrierless region corresponding to $\alpha_{ex} \approx 1$, and levels above the Fermi level dominate here. For this region to prevail for cathodic currents it is necessary that $k_B T \ln(C_R/C_O) < -E_r$, or for $E_r \approx 1$ ev, $C_R \approx 10^{-18} C_O$. Barrierless processes are therefore hardly observable for simple redox processes for which the reverse process would anyway immediately follow in an activationless fashion, but possibly for processes in which the discharged ion is weakly adsorbed and rapidly removed in subsequent steps (29).

(3) For $e\eta < -E_r - k\ T\ln(C_R/C_0)$, $n(\varepsilon^*) \gtrsim 1$, $\varepsilon^* - \varepsilon_F \approx E_r - k_B T\ln(C_R/C_0) + e\eta$, and $\alpha_{ex} \approx \alpha(\varepsilon^*) \approx 0$. This gives for the current density

$$i \approx 2eC|V|^2 \pi\hbar^{-1} \rho(\varepsilon^*) \tag{8.30}$$

This is the activationless region in which levels below the Fermi level contribute most to the current, corresponding to zero activation energy.

8.2.2 Semiconductor electrodes.

The expression for the total current at semiconductor electrodes is formally identical to eq.(8.23) (336). However, if we now view the electron and hole components of this current separately, then $i = i_e + i_h$, where i_e and i_h are given explicitly by the equations (336)

$$i_e = eC|V|^2 (\pi/k_B T\hbar^2 E_r)^{\frac{1}{2}} \int_{\varepsilon_F+\Delta_c+e(\eta-\eta_c)+e\Delta\phi_o}^{\infty} \rho_c(\varepsilon + e\eta - e\eta_c)\, n(\varepsilon) \exp[-E_A(\varepsilon,\eta)/k_B T]\, d\varepsilon \tag{8.31}$$

$$i_h = eC|V|^2 (\pi/k_B T\hbar^2 E_r)^{\frac{1}{2}} \int_{-\infty}^{\varepsilon_F-\Delta_v+e(\eta-\eta_v)+e\Delta\phi_o} \rho_v(\varepsilon - e\eta + e\eta_v)\, n(\varepsilon) \exp[-E_A(\varepsilon,\eta)/k_B T]\, d\varepsilon \tag{8.32}$$

where

$$E_A = [E_r - k_B T\ln(C_R/C_o) - (\varepsilon - \varepsilon_F) + e\eta]^2 / 4E_r \tag{8.33}$$

(cf.eqs.(8.17), (8.18), and (8.23)). We consider furthermore only semiconductor electrodes of low charge concentrations ($\partial\eta_c/\partial\eta \approx 0$), and we shall assume that the inequalities $E_r > \Delta_{c,v} - k_B T \ln(C_R/C_O) + e\eta$ and $|e\eta| < \Delta_{c,v}$, are valid, i.e. the reorganization energy is larger and the overvoltage smaller than the width of the band gap. It is then convenient firstly to derive the expressions for the exchange currents, i_e^o and i_h^o, i.e. the currents for zero overvoltage. The integration of eq.(8.31) gives, if we use the asymptotic formula $\int_t^\infty \exp(-u^2)du \approx \exp(-t^2)/2t$ $(t > 2)$ (336)

$$i_e^o \approx e(C_R C_O)^{\frac{1}{2}}|V|^2(\pi/k_B T\hbar^2 E_r)^{\frac{1}{2}}\rho(\varepsilon^*)\Delta\varepsilon^* \cdot \exp\left\{-[E_r + \Delta_c + e\Delta\phi_o]^2/4E_r k_B T\right\} \tag{8.34}$$

where $\varepsilon^* \approx \varepsilon_F + \Delta_c + e\Delta\phi_o$, and $\Delta\varepsilon^* \approx 2k_B T$, i.e. the levels at the lower edge of the conduction band primarily contribute. Similarly,

$$i_h^o \approx e(C_R C_O)^{\frac{1}{2}}|V|^2(\pi/k_B T\hbar^2 E_r)^{\frac{1}{2}}\rho(\varepsilon^*)\Delta\varepsilon^* \times \exp\left\{-[E_r + \Delta_v - e\Delta\phi_o]^2/4E_r k_B T\right\} \tag{8.35}$$

If $\Delta_c = \Delta_v = \Delta$, then

$$i_e^o/i_h^o \approx \exp[-e\Delta\phi_o(1 + \Delta E_r^{-1})/k_B T] \tag{8.36}$$

i.e. depending on the sign of the band bending, $e\Delta\phi_o$, the exchange current may be dominated by either electron or hole currents (336).

When a cathodic electrochemical reaction proceeds, similar integrations of eqs.(8.31) and (8.32) with the appropriate overpotential dependent integration limits give

$$i_e \approx i_e^o \exp(e\eta_c/2k_BT)\exp(-e\eta/k_BT) \qquad (8.37)$$

$$i_h \approx i_h^o \exp(-e\eta_c/2k_BT) \qquad (8.38)$$

i.e. $\alpha_e \approx 1$ and $\alpha_h \approx 0$ for the cathodic currents if the overpotential at the contact depends slowly on the bulk overpotential. The general expressions for anodic current densities are analogous to those of the cathodic currents, except that the behaviour of the electron and hole components are now inverted, i.e. $\alpha_e \approx 0$ and $\alpha_h \approx 1$.

8.3 Relation to Experimental Data.

We shall now proceed to a consideration of some experimental data on electrochemical processes in order to illustrate some of the implications of the formalism outlined in sections 8.1 and 8.2:

8.3.1 The current-voltage relationship.

Considerable attention has recently been devoted to attemots to find simple electrochemical systems which display nonlinear current-overvoltage characteristics. For a simple ET process at metal electrodes and involving no specific adoption or reorganization of the molecular structure a parabolic current-overvoltage is expected (cf. eq.(8.23)). However, although a possible discovery of this effect would of course present a substantial interest, we should recall three reservations. Firstly, the appearance of a smooth Tafel relationship is interpretable in terms of any pair of intersecting potential energy surfaces and not in itself indicative of a particular molecular mechanism. Secondly, the parabolic relationship is expected when the linear medium modes and/or a set of low-frequency intramolecular harmonic modes are subject to equilibrium coordinate shift only. If frequency shift, anharmonicity, reorganization in high-frequency modes etc. occur as well, less transparent current-overvoltage relationships are expected, and proper account of these effects is essential even for qualitative considerations. In particular, we have seen that anharmonicity cause a 'flattening' of the Tafel relationship. Finally, with reference to chapter 5 and 6 we conclude that curvature of the Tafel relationship cannot itself distinguish between several possible mechanisms such as inner and outer sphere ET processes.

In addition, important experimental problems commonly arise in these investigations. A substantial deviation from linearity is thus expected as $e\eta$ becomes appreciable compared with E_r. At these high overvoltages the electrochemical reaction rates are rapid, and disentanglement from diffusion and other effects is commonly insufficiently accurate. More important are the difficulties of estimating the double layer effects. If we include the double layer effects explicitly, the activation energy for a cathodic one-electron process takes the form

$$E_A = Ze\psi_1 + [E_r - k_BT\ln(C_R/C_o) + e\eta - e\psi_1]^2/4E_r \qquad (8.39)$$

where ψ_1 is the potential at the distance of discharge of the depolarizer ion. ψ_1 thus determines the work in bringing the depolarizer ion from the bulk of the solution to the point of discharge, and it appears because at this location the depolarizer concentration differs from the bulk value, and because the 'effective' electrode potential is not φ_m but $\varphi_m - \psi_1$.

The fundamental importance of the ψ_1 -potential was first recognized by Frumkin (337). Corrections of electrochemical rate data for the ψ_1 -potential usually requires either independent measurements of the double layer capacity, a suitable theoretical model for the double layer, or in certain cases it can be determined from the kinetic data themselves. Moreover, for multiply charged ions the double layer corrections may be appreciable, and since they are in principle potential dependent, they may conceal a potential dependence of the transfer coeffidient.

With these reservations we then notice that an approximately parabolic current-voltage dependence for both the cathodic and anodic branches of the $[Fe(H_2O)_6]^{3+/2+}$ and $[Fe(CN)_6]^{3-/4-}$ couples at electrodes of gold, platinum, and mercury, have been reported (338-340). However, the medium for the $[Fe(H_2O)_6]^{3+/2+}$ couple contained sulphuric acid which means that several labile sulphate complexes must be present in the solution, and the $[Fe(CN)_6]^{3-/4-}$ ions are also known to form several ion pairs with the alkali ions present in the solution (cf. section 7.). Very much information is therefore contained in the current-voltage data, and it is not surprising that the consistency of the values of E_r, which determine the curvature, was poor.

No detectable curvature was observed for the electrochemical reduction of complexes of the form $[Cr(H_2O)_5X]$ ($X = F^-$, H_2O, SO_4^{2-}) at a mercury electrode over a 600 mv potential range (341). However, in view of the fact that the Cr(III)/Cr(II) couples are subject to large molecular reorganization, the result of this observation is perhaps not surprising. On the other hand, a curvature, corresponding to E_r = 0.8-1 ev, was observed for the reduction of tert-nitrobutane, nitrodurene, and nitromesitylene in acetonitrile and dimethylformamide over nearly 1 v (72). These systems involve electrically neutral reactants which makes the double layer corrections less critical. A pronounced (parabolic) curvature, even going through a maximum and corresponding to $E_r \approx 1$ v, was finally observed for the two-electron reduction of Hg_2^{2+} to elemental Hg at carbon paste electrodes (73). Even though this process may also involve bond reorganization when the Hg is eventually incorporated in the mercury drop, it is still expected to exhibit the basic features predicted by the theory. We then conclude that although the ascending branch of the Tafel relationship and the linear dependence of the transfer coefficient on the overpotential is fully compatible with the theory, the decrease of the current with increasing overvoltage for large values of the latter, is at variance with expectations form the theory.

ET at semiconductor electrodes can occur with an appreciable velocity only if the redox potential of the depolarizer couple is located sufficiently close to the lower edge of the conduction band or the top edge of the valence band. Couples with high redox potentials are then generally expected to participate in hole transfer, i.e. ET via the valence band, whereas couples of low redox potentials exchange electrons with the conduction band (332). ET reactions at semiconductor electrodes are, however, frequently complicated by other effects such as decomposition of the electrode, changes of the surface properties or participation of surface states, and quantitative data to illustrate the theoretical framework are relatively scarce.

Examples of reasonable unambiguous illustrations of the theory are the reduction of $[Fe(CN)_6]^{3-}$ and $[Fe(H_2O)_6]^{3+}$ at CdS electrodes (342). The cathodic branches do show a transfer coefficient close to unity corresponding to ET via the conduction band, and the whole electrode-solution potential drop in the electrode. An example of an anodic process is the oxidation of V^{2+} at a germanium electrode the rate of which is approximately independent of the electrode potential over nearly 1 v, corresponding in a similar way to ET from the depolarizer ion to the conduction band (332).

8.3.2 The nature of the substrate electrode.

The dependence of the current density on the properties of the metal electrode is seen to be solely reflected in the electronic exchange integral and in the density of states of the electrode, provided that double layer and adsorption effects can be ignored. This result is in a sense a reformulation of the general result that the rate of simple ET processes at metal electrodes is independent of the work function of the metal (343). With reference to eq.(8.8) this is associated with the fact that by changing the metal, φ_m^o and ε_F are subject to similar shifts, and that energy levels close to the Fermi level usually give the dominating contributions to the current density. The exchange current is then expected to vary relatively little at different metal electrodes. This was confirmed by Capon and Parsons (344) for the benzoquinone anion radical couple for which the exchange rate only varies by a factor of two at six different metal electrodes. These reactants have low electrical charges giving small double layer corrections. Vojnović and Šepa (345) studied the ET reactions of the $[Fe(CN)_6]^{3-/4-}$ couple at tungsten bronze electrodes at several different compositions, i.e. by systematically varving the electronic density of states. By ignoring double layer effects they found an approximately linear relationship between the exchange current and the density of states of the electrodes. However, in view of the fact that the surface composition of tungsten bronzes may differ from the bulk composition of these materials, and of the difficulties about unambiguous predictions of the double layer effects for such strongly charged ions even at the high electrolyte concentrations applied, the results of Vojnović and Šepa hardly possess sufficient unambiguity to be taken as evidence for the expected relationship.

The exchange current density of the $[Fe(H_2O)_6]^{3+/2+}$ couple at several different metal electrodes (platinum metals, Au, Ni, and

Ta) has also been reported (346). While the activation energy is approximately independent of the metal, the current density shows a strong (exponential) dependence on the work function and an approximately linear dependence on the electronic density of states of the metal. The exponential dependence on the work function is qualitatively understandable as a double layer effect (346), but quantitative agreement with the experimental data can only be achieved by invoking several assumptions for which there is so far no independent experimental check. The conclusion is then that an experimental disentanglement of the various properties of the electrode material which might affect the current density is at present hardly available.

8.3.3 The electrochemical hydrogen evolution reaction (her).

Since the early days of electrochemistry this has been one of the most comprehensively studied electrochemical processes. The her thus provided the basis for the first report of the Tafel law at the beginning of this century, and the first approaches to a theoretical description of electrochemical processes also referred to the her in particular. The mechanism in terms of the sequence and nature of the elementary steps constituting the overall process at many different substrate electrodes is now generally well known, even though some features remain obscure. The first step of the her at metal electrodes is generally believed to be the proton discharge

$$H_3O^+ + e \rightarrow H_a + H_2O \qquad (8.40)$$

or

$$H_2O + e \rightarrow H_a + OH^- \qquad (8.41)$$

in acid and alkaline solution, respectively (347). H_a refers to a hydrogen atom adsorbed to the electrode surface. The subsequent step is either the electrochemical desorption, i.e.

$$H_3O^+ + H_a + e \rightarrow H_2 + H_2O \quad (8.42)$$

$$H_2O + H_a + e \rightarrow H_2 + OH^- \quad (8.43)$$

in acid and alkaline solution, respectively, providing gaseous molecular dihydrogen. Alternatively, the second step may be 'catalytic' recombination of adsorbed hydrogen, i.e.

$$2\,H_a \rightarrow H_2 \quad (8.44)$$

The latter mechanism requires a high coverage of the electrode by hydrogen atoms. On the other hand, the hydrogen atoms must be 'mobile' and the recombination step therefore occurs at electrodes of moderately high adsorption energies, e.g. platinum metals. If the adsorption energy is small, such as for Hg, Ag, and other metals, the second step is rapid electrochemical desorption. The second step for high adsorption energies (W, Ta) is also electrochemical desorption, but now this step is rate determing (347).

No other electrochemical process has provided as many experimental results appropriate for the theory of rate processes as the her. Since furthermore some of these data have appeared to be particularly diagnostic with respect to a possible distinction between the quantum mechanical theory of rate processes and the semiclassical theories, we shall summarize the most important of these data below. Since, however, the concept of barrierless processes is of crucial importance in the interpretation of these data, we recall at first a few features of this important class of processes.

Barrierless processes may be observed if the particular, strongly endothermic process is followed by a subsequent rapid step to prevent the system from returning to the initial state. The rate determining proton discharge step of the her may therefore occur under barrierless conditions at metal electrodes of low adsorption energy for hydrogen atoms. Observation of these effects requires, however, that the equilibrium potential for the overall her is located at values lower than the transition region between barrierless and normal discharge. These expectations were corroborated in particular by experimental studies of Krishtalik and his associates (347,348), who found that the current-overvoltage curves for the her at Hg electrodes in strongly acid solutions generally display two approximately linear branches of slope about 60 mv and 120 mv at low and high overvoltages, respectively ($\alpha \approx 1$, and $\alpha \approx 0.5$). The transition region was quite narrow (from a few to about 30 mv) and located sufficiently far from the potential of zero charge that any changes of the double layer structure could be responsible for the observed changes of the slope. Finally, while the high-overvoltage ('normal') branch displays pronounced shifts when the concentration of either depolarizer (H_3O^+) or the electrolyte was changed, indicative of a strong influence of the double layer structure, this effect was almost absent in the low-overvoltage region (348).

Similar effects have been found for the her at Ag, Mo, W, and Nb electrodes, for the chlorine evolution reaction at graphite electrodes, for the anodic oxidation of azide ions to molecular dinitrogen, and for certain other processes. All these effects are compatible with the assumption that the low-overvoltage branch of the polarization curves corresponds to a barrierless proton discharge and very difficult to reconcile with any other conceivable effects.

The proton discharge and the electrochemical desorption steps have been the subject of extensive theoretical treatment in terms of both the semiclassical rate theory (37-41) and the quantum mechanical theory (95,197,259). We restrict here ourselves to the discharge step and recall that according to the semiclassical theory the proton transfer is essentially viewed as classical motion along a proton stretching mode (which can be generalized to a three-dimensional trajectory), but corrected for proton tunnelling near the barrier top. Motion along this mode thus provides both the activation energy and possible tunnel corrections. In contrast, according to the multiphonon approach outlined in chapter 6, all vibrational molecular and medium modes are explicitly incorporated and their role viewed differently, according to the values of their vibration frequencies. The proton discharge is thus represented by the following general equation (in the nonadiabatic limit)

$$i = e\, C_{H^+} (\pi / k_B T \hbar^2 E_r)^{\frac{1}{2}} \sum_{v_j} \sum_{w_j} \int_{-\infty}^{\infty} n(\varepsilon) \rho(\varepsilon) \exp[-\sum_{v_j} \varepsilon_{v_j} / k_B T] \tag{8.45}$$

$$\prod_j S_{v_j, w_j} \exp \left\{ -[E_r + \sum_{w_j} \varepsilon_{w_j} - \sum_{v_j} \varepsilon_{v_j} - (\varepsilon - \varepsilon_F) + e\eta - \frac{1}{2} D_{H_2} + E_H]^2 / 4 E_r k_B T \right\}$$

where v_j and w_j are the quantum numbers of the j'th high-frequency mode in the initial and final state, respectively, ε_{v_j} and ε_{w_j} the corresponding vibrational energies, D_{H_2} the dissociation energy of the hydrogen molecule, E_H the adsorption energy of the hydrogen atom, and the other symbols have been defined previously. E_r thus includes not only the solvent modes but also other low-frequency vibrational modes such as hindered translation and rotation of water molecules etc. (197). Similarly, in addition to proton stretching j incorporates also all other high-frequency modes subject to reorganization.

Even though preference to one or the other of the two approaches may be given on the basis of the physical and chemical properties of the system, their validity must ultimately be evaluated from experimental data. We recall, however, that most of the fundamental features of proton transfer reactions such as Bronsted or Tafel relationships, kinetic isotope effect etc. are not sufficiently diagnostic in this respect. On the other hand, new experiments were recently designed by Krishtalik and his associates and have appeared to be particularly illuminating on this point. Below we summarize this new evidence in the following points (259);

(A) Barrierless processes. The rate of the discharge process was written by Krishtalik in the general form of the theory of absolute rates

$$i = \varkappa \frac{k_B T}{h} e\, C_p^s \exp(S_A/k_B) \exp\left\{ -[E_A^o + \alpha(\varphi_m - \psi_1)e]/k_B T \right\} \tag{8.46}$$

where S_A and E_A^o are the entropy and energy of activation (at zero potential φ_m), ψ_1 the potential at the plane of discharge, C_p^s the concentration of depolarizer ions H_3O^+) at this plane. Introducing the surface concentration of water, C_w^s (mol cm^{-2}), and the surface mole fraction of H_3O^+, $X_p^s \approx C_p^s\, C_w^{s\,-1}$ we have

$$X_p^s = X_p \exp(\Delta S_a/k_B) \exp[-(\Delta H_a + e\psi_1)/k_B T] \tag{8.47}$$

where X_p is the mole fraction of H_3O^+ in the bulk solution, and ΔS_a and ΔH_a the entropy and enthalpy of adsorption of H_3O^+ from the bulk solution. Introducing further the overvoltage $\eta = \varphi_m - \varphi_m^o$, where $\varphi_m^o = \varphi_e + (k_B T/e)\ln X_p$ and φ_e the standard equilibrium potential, then

$$i = \varkappa \frac{k_B T}{h} e\, C_w^s\, X_p^{1-\alpha} \exp[(S_A + \Delta S_a)/k_B] \cdot \tag{8.48}$$

$$\exp[-(E_A^\circ + \Delta H_{ad} + e\phi_e)/k_BT]\exp(\alpha e\eta/k_BT)\exp[(1-\alpha)e\psi_1/k_BT]$$

For barrierless processes $\alpha = 1$. If ϕ_e is decomposed into enthalpy (ΔH_e) and entropy (ΔS_e°) contributions, eq.(8.48) can be rewritten in the form

$$i^b = \varkappa^b \frac{k_BT}{h} e C_w^s \exp[(S_A^b + \Delta S_a + \Delta S_e^\circ)/k_B] \qquad (8.49)$$

$$\exp[-(\Delta H_a + E_A^{ob} + \Delta H_e)/k_BT] \exp(e\eta/k_BT)$$

where the superscript 'b' refers to the barrierless process. Furthermore, $\Delta H_e = H_p + H_e - H_w - \frac{1}{2}H_{H_2}$ and $\Delta S_e^\circ = S_p^\circ + S_e - S_w^\circ - \frac{1}{2}S_{H_2}^\circ$, where the subscripts refer to the appropriate molecular species. Since the transition state coincides with the final state for a barrierless process, then $S_A^b = S_{H_a}^\circ + S_w^{so} - S_p^{so} - S_e$, and the sum of all the entropy terms becomes $S_{H_a}^\circ + \Delta S_w - \frac{1}{2}S_{H_2}$. For barrierless processes difficult estimates of double layer effects and activation entropies are thus avoided. Moreover, the entropy terms appearing in eq.(8.49) can be estimated reasonably accurately. The adsorption entropy of water, $\Delta S_w (= S_w^{so} - S_w^\circ)$, can be found from the temperature dependence of the surface tensions of mercury/water and mercury/air interphases (-1.15 cal K^{-1}). $S_{H_a}^\circ$ primarily contains vibrational contributions from the Hg-H bond (1.98 cal K^{-1}), while $S_{H_2}^\circ$ is available from tables of thermodynamic quantities of elements. Inserting these values in eq.(8.49) and comparison with experimental values of the overall pre-exponential factor gives a value of $\varkappa^b \approx 10^{-3} - 10^{-2}$. This conclusion is not modified if the estimates of $S_{H_a}^\circ$ and ΔS_w are wrong by an amount comparable to their absolute values, and the experimental accuracy is sufficient to maintain this small value of $\varkappa^b$.

This result shows that a substantial proton tunnelling is likely to occur even under conditions where the process is barrierless.

This is understandable on the basis of the quantum theory of proton transfer reactions, according to which the barrier for proton transfer differs from the barrier which provides the activation energy. However, if the proton stretching mode is the only one considered, proton tunnelling is expected to vanish when the barrier vanishes.

(B) Isotope effects. The isotope separation factor, $S_{H/L}$ (L is deuterium or tritium) is defined as

$$S_{H/L} = i_H C_L / i_L C_H \tag{8.50}$$

where $C_{H,L}$ are the appropriate total isotope concentrations, and $i_{H,L}$ refers to the total currents of isotope transfer from the solution to the gas phase. When $C_L << C_H$, $S_{H/L}$ takes the form (39b,259)

$$S_{H/L} = 2 S_{di} S_{de} / (S_{di} + S_{de}) \tag{8.51}$$

where S_{di} and S_{de} are the separation factors for the discharge and desorption steps, respectively. It is, however, convenient to express the separation factors in terms of the ratio between the rates of the appropriate isotope transfers, i.e. ξ_{di} and ξ_{de} (cf. chapter 6). Since the dominating depolarizer for the discharge process in acid solution is H_3O^+, S_{di} should be corrected by the equilibrium constant for the isotope distribution between H_3O^+ and H_2O, i.e. $\xi_{di} = S_{di}/K_{H/L}$, where $K_{H/L} = C'_H C_L/C'_L C_H$, $C'_{H,L}$ the isotope concentration in H_3O^+ or H_2LO^+, and $C_{H,L}$ the concentration in H_2O or HLO. (This correction is not necessary for the desorption step since the adsorbed hydrogen atom here reacts with H_2O rather than with H_3O^+). In terms of kinetic isotope effects, the overall separation factor can then be written as

$$S_{H/L} = 2 K_{H/L} \xi_{di} \xi_{de} / (K_{H/L} \xi_{di} + \xi_{de}) \tag{8.52}$$

There is now a considerable amount of evidence that the electrochemical desorption step occurs in an activationless fashion. This evidence partly comes from photoemission experiments, in which a photoemitted electron reacts with H_3O^+ in solution giving hydrogen atoms which subsequently diffuse to the electrode surface where they react thermally, either to reionize to H_3O^+ or to molecular H_2 by the electrochemical desorption (349). However, the strongly exothermic nature of the electrochemical desorption can also be inferred in a less direct way. Thus, at the equilibrium potential of the overall process the extrapolated activation energy for the barrierless process is 22.9 kcal. Since the system is at equilibrium both the reverse process, and the formation of molecular hydrogen are exothermic by this amount at the equilibrium potential. Since furthermore both of the latter processes involve approximately similar structural reorganization they are also both likely to proceed in an activationless fashion. For the desorption the exothermicity will be even more pronounced in the overvoltage region (300-500 mv) where the normal her begins.

If only proton motion is considered we must therefore expect that $\xi_{de} = 1$, and $S_{H/L} \lesssim 2$. However, experimental values can be substantially higher (3-16 for $S_{H/T}$ at mercury and gallium), which is only understandable if the electrochemical desorption also shows an isotope effect. On the other hand, this is compatible with the quantum theory for which the activationless nature refers to classical modes rather than the proton modes.

(C) The nature of the metal electrode. Since the adsorbed hydrogen atom constitutes an intermediate state in the her, the nature of the metal, M, is expected to affect the rate of the proton discharge. For a given overpotential a higher M-H bond energy, E_H , thus provides a lower barrier for the proton motion. In terms of the semiclassical theory this would give both a lower activation energy and a lower barrier for tunnelling, with increasing bond energy. This effect might be partially compensated, however, by an increased proton vibration frequency of the M-H bond with increasing bond energy.

In terms of the quantum theory the proton transfer probability is determined by overlap integrals of the proton wave functions. For harmonic potentials and proton transfer between ground vibrational levels the overlap integrals have the form $\exp[-m_H \Omega_i \Omega_f (\Delta R_H)^2/(\Omega_i+\Omega_f)]$ where m_H is the mass of the proton, ΔR_H its transfer distance, and Ω_i and Ω_f the frequencies in the initial and final states. Since $\Omega_f \approx (0.3\text{-}0.5)\Omega_i$ for the proton discharge, we would expect a decreasing pre-exponential factor with increasing Ω_f or E_H. Experimental data show an approximately exponential decrease, while the activation energy does not exhibit a systematic dependence on E_H. This also corroborates the predictions of the quantum theory.

(D) Effect of the solvent. Further illumination of the role of the nature of the proton donor and the solvent might be provided either if the discharge of the same donor molecule in different solvents, or different donor molecules of approximately similar geometrv in the same solvent, could be investigated. For homogeneous processes a change of the solvent causes a change of both the solvation energies of the reactants, i.e. a change of the heat of reaction, and a change of the solvent reorganization energy, and disentanglement of these two effects is generally difficult. On the other hand, for electrochemical processes, and provided that specific adsorption effects can be ignored, the energies of the initial and final states are still equal at the equilibrium potential, also when the solvent is changed. This unique feature of electrochemical processes means that a comparison of the kinetic parameters for different solvents at a given overpotential reflects different reorganization energies, proton transfer distances etc. and not directly different solvation energies.

Such data for the her at Hg in acid perchloric aqueous and acetonitrile solutions were also provided by Krishtalik and his associates. At very small concentrations of water in this solvent ($< 5.10^{-4}$ mol dm^{-3}) the dominating depolarizer species is protonated acetonitrile CH_3CNH^+. At concentrations in the region 0.08-0.1 mol dm^{-3} the proton is primarily solvated by water forming mono- and dihydrates. In this concentration range the depolarizer species is thus (hydrated) hydroxonium ion, while the surrounding solvent is primarily acetonitrile. At still higher water concentrations the solvent sphere gradually becomes more similar to that prevailing in aqueous solution.

The activation energies for the proton discharge of CH_3CNH^+ and H_3O^+ in acetonitrile solution were both found to be about 18 kcal, whereas the pre-exponential factor for H_3O^+ was about 10 times higher than for CH_3CNH^+. The activation energy for H_3O^+ discharge in aqueous solution is 21.7 kcal and the pre-exponential factor about six times higher than for H_3O^+ in acetonitrile. The activation energy is thus primarily determined by the nature of the solvent, whereas the different nature of the depolarizer molecules, although less unambiguously, is reflected in the pre-exponential factors.

In qualitative terms the data summarized above are all compatible with predictions of the theory of proton transfer reactions as a multiphonon process with strong coupling to the medium, whereas they would be hard to reconcile with the assumption that the classical proton motion with tunnel corrections near the barrier top is the only mode involved in the reorganization process. These new data are thus a strong indication at least that (1) the proton mode is not the only reaction coordinate of the discharge step of the her. (2) the barrier for proton tunnelling is not directly associated with the barrier which provides the activation energy, and (3) solvent reorganization strongly contributes to the activation energy.

8.4 Electrode Processes at Film Covered Electrodes.

ET processes at metal electrodes covered by insulating or semiconducting films is of great practical and theoretical importance in such phenomena as current flow across metal-insulator-metal junctions (tunnel diodes), heterogeneous gas phase chemical processes, passivation and electrocatalytic phenomena, and possibly electronic conduction across biological membranes. The actual electron transport across the film is expected to proceed by analogous mechanisms in the metal-film-metal, metal-film-electrolyte, and membrane systems, and the three kinds of systems to display interesting common features. However, the initial and final states in the solid-state junctions are both delocalized band states, whereas one or both states are localized on a molecular reactant in the electrochemical and membrane systems, respectively.

The analogies and differences between the three classes were recognized recently in attempts to formulate a theory of ET at film covered metal electrodes (350,351), and we conclude the present chapter by a summary of these analogies and their implications for electrochemical systems.

We consider only electron transfer mechanisms (i.e. in contrast to ionic conduction). The ET mechanisms are then expected to be qualitatively different for thin and for thick films. For thin films ET occurs from an energetically low-lying level (around the Fermi level of the metal) by tunnelling through the barrier. For thick barriers the ET proceeds either through the conduction band of the film or by 'hopping' via localized states in the band gap. Semiquantitative criteria for the prevalence of either tunnelling or band conduction can be obtained by considerations almost identical to those given in chapter 6 in our estimate of the relative importance of tunnelling and thermally activated barrier passage in nuclear motion. Thus, tunnelling is favoured if the following inequality is valid

$$(\Delta\varepsilon)_{min} > k_B T \qquad (8.53)$$

where $(\Delta\varepsilon)_{min}$ is the smallest electronic energy interval between levels in the potential well which arises when the barrier is inverted relative to the axis defined by the distance perpendicular to the electrode. Similarly, if the inequality

$$(\Delta\varepsilon)_{max} < k_B T \qquad (8.54)$$

is valid, ET occurs in a thermally activated fashion via the conduction band. For barrier heights of about 1 ev the critical width is found to be about 40 A.

It is now convenient to consider the two classes separately:

8.4.1 Tunnelling mechanisms

Direct elastic tunnelling involves ET trough a uniform barrier with conservation of both the energy of the electron and its momentum parallel to the surface, $\vec{k}_{\parallel}$. Within the single-particle approximation the analogy between ET in the solid-state junctions and at the electrode surface emerges from the expressions for the appropriate current-voltage relationships for the two kinds of systems (351a,352)

$$i_{ss} = e/(\pi\hbar) \int_{-\infty}^{\infty} d\varepsilon\, n_L(\varepsilon)[1-n(\varepsilon)] \int_{\vec{k}_{\parallel}<2m_e\varepsilon/\hbar^2} (2\pi)^{-2} d\vec{k}_{\parallel}\, T_{ss}(\varepsilon,\vec{k},\eta) \qquad (8.55)$$

$$i_{SL} = (m_e^{\frac{1}{2}} e/\pi\hbar^2)(\pi/k_B T E_r)^{\frac{1}{2}} \int_{-\infty}^{\infty} d\epsilon\, n(\epsilon) \exp[-E_A(\epsilon,\eta)/k_B T]$$

$$\int_{\vec{k} < 2m_e \epsilon/\hbar^2} (2\pi)^{-2} d\vec{k}_{\parallel} (\epsilon - \hbar^2 k_{\parallel}^2/2m_e)^{-\frac{1}{2}}\, T_{SL}(\epsilon, \vec{k}_{\parallel}, \eta) \qquad (8.56)$$

i_{ss} is the density of current flow across the solid-state junction, $n_L(\epsilon)$ and $n_R(\epsilon)$ the Fermi functions of the two metals, $T_{ss}(\epsilon, \vec{k}_{\parallel}, \eta)$ the tunnel probability at given ϵ, $\vec{k}_{\parallel}$ and bias voltage η. Similarly, i_{SL} is the cathodic current density at the solid-liquid interphase with $n(\epsilon)$ and T_{SL} having analogous meaning. $E_A(\epsilon, \eta)$ is, however, now the activation energy for the solvent and molecular modes of the depolarizer molecule.

The main qualitative result of analysis of expressions such as eq.(8.56) is that the current distribution over the energy levels is much wider than for uncovered metal electrodes (351a). Thus, when the electronic energy of the metal donor level increases, the microscopic activation energy decreases, whereas the Fermi function decreases. As a result electrons near the Fermi level contribute most. At film covered electrodes the electrons also have to pass a barrier, the height and width of which decrease with increasing energy. Electrons from higher levels than the Fermi level will then also contribute and more so the lower and wider the barrier.

A consequence of this is that even though the transfer coefficients do show some characteristic variation with overpotential and barrier characteristics, they are probably generally too featureless to be of unambiguous diagnostic value. i_{SL} shows, however, an exponentially decreasing exchange current density with increasing film thickness. This has also been observed for several redox reactions at oxide covered Pt electrodes of film thickness in the range 4-10 A. (353).

More information is contained in the current-voltage relationship for ET via discrete electronic states located in the band gap sufficiently far from the valence and conduction bands (351b,c). This impurity-assisted tunnelling is closely similar to the concept of second order processes which we discussed in some detail in chapter 7. The characteristics for current flow in solid-state junctions (355) and for (cathodic) electrochemical processes (351b,355) are now represented by the equations

$$i_{ss} \propto \int_{-\infty}^{\infty} d\varepsilon |T_{ss}(\varepsilon,\eta)|^2 \rho_L(\varepsilon)\rho_R(\varepsilon)n_L(\varepsilon)[1-n_R(\varepsilon)] \tag{8.57}$$

$$i_{SL} \propto \int_{-\infty}^{\infty} d\varepsilon |T_{SL}(\varepsilon,\eta)|^2 \rho(\varepsilon)n(\varepsilon)\exp[-E_A(\varepsilon,\eta)/k_B T \tag{8.58}$$

where ρ is the level densities in the metal electrodes, and the electronic factors, T_{ss} and T_{sL} are now of second order (cf. chapter 7). If the electronic-vibrational coupling in the intermediate impurity state can be ignored, elastic resonance tunnelling prevails, in which the electron at first tunnels from the donor (metal or depolarizer) to the impurity level and subsequently to the acceptor level (351b,c,354,355). A closer analysis then shows that due to the rapid dependence of the first order electronic coupling factors on the ET distance, impurity levels near the centre of the film are the primary contributors to the resonance current. Moreover, levels close to the Fermi level of the metal contribute perferentially. It is therefore conceivable that when the intermediate level is off-resonance with respect to the Fermi level at zero overvoltage it will coincide energetically with the Fermi level at some value of this potential. This is typicially manifested as changes (i.e. within about 100 mv) in the apparent transfer coefficient at this potential. Such effects have been observed both at solid-state contacts (355) and for redox processes at oxide covered iron and niobium electrodes (301b,356).

If the vibronic coupling in the impurity state is strong, the ET mechanism is essentially identical with the one discussed in chapter 7 for reactions in homogeneous solution (106,350,351a). Depending on the localization of the impurity level, the potential distribution in the film region etc., different phenomenological current-overvoltage relationships may be envisaged for this inelastic tunnelling process. In particular, 'sudden' changes in the transfer coeffient are predicted, associated with the change of the relative energies of the intersection regions of the initial-intermediate and intermediate-final state potential energy surfaces at certain potentials.

An analysis of the electrochemical dioxygen reduction at carbon-supported metal phthalocyanines along these lines was in fact recently attempted (300). Since electrochemically generated oxide layers are polar materials, the impurity-assisted ET processes referred to above, are also most likely more adequately viewed as inelastic tunnelling than as resonance tunnelling.

Inelastic tunnelling in solid-state junctions has been developed into a sensitive tool in the study of the nature of 'impurity' molecules in the tunnelling barrier at solid-state junctions (357,358). The interaction between the electron and the molecular centres is manifested as peaks in the second derivative of the current-voltage relationship of the junction at potentials corresponding to the appropriate vibrational energies. Detailed information about vibration frequencies, isotope effects etc. of molecules in monolayer quantities can be obtained in this way. Furthermore, since the solid-state oxides can normally be produced with better defined properties than the electrochemical oxide layers, important predictions of the theory of intermediate states in ET processes may eventually be more conveniently studied by means of the solid-state junctions.

8.4.2 Mobility mechanisms

Two mechanisms are particularly important for electron transport across thick films:

(A) If the semiconducting film is crystalline, and coupling to the nuclear modes is unimportant, the electron is transferred through the conduction or valence band. The expression for the current density is then formally that for semiconductor electrodes (359). However, the potential distribution in the film region is now determined by the metal potential, φ_m, in a way which depends on the dielectric properties and the thickness of the film. For example, if the potential distribution can be considered linear the following relation between φ_m and the film-electrolyte contact potential, φ_c is found (359)

$$\varphi_c = \varphi_m (1+\varepsilon_s^{sol} d/\varepsilon_s^{sc} \delta)^{-1} \tag{8.59}$$

where d and δ are the film thickness and the extension of the Helmholtz layer in the solution, and ε_s^{sc} and ε_s^{sol} the corresponding dielectric constants. This gives for the cathodic transfer coeffients of the electron and hole currents for small overvoltage

$$\alpha_e = 1-\alpha_h \tag{8.60}$$

$$\alpha_h = \frac{1}{2} (1 + \varepsilon_s^{sol} d/\varepsilon_s^{sc} \delta)^{-1} \tag{8.61}$$

respectively. We see that as $d \to \infty$, then $\alpha_h \to 0$ and $\alpha_e \to 1$, i.e. the behaviour expected for a bulk semiconductor electrode. As $d \to 0$, both α_e and α_h become 0.5 as for metal electrodes. However, as $d \to 0$ tunnel effects predominate, and the band description is also invalid. Furthermore, the dependence of the current density on d is only reflected in eq.(8.60) which means that band conduction, in contrast to both tunnelling and hopping, depends relatively weakly on the thickness of the film

(B) In disordered solid materials the band structure is believed to be basically preserved, although the density of states remains finite in the band gap. These low-density states provide a new mobility mechanism by electron 'hops' between the states (360). Since the disorder also causes a spread in energy of the localized states, hopping is generally a thermally activated process which, in the limit of strong coupling to the phonon modes is equivalent to thermal ET as discussed previously.

We shall illustrate the hopping mechanism by reference to a particular system studied experimentally, i.e. simple electrochemical processes at steel- and Pt-supported polyacrylonitrile layers in the thickness range 40-500 A. (302). In contrast to other powerful organic electrocatalysts (such as metal phthalocyanines) these materials are both mechanically stable and resistent to corrosion, being thus possible effective catalysts for electrochemical energy conversion. Moreover, both the thickness and the density of electronic states can be varied, and both solid-state junctions and metal-film electrolyte systems of the film can be studied. The experimental data for these systems can be summarized in the following way:

(a) The polymers contain large numbers of localized reversible redox centres e.g. of the quinone type, which can transmit the electrons.

(b) The current-voltage curves of the solid-state junctions are linear over up to several hundred millivolts, and the exchange current density decreases strongly (approximately exponentially) with increasing thickness and depends exponentially on ($-T^{\frac{1}{4}}$). These features are consistent with a variable range hopping mechanism. A closer analysis provided impurity concentrations of 10^{18}-10^{19} cm^{-3} and optimum hopping distances of 20-30 A.

Cathodic and anodic current-voltage curves of the $[Fe(CN)_6]^{3-/4-}$ and $[Cu(NH_3)_4]^{+/2+}$ couples participating in electrochemical processes at the metal-supported polymer films furthermore showed:

(c) Platinum and steel supported layers give similar results showing that polymer properties are in fact being measured.

(d) The exchange current density strongly decreases when the density of states is decreased by dilution of the acrylonitrile polymer with hexafluoropropene which does not form low-lying impurity states.

(e) The exchange current density decreases approximately exponentially with increasing thickness.

(f) For a layer thickness in the range 150-800 A linear Tafel plots with transfer coefficients in the range 0.16-0.25 were found for both cathodic and anodic processes. The curves crossed at zero overvoltage, and there were no indications of rectification effects. These values of the transfer coefficient correspond to potential drops between 30 % and 50 % of the total metal-electrolyte potential drop.

All these data seem consistent with a hopping mechanism, whereas they would be hard to reconcile with any of the other mechanisms discussed.

9 APPLICATION OF THE RATE THEORY TO BIOLOGICAL SYSTEMS

9.1 General

ET and AT are the fundamental elementary chemical steps in many biological processes. Consecutive or concerted series of ET reactions through redox centres of transition metal complexes (iron, copper, molybdenum) or organic redox couples (quinones, thiols) thus represent the key steps of photosynthesis (361-363) and the respiratory chain (364). Proton transfer is one of the most important elementary steps in the action of hydrolytic enzymes (361,365) and in the primary photochemical processes in some visual pigments (366). Ligand substitution is important in the binding of substrate molecules to the metal centres in hydrolytic metalloenzymes (361,367), and transfer of heavy molecular groups is finally the dominant elementary process in the reversible uptake of dioxygen, carbon monoxide and other small molecules by myoglobin (mb), hemoglobin (hb), and related compounds (273) (cf. chapter 6). It is therefore of substantial interest to investigate to what extent the theory of elementary rate processes is applicable to biological systems. The systems where the ET chains have been resolved with a reasonable certainty into elementary steps is, however, restricted to ET in the mitochondrial membranes and the primary processes of photosynthesis, and the discussion in the following therefore refers to these systems in particular.

Several mechanisms have been suggested for ET in biological systems: (a) Thermal ionization and subsequent free motion of electrons to the acceptor site in delocalized conduction bands formed by an assumed more or less regular protein structure (368,371). (b) Thermally activated hopping of an electron along a chain of trapping sites from the donor to the acceptor site

(371). (c) Modification of mechanisms (a) and (b) by electron-lattice interactions leading to a polaron or conformon motion in the protein structures (371). (d) Quantum mechanical tunnelling between donor and acceptor centres of fixed nuclear configuration (362,372). (e) Multiphonon radiationless transitions induced by coupling to the 'intramolecular' and medium modes of the system analogous to thermal ET between ions in low-molecular weight solvents (109,374-76). (f) Second order multiphonon transitions analogous to (e) where the intermediate state ensures an efficient overlap with donor and acceptor centres (110,112).

The basis of mechanism (a) was the observation of electronic semiconductivity and a mobility activation energy in crystalline samples of various biological materials. However, biological materials contain numerous centres with donor and acceptor properties, and the observation of an exponential temperature dependence of the conduction is characteristic of a variety of processes of different mechanisms. Also, a thermal ionization of the electron into the conduction band requires a low ionization energy of the donor, whereas electron injections into a medium far from any traps usually requires a considerable energy (30), in contrast to the ease with which many biological processes proceed. We shall, however, in section 9.3 discuss the features of certain membrane systems where quasicontinuous electronic spectra may contribute to the observed ET processes.

The difficulties of electron transport by a band mechanism is also partially met by a hopping or a polaron/conformon mechanism, i.e. the high injection energy from the donor level. Electron transport by either of these mechanisms could therefore only be expected in light-induced processes, where, however, the polaron states are likely to be shallow due to the apolar nature of the medium.

Electron tunnelling between fixed nuclear configurations of the donor and acceptor centres (mechanism(d)) has been suggested for ET in biological systems as an explanation that biological ET is activationless at low temperatures and an activated process at higher temperatures (362,372). However, this approach ignores the coupling between electronic and nuclear motion and the nature of the ET as a radiationless transition being determined by the Franck Condon overlap integrals. The central role of the Franck Condon principle has only recently been recognized for biological systems (110,112,279,374-76). Thus, several authors (373) pointed out that the ET probability is determined by the coupling with the nuclear modes as expressed by the vibrational overlap integrals. Hopfield (375) has emphasized the analogy between ET in biological systems and electronic energy transfer in condensed phases and derived a semiclassical rate equation. However, both of these multiphonon formalisms are incomplete since the former were able to provide only qualitative results, while the latter is inadequate to account for ET at low temperatures. Vol'kenstein, Dogonadze and their associates (110,112,374), and Jortner (108) applied the theory of ET between 'simple' ions in low molecular weight solvents to biological ET reactions. They noted especially the role of conformational states (110,374), of intermediate states (112), and the possibility of inner sphere ET (112).

Our analysis of biological systems will be based on the view that the elementary rate processes in biological systems are multiphonon radiationless processes analogous to elementary homogeneous and heterogeneous processes in low-molecular weight solvents. The theoretical formalism outlined in previous chapters must then incorporate the following features specific to biological processes:

(1) The biological redox centres are frequently fixed at given positions in membrane structures. The fixation of translational and rotational modes has been considered a necessary condition for enzymatic activity both in classical theories of enzyme catalysis (238,377) (according to which the enzymes function as 'entropy traps',) and in more recent theories according to which

the enzyme activity is governed by orbital symmetry adaption (377) or 'orbital steering' (378). The fixation means that only the direct transition probability of unimolecular process of the biological macromolecules needs to be calculated, and complications due to the diffusion of the reactants can be ignored.

(2) X-ray investigations of enzyme-substrate complexes show that when this complex is formed the substrate is transferred from surroundings which are largely aqueous to a medium in which it is at least partially surrounded by amino acid residues of predominantly hydrophobic character (380). In enzyme reactions the biological macromolecules are therefore not only reactants but also to a considerable extent the reaction medium.

(3) Electronic properties (optical, magnetic, EPR), redox potentials, ligand binding properties, etc.) of the redox centres are commonly different from those of better characterized model complexes (361,381). This implies that both the active centres and the protein part of the macromolecules determine the enzyme activity. Apart from being a reaction medium the protein also imposes certain geometric restrictions on the ligand sphere around the metal centre which in turn directs the conformational stability of the protein. The atypical properties of the metallic centres are then a result of the metal-protein interaction and these properties have no analogue in simple model compounds. The concept of an 'entatic' state has been introduced in order to describe this situation (381). In terms of ET theory the process is facilitated due to a low 'intramolecular' reorganization energy.

(4) The main polar component of the globular protein medium in which the biological redox centres are located are peptide units, the neighbouring dipoles of which are largely oriented in opposite directions. The medium reorganization energy of ET reactions in a protein medium is then strongly dependent on the ET distance and direction with respect to the position of the dipoles (382). If the electron is transferred over a short dis-

tance the activation energy may be high due to the large amplitude of the dipole reorientation. On the other hand, if the electron is transferred over such a large distance that the final state corresponds to a favorable interaction with the neighbouring dipole the activation energy may be lower due to a lower medium reorganization energy. The special nature of the protein medium, may thus favour long-distance ET in contrast to ET in low molecular weight solvents for which the medium reorganization energy always increases with increasing ET distance.

(5) It is still an open question wether ET in low-molecular weight solvents are commonly adiabatic or nonadiabatic. However, the small values of the electronic exchange integral found for biological ET processes (in the few cases where such an analysis is possible) characterize these processes as strongly nonadiabatic. This implies that a second order mechanism, i.e. ET through intermediate states may compete with the direct ET from the donor to the acceptor.

(6) Most current ideas on enzyme reactions are based on Koshland's theory of induced conformational changes in the enzyme-substrate complex (377,379,383). The key point of this theory is that by 'orbital steering' the electronic interactions induce conformational changes in the enzyme-substrate complex, and that these changes in turn affect the electronic structure around the active centre to ensure optimum relative orientation. However, the biological macromolecules are not conformationally static, such as implied by Koshland's theory, but subject to conformational fluctuations during an elementary process (383,384). Electronic-conformational interactions are in fact the basis of recent reformulations of the theory of elementary biological processes. Thus, according to Blumenfel'd (385), it is appropriate to redefine the elementary act of enzyme catalysis as the conformational changes of the enzyme-substrate complex, which determine the overall rate of conversion of substrate into products. The basis of this reformulation is that the relaxation

times of the conformational changes following the formation of the enzyme-substrate complex are typically long compared to both electronic and vibrational relaxation times (10^{-1}-10^{-8} s). The macromolecules in biological ET reactions therefore pass through nonequilibrium conformations, which only slowly relax into new equilibrium values.

9.2 Specific Biological Electron Transfer Systems

9.2.1 Primary Photosynthetic Events

Elementary biological ET steps are organized in consecutive or coupled series the nature of which has been elucidated in considerable detail for the mitochondrial ET chain leading ultimately to the reduction of molecular dioxygen, and for the primary steps in bacterial photosynthesis. We shall at present deal with the latter only.

Studies of the primary photo-induced ET processes in bacterial photosynthesis have provided some of the most remarkable results in relation to the theory of rate processes. Rate data are now available in the whole temperature range from liquid helium to room temperature for some of these processes which proceed with a high velocity even at cryogenic temperatures. One of the most notable examples is the thermal oxidation of cytochrome c (cyt c) by photooxidized bacteriochlorophyll (bch) in the bacterium Chromatium. This process was studied by DeVault and Chance (361), and others (386), and displayed an Arrhenius temperature dependence of activation energy 3.3 kcal from room temperature down to about 100° K, and a temperature independent rate corresponding to a half life $\tau_{1/2}$ = 2.3 ms at lower temperatures. Similar data are now available for several other photosynthetic bacteria (386).

While the thermal oxidation of cyt c involves processes with half-lives in the milli- and microsecond ranges, basic new information has recently been provided by picosecond absorption techniques. This has now resulted in the following picture of the primary processes in bacterial photosynthesis which is likely to be representative for a variety of different bacterial species (363,386).

The ET processes occur in reaction centres located in membranes of chromatophores, i.e. vesicles of average diameter 500 A. The chromatophore membrane contains the light-harvesting pigments necessary to transfer the absorbed light energy to the reaction centres. The latter contain a bch dimer, $(bch)_2$ located at the inner side of the membrane, two bacteriopheophytins, bph, i.e. chlorophylls in which magnesium is replaced by hydrogen, a ubiquinone molecule associated with iron (UQFe) and located nearer the outer side of the membrane, several quinones (Q), and several water-soluble cytochromes possibly adsorbed at the inner membrane surface. Initially the reaction centre can be written cyt(II) c$[(bch)_2$(bph)]UQFe. After photoexcitation of $(bch)_2$ the excited$(bch)_2^*$ is a strong reducing agent (the redox potential changes from 0.5 v to -0.7 on photooxidation) and within approximately 10 ps the electron is transferred to bph forming the radical $[(bch)_2^+(bph)^-]$ (the redox potential of the (bph)/$(bph)^-$ couple is -0.4 v), and subsequently, within 100-200 ps, transmitted to the UQFe complex (redox potential -0.15 v) (363). The life time of the $[(bch)_2^+(bph)](UQFe^-)$ (before ET to the quinones) is suitably in excess (100 μs) of the time for ET from cyt(II) c to the $(bch)_2^+$ radical (about 1 μs). Within althogether 100 μs the cyt(II)c$[(bch)_2(bph)](UQFe)^-$ state is therefore formed causing both ET across the membrane, a conversion of light energy to chemical energy, and a charge separation across the membrane, probably associated with changes in the membrane structure. The photosynthetic cycle is completed with subsequent slow ET steps (in the millisecond range) which are less well known. They may involve ET from the quinones to cytochromes embedded in the membrane and eventually back to the oxidized cyt c. The energy liberated in these exothermic ET steps is exploited in other cycles involving phosphorylation, reduction of CO_2 etc., i.e. this energy is not dissipated but 'recuperated' for the subsequent endothermic processes (cf. section 9.4).

The first two steps, i.e.

$$[(bch)_2^{*}(bph)]UQFe \rightarrow [(bch)_2^{+}(bph)^{-}]UQFe \quad \tau_{1/2} \approx 10ps \qquad (9.1)$$

$$[(bch)_2^{+}(bph)^{-}]UQFe \rightarrow [(bch)_2^{+}(bph)](UQFe)^{-} \quad \tau_{1/2} \approx 150ps \qquad (9.2)$$

where recently investigated and found to be temperature independent in the whole range from liquid helium to room temperature (363). Considering the strongly exothermic nature of both reactions ($\Delta E \approx -0.4$ v in both cases) a plausible reason for this is that they are both activationless, i.e. the energy gap is of the same order as the total medium and intramolecular reorganization energy, which must then have the plausible value of approximately 0.4 ev.

We now reconsider the thermal ET between cyt c and $(bch)_2^{+}$ in Chromatium (362). This reaction has been analyzed by Hopfield (375) and Jortner (108) in terms of multiphonon ET theory. However, Hopfield's approach is inadequate in two respects. Firstly, as noted by several people (104,108), the mathematical technique used is inappropriate except when the process is strongly exothermic, i.e. when the energy gap approaches the total reorganization energy. Secondly, the rate constant is written in the form

$$W \propto \int D_D(E)\, D_A(E)\, dE \qquad (9.3)$$

where D_D and D_A are energy distribution functions for the donor and acceptor caused by the coupling of the electron to nuclear modes in the surroundings. This approach is analogous to the treatment of both electronic energy transfer (122), and to Gerischer's treatment of ET processes (142). However, even though the resulting rate expression in the high-temperature limit is formally identical to the one derived in chapter 3, there is a conceptual shortcoming in expressions such as eq.(9.3) when applied to chemical processes. Thus, eq.(9.3) implies that the donor and acceptor are coupled to independent sets of nuclear modes. This is appropriate if the modes are of local nature, but for medium modes only when the donor and acceptor molecules are located far from eachother.

Jortner's analysis incorporated both intramolecular and medium modes, the latter being represented by a single mode of a given average frequency, which for solid media is expected to be 10-100 cm^{-1}. Since coupling to modes in this frequency range would be strongly thermally excited at temperatures below 100° K and provide an activation energy in this range - in contrast to experimental observations - it was concluded that coupling to such modes can be ignored. An analysis almost identical to the one provided in chapter 6 in our discussion of CO-hb system then showed that a single harmonic mode, of frequency about 400 cm^{-1} and coordinate displacement S = 15-20 could reproduce the data in the whole temperature range. Moreover, the electronic factor V could be estimated to be about 10^{-5} ev (or 0.1 cm^{-1}) emphasizing the strongly nonadiabatic character of the process.

This analysis is also subject to modifications (109,111). Firstly, the cyt c donor is located in the aqueous phase at the membrane surface. Cyt c contains a heme group with histidine (donor atom nitrogen) and methionone (donor atom sulphur) as the axial ligands bound to the protein residue (361). The latter is wound up around the heme group leaving the edge exposed to the external medium. This edge probably points towards the membrane to ensure optimum electronic overlap between donor and acceptor. Even though this would cause shielding of the reaction centre (the heme group) from the medium, some electrostatic interaction is still expected to give a finite medium reorganization energy. Secondly, the medium is disordered and must be subject to a broad frequency dispersion. We recall (section 3.4) that this will cause an apparent temperature independence in the low-temperature region, even when the medium reorganization energy is large. This is related to the fact that the number of excitable modes, and therefore the activation energy, increases monotonously with increasing temperature in the low-temperature region thus concealing a finite temperature effect in the Arrhenius relationship.

It is therefore more appropriate to apply the theory of low-temperature ET processes for systems subject to medium frequency dispersion (88,109,111), i.e. in terms of the following rate expression (cf. section 3.4)

$$W = Z_{int}^{-1} \sum_{v,w} \exp(-\varepsilon_v / k_B T)\, S_{v,w}\, W_m(\Delta E + \varepsilon_w - \varepsilon_v) \qquad (9.4)$$

where $W_m(\Delta E)$ is given by eqs.(3.83) and (3.84), and all other symbols have been defined previously. In order to keep the number of parameters small, and since we have no accurate knowledge of the nature of the frequency dispersion we shall represent it by a Debye distribution function of the form of eq.(2.14). For given values of $\Delta E, \Omega_D$, and the medium reorganization energy E_r^m it is then possible to calculate W_m and W numerically for all temperatures. Examples of the resulting dependence of W on T were shown in fig.(3.1), where we also noticed that values of E_r^m in the range 0.2-0.4 ev give a practically temperature independent rate up to 100 °K for Debye frequency values in the range 100-200 cm^{-1}. In order to estimate the remaining parameter values we proceed as follows:

(1) We notice that a single Debye frequency distribution cannot reproduce the rate data in the whole temperature range. The transition from the activationless to the activated region becomes too smooth.

(2) A single (harmonic) mode of frequency about 400 cm^{-1} can reproduce the sharp transition between the high- and low-temperature regions (cf. Jortner's analysis). This mode is plausibly interpreted as a metal-ligand stretching mode of one of the axial ligands, although structural data to confirm this are not available at present. At high temperatures the behaviour of this mode is practically classical, and the experimental high-temperature activation energy can therefore provide the total reorganization energy from the equation

$$E_A = (E_r^m + E_r^c + \Delta E)^2/4(E_r^m + E_r^c) \qquad (9.5)$$

where E_r^c refers to the high-frequency modes. We should here take ΔE = -0.45 ev as opposed to the value of -0.1 ev used previously (108,109,375). This is in accordance with the common belief (387) that the electron donor is the 'low-potential' cyt c (redox potential 0 v) and not the 'high-potential' cyt c (redox potential +0.38 v).

(3) Taking $\Omega_D \approx 150$ cm^{-1} ($\approx 2k_BT_{cr}$ corresponding to the transition temperature, T_{cr}, between the two branches), and E_r^m = (0.9 ± 0.1) ev from the low-temperature branch, then E_r^c = (0.4 ± 0.05) ev and the coupling factor $S = E_r^c/\hbar\omega_c \approx 9 \pm 1$.

(4) Finally, the low-temperature form of eq.(9.4) is (cf. eqs.(3.32))

$$W = (|V|^2/\hbar k_BT)S_{o,o}\,(E_r^m/\hbar\Omega_D)^{\frac{1}{2}} \cdot \qquad (9.6)$$

$$(\pi e|\Delta E|k_BT/\gamma E_r^m)^{2E_r^m/\pi\hbar\Omega_D}$$

where all the symbols were defined in section 3.4. From the absolute value of $W = 4 \cdot 10^2\ s^{-1}$ as $T \rightarrow 0$ we then find the exchange integral $|V| = (1-10) \cdot 10^{-5}$ ev. In line with the conclusions of Hopfield and Jortner this emphasizes the strongly nonadiabatic character of the process. In fact, if the transmission coefficient, κ, is calculated from eq.(5.15) using this value of V, then $\kappa \simeq 10^{-4}$ corresponding to a value of about $10^9\ s^{-1}$ for the pre-exponential Arrhenius factor. This is in good agreement with the experimental value of $7.8 \cdot 10^8\ s^{-1}$ obtained by DeVault and Chance.

As for the CO/hb system this analysis therefore shows that by means of a consistent set of parameters compatible with the physical and chemical properties available or expected for the photosynthetic system, the theory is able to reproduce the rate data in the whole temperature range.

9.2.2 Bioinorganic ET Reactions

There has been much recent interest in the kinetics and mechanism of ET reactions involving 'isolated' biological redox components a few of which we shall briefly consider.

<u>(A) Reactions involving cytochrome c.</u> ET processes between cyt c and small mobile ions in aqueous solution (e.g. oxidation of cyt(II) c with $[Co(phen)_3]^{3+}$ or reduction of cyt(III) c with $[Fe(EDTA)]^-$ or $[Ru(NH_3)_6]^{2+}$ apparently do not exhibit any particular features different from 'normal' outer sphere ET reactions (388). These reactions are therefore commonly viewed as ET via the exposed edge of the heme group, the latter being accessible from the external solvent, and not as a more involved ET pathway through the protein structure. Moreover, the self-exchange rate between cyt(II) c and cyt(III) c calculated from Marcus' cross relationship with reference to different counterreactants gener-

ally agrees well with the experimental value (10^3-10^4 $dm^3 mol^{-1} s^{-1}$).

The fact that the self-exchange rate constant also does not differ significantly from the self exchange rate constants of dipyridyl or phenanthroline complexes of iron, when corrections for the small 'effective' area on the molecular surface are made is perhaps surprising. The heme edge is thus assumed to be located at the bottom of a crevice and the electronic overlap between oxidized and reduced cyt c would be sterically hindered by the protein part, unless the latter is modified by conformational reorganization. Such changes are in fact suggested by the relatively large activation energy of the self-exchange process (12.4 kcal) compared to that for small molecules, in spite of the much larger size of the cytochrome molecules. Other evidence for conformational lability in solution comes from the strong dependence of the ET rate of the reactions of both cyt(II) and cyt(III) c with several outer sphere reagents ($[Ru(NH_3)_6]^{2+}$, $[Co(phen)_3]^{3+}$, $[Fe(EDTA)]^-$) on pH (caused by pH-induced conformational changes), from the observation of a relaxation process involving cyt(II) c subsequent to a fast reaction of cyt(III) c with hydrated electrons (389), and from the observation of a first order reaction prior to the reduction with Cr(II) (indicative of a 'crevice-opening' step (388)). However, in contrast to earlier beliefs (370b), there is now crystallographic evidence for only minor conformational differences between the oxidized and reduced forms of cyt c (390).

When conformational reorganization is important, certain modifications of the theory are required. The conformational subsystem thus represents a classical nuclear subsystem coupled to the reaction centre in a way somewhat similar to the solvent. This also suggests a way of incorporating the conformational dynamics (see further section 9.4).

<u>(B) Reactions involving iron-sulphur proteins.</u> Data are available for the reaction of the bacterial ET mediators rubredoxin

(with aqua ions of V(II) and Cr(II) and with $[Ru(NH_3)_6]^{2+}$ (391)), and the $Fe_4S_4S_4^*$ cluster (S and S* refer to donor atoms in amino acids and in the sulphur bridges, respectively) of Chromatium high potential iron-sulphur protein HiPIP) with $[Fe(EDTA)]^-$ and $[Co(phen)_3]^{3+}$ (81b,392). The metal centre in rubredoxin consists of a single iron atom tetrahedrally coordinated to four cystein amino acids via sulphur atoms and is subject to very little structural change when the oxidation state of the metal atom changes. The reactions all showed rate constants in the range 10^3-10^4 $dm^3\ mol^{-1}\ s^{-1}$, an almost vanishing activation energy, and large negative activation entropies. (about -40 cal K^{-1}). The small activation energies are understandable in view of the small intramolecular changes or large exothermicities (V(II), Cr(II)) and of the partial screening of the aqueous solvent by the protein part of rubredoxin. The large negative activation entropies most likely reflect a small 'effective' area of the rubredoxin molecule rather than a nonadiabatic character of the process, since crystallographic data show that the FeS_4 group is located close to the surface. The HiPIP shows a quite different behaviour in the reaction of its oxidized form with $[Fe(EDTA)]^-$ and its reduced form with $[Co(phen)_3]^{3+}$. The former reaction has a low activation energy ($\approx$ 1 kcal) and a large negative activation entropy (-41 cal K^{-1}), while the values for the second process are 15.5 kcal and 7 cal K^{-1}, respectively. This is indicative of a nonadiabatic reaction path for the former, compatible with structural data which show that the iron centre is located 'inside' the molecule. The activation parameters for the second reaction rather suggest a direct access of $[Co(phen)_3]^{3+}$ to the iron centre, which must involve conformational reorganization of the HiPIP molecule.

(C) Reactions involving copper proteins. Copper proteins occur in many plant, bacterial, and animal materials as reversible electron carriers and in more involved oxidation reduction reactions. Illuminating in the context of the ET theory in particular are the 'blue' copper proteins (referring to their colour), such as plastocyanin, azurin, and stellacyanin, i.e. monomeric units which serve as electron mediators between membrane bound redox systems, being thus reminiscent of cyt c (81b).

Spectroscopic data have suggested that the coordination geometry of a blue copper site is distorted tetrahedral. One ligand is cystein and the others probably histidine and a nitrogen or oxygen coordinated peptide link. Being intermediate between tetrahedral and square planar, characteristic for small Cu(I) and Cu(II) complexes, respectively, the geometry seems to ensure a minimum amount of structural reorganization associated with the ET processes. Moreover, at least some of the copper centres are likely to be located inside the protein.

The reactions of the oxidized stellacyanin, azurin, and plastocyanin with $[Fe(EDTA)]^-$ have small activation energies (2-3 kcal) and substantial negative activation entropies (-21, -37, and -29 cal K^{-1} for the three compounds respectively. The data suggest that little reorganization occures but the large negative activation entropies for plastocyanin and azurin are indicative of nonadiabaticity effects compatible with the localization of the copper centre inside the protein.

The calculated self-exchange rate for stellacyanin is consistent with the value calculated from the reaction of its reduced form with $[Co(phen)_3]^{3+}$, the activation parameters of which are furthermore similar to those of the reaction of the oxidized form with $[Fe(EDTA)]^-$ (E = 6.9 kcal, S = -13 cal K^{-1}). The copper centre in this compound therefore exhibits properties similar to those of 'normal' outer sphere adiabatic ET reactions. The self-exchange rates of both plastocyanin and azurin calculated from reactions with $[Fe(EDTA)]^-$, $[Co(phen)_3]^{3+}$, and cyt c 551 (which is believed to be the physiological redox partner of azurin) are widely different. In addition, the activation parameters for the azurin reactions with $[Co(phen)_3]^{3+}$ and cyt c 551 (E = 14.9 and 11.2 kcal; S = 7 and 8.6 cal K^{-1}) suggest that these reactions, in contrast to those of stellacyanin, involve conformational reorganization of the protein leading to a more easy access to the reaction centre from the external solvent.

9.3 Electronic Conduction in Biological Systems

Jordan and Szent-Gyorgyi (368) were the first to suggest that ET in biological systems may proceed by electronic conduction via delocalized band states arising from the regular structural order in the protein. Much effort has been devoted to investigations which might illuminate this suggestion, but although isolated biological samples do display electronic conductivity, there is at present no indication that this should be the generally perferred thermal electron transfer pathway.

Electronic semiconductivity is displayed by molecular crystals of a variety of organic materials such as aromatic hydrocarbons, metalloorganic compounds (in particular metal phthalocyanines), charge transfer couples, polymers, and by samples of biological materials (369) (DNA, proteins etc.). However, with a few exceptions the conductivities have activation energies in the range 1-3 ev, in contrast to the low values characteristic e.g. for the primary photosynthetic and mitochondrial processes. In addition, the apparent thermally activated conductivity may arise from other elementary processes such as hopping (cf. section 8.4).

The redox species in biological redox processes (cytochromes, copper proteins) are commonly located at the surfaces of membranes. The latter consist essentially of an ordered bilayer of lipid molecules the hydrophobic ends of which point towards eachother, while the hydrophilic ends ensure that the bilayer is 'sandwiched' between two layers of protein (393) providing an overall width of 50-100 A . The whole structure separates two aqueous phases with the ultimate electron donor and acceptor located on each side of the membrane. In addition it contains 'impurities' having different important functions, such as chlorophyll, carotenes (light-harvesting pigments), enzymes which couple elementary endothermic processes (primary steps in the conversion of ADP to ATP) to the exothermic primary ET processes, etc.

We have seen that many of the membrane-bound redox components have been identified, and that the chains of ET processes are adequately viewed as successive or coupled ET steps between discrete molecular entities. Electronic semiconductivity may, however, be important in the following two contexts. Firstly, in the primary light absorption in the 'antenna' chlorophyll or carotene molecules and the subsequent transfer of the excitation energy to the reaction centre. Many, presumably structurally organized chlorophyll molecules are involved in these steps preceding the thermal processes, and the extreme efficiency may suggest band conductivity of the radiatively excited electron as a plausible mechanism. Secondly, artificially produced biplid layers, separating two aqueous solutions, display several features which are all compatible with electronic 'conduction'. Thus (393):

(1) Artificial bilayer membranes exhibit thermally activated conductivity.

(2) Addition of iodine to the aqueous solution raises the normally very low conductivity of certain membranes by several orders of magnitude. This could be caused by a dissociation of I_2 to form a charge transfer complex and an I^- ion which could then move through the membrane under the influence of the electric field. Alternatively, an electron may be conducted across the membrane and an iodine molecule reduced on the opposite side, analogous to the hole injection into anthracene crystals from aqueous solutions of I_3^- and other redox agents (332). The increased conductivity is, however only accompanied by a very

small flux of I^{131} across the membrane which is indicative of the electronic conduction mechanism.

(3) Redox processes can occur at the membranes, for example, by adding mercuric, platinum or silver salts to one of the aqueous solutions and I_3^- to the solution on the other side of the membrane, and applying a small potential difference across the membrane with the metal side positive a metallic film appears on the membrane.

(4) Bilayer membranes sensitized with pigment molecules (e.g. chlorophyll) induce photoredox reactions across the layer in which electrons are transferred from a donor molecule on one side of the membrane to an acceptor on the other side.

There is thus evidence that electronic processes across bilipid membranes do occur. We shall therefore conclude the present section by outlining an extension of the rate theory to incorporate certain of these electronic effects. We consider the model system shown in fig.(9.1), i.e. ET from a donor (D) to an acceptor (A) via several possible intermediate (M) electronic states. We shall treat this process as a second order process which now incorporates excited electronic states in the intermediate channel (membrane) in principle corresponding to an arbitrary distribution of levels, although we shall largely consider the limit of a continuous distribution. In this limit the model bears a superficial similarity to a previous phenomenological description of enzyme reactions formulated by Cope (371). In Cope's theory the enzyme was assumed to be an electronically conducting macromolecule providing separate noninteracting sites for electron exchange. However, this gives the rate in terms of a macroscopic enzyme resistance and no molecular properties of the reactants were considered.

In order that several electronic levels may contribute to the reaction the level spacing should not greatly exceed k_BT. Using for a rough estimate the model of an electron in a box of length

L gives a level spacing $\hbar^2/2m_e L < k_B T$, or L > 15-20 A which is therefore the approximate lower limit of the size of the local structural entity (membrane or dopant) which can provide several closely spaced electronic levels.

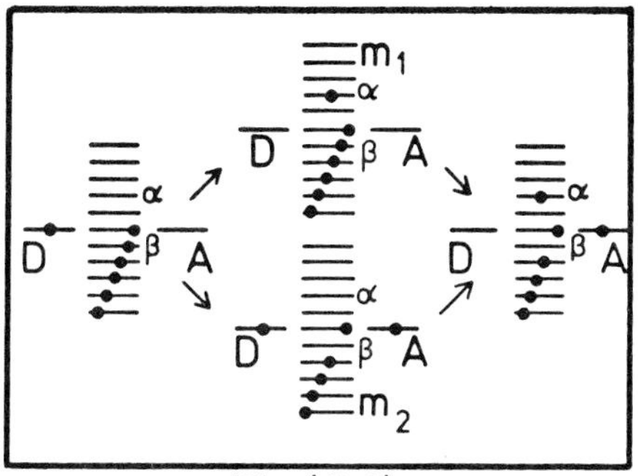

Fig (9.1)

Shcematic view of transition from initial (i) to final (f) state by push pull (m) and pull push (m) mechanism. D: donor. A: acceptor.

Since we shall only consider the nonadiabatic limit, the lower limit of the level spacing is approximately the exchange integrals coupling the initial and intermediate, and the intermediate and final states. Finally, for the sake of simplicity we shall represent the process by a single classical nuclear coordinate q, but the formalism can be extended to incorporate an arbitrary number of classical and high-frequency modes.

Two reaction schemes may be envisaged. An electron may be transferred from the donor to the membrane and this followed by a second ET from the membrane to the acceptor (the push pull mechanism). Alternatively, the ET steps may proceed in the reverse order (the pull push mechanism). If the membrane levels are sufficiently closely spaced, the donor and acceptor membrane levels will furthermore not normally coincide. These two mechanisms correspond to different intermediate states, and for low-lying (energy-conserving) states the two ET paths will have different activation energies, and only occasionally occur in parallel. The following discussion therefore refers explicitly to the push pull mechanism via low-lying electronic states (in the sense discussed in chapter 7).

The electronic coupling factors can be written in second quantized form

$$V = \sum_{s,t} (L_{ds} a_s^+ a_d + L_{at} a_a^+ a_t + L^*_{ds} a_d^+ a_s + L^*_{at} a_t^+ a_a) \qquad (9.7)$$

where L_{ds} and L_{at} are the matrix elements with respect to the electronic wave functions of the donor (d) and a given membrane acceptor level, s, and to the donating membrane (t) and the acceptor level (a), respectively, '*' refers to the complex conjugate, and the a^+ 's and a's are the creation and annihilation operators. Separating the nuclear and electronic motion, the coupling factors between the total vibronic wave functions Ψ_n (n = i,m,f) become

$$(\langle \Psi_m | V | \Psi_i \rangle) = (\chi_{N_m}, \chi_{N_i}) \sum_{\alpha} L_{\alpha d} (1-\nu^i_\alpha) \delta_{o,\nu^i_\alpha} \prod_{s \neq \alpha} \delta_{\nu^i_s, \nu^m_s} \qquad (9.8)$$

$$(\langle \Psi_f | V | \Psi_m \rangle) = (\chi_{N_f}, \chi_{N_m}) \sum_{\beta} L_{a\beta} \nu^m_\beta \delta_{1,\nu^m_\beta} \prod_{t \neq \beta} \delta_{\nu^m_t, \nu^f_t} \qquad (9.9)$$

where <...> and (...) refer to integration with respect to the electronic and nuclear coordinates, respectively. ν^n is the occupation number of the membrane level indicated by the subscript (α and β represent a given pair of donor and acceptor levels), N_n the nuclear quantum numbers, and the delta symbols indicate that a vacant acceptor level α and an occupied donor level β are available, while the population of all other levels is unaffected by the process. Insertion of eqs.(9.8) and (9.9) in the averaged rate expression for the second order process (eq.(7.11)) then gives

$$W = W_I + W_{II} \tag{9.10}$$

where

$$W_I = \frac{2\pi}{\hbar} Av_{N_i, \nu^i_\alpha, \nu^i_\beta} \sum_{N_f} \delta(E^o_i - E^o_f) \times \tag{9.11}$$

$$\left| \sum_{N_m} \frac{(\chi_{N_f}, \chi_{N_m})(\chi_{N_m}, \chi_{N_i})(1-\nu^i_\alpha) L_{d\alpha} L_{\beta a}}{E^o_i - E^o_m + i\gamma} \right|^2$$

corresponding to $\alpha \neq \beta$, and

$$W_{II} = \frac{2\pi}{\hbar} Av_{N_i, \nu^i_\alpha} \sum_{N_f} \delta(E^o_i - E^o_f) \times \tag{9.12}$$

$$\left| \sum_{N_m, \nu^m_\alpha} \frac{(\chi_{N_f}, \chi_{N_m})(\chi_{N_m}, \chi_{N_i})(1-\nu^i_\alpha) L_{d\alpha} L_{\alpha a}}{E^o_i - E^o_m + i\gamma} \right|^2$$

corresponding to $\alpha = \beta$. E^o_i, E^o_m, and E^o_f now contain both vibrational and electronic energy, and Av denotes averaging with respect to the subsystems the quantum numbers of which are indicated as subscripts. In this context we notice that when $\alpha \neq \beta$ a given initial and final electronic state (i.e. given α and β)

unambiguously define the intermediate electronic state. For this reason the summation over the intermediate states is only over the nuclear subsystem. Moreover, the averaging with respect to ν_α^i and ν_β^i includes the necessary summations over the final electronic states, and the additional summation over the states of the latter is therefore also restricted to vibrational levels. The separate averaging over nuclear and electronic levels then leads to the following plausible form of W_I

$$W_I = \sum_\alpha \sum_\beta [1-n(\varepsilon_\alpha)]n(\varepsilon_\beta)\, W(\varepsilon_\alpha,\varepsilon_\beta) \qquad (9.13)$$

where

$$W(\varepsilon_\alpha,\varepsilon_\beta) = \frac{2\pi}{\hbar} Av_{N_i} \sum_{N_f} \delta(E_i^o - E_f^o) \times \qquad (9.14)$$

$$\left| \sum_{N_m} \frac{(\chi_{N_f},\chi_{N_m})(\chi_{N_m},\chi_{N_i})\, L_{d\alpha} L_{\beta a}}{E_i^o(\varepsilon_\alpha,\varepsilon_\beta) - E_m^o(\varepsilon_\alpha,\varepsilon_\beta) + i\gamma} \right|^2$$

and $n(\varepsilon_{\alpha,\beta})$ is the Fermi function of the membrane. $W(\varepsilon_\alpha,\varepsilon_\beta)$ has the same form as eq.(7.11), i.e. it is the microscopic reaction probability for a given set of membrane levels ε_α and ε_β. W is therefore the microscopic probability averaged with respect to the probability of finding a vacant level ε_α and an occupied level ε_β. The summation over the intermediate states in the expression for W_{II} (eq.(9.12)) also includes the electronic states. However, in the limit when $\Delta\varepsilon_{\alpha,\beta} < k_B T$ a proof can be given (112) that $W_{II} \ll W_I$. This is associated with the fact that for sufficiently dense level spacing the number of effectively contributing levels is much larger when $\alpha \neq \beta$ than if the additional restriction $\alpha = \beta$ is invoked. If, on the other hand the membrane level which transmits the electron is a localized level, then the inverse inequality, i.e. $W_I << W_{II}$ may be valid, by a more favourable activation energy and/or electronic overlap than ET via the quasicontinuous levels. In this case the process is, however, most conveniently viewed either as discrete ET from donor to membrane level (cf. the cyt c-(bch) reaction) or as hopping or inelastic tunnelling analogous to the mechanisms discussed in section 8.4. In view of this we shall take W_I (eq.(9.13)) as the rate expression for the process.

The summation over ε_α and ε_β covers in principle the whole electronic distribution of the membrane. As an illustration we shall derive the result for the special cases when the electronic distribution is continuous and the density of states not limited by mobility edges, i.e. the electronic distribution is similar to that of a metal. In this limit eq.(9.13) can be rewritten as

$$W = \int_{-\infty}^{\infty} d\varepsilon_\alpha \int_{-\infty}^{\infty} d\varepsilon_\beta \rho(\varepsilon_\alpha)\rho(\varepsilon_\beta)[1-n(\varepsilon_\alpha)]n(\varepsilon_\beta)W(\varepsilon_\alpha,\varepsilon_\beta) \quad (9.15)$$

where $\rho(\varepsilon_{\alpha,\beta})$ is the density of levels for the membrane. We also assume that $\rho(\varepsilon_{\alpha,\beta})$ is a slowly varying function of $\varepsilon_{\alpha,\beta}$ compared to $n(\varepsilon_{\alpha,\beta})$ and $W(\varepsilon_\alpha,\varepsilon_\beta)$. The following equations are therefore appropriate for electron transfer either via conduction or valence bands, or via groups of levels for which ρ $(\alpha \neq \beta)$ is approximately constant. We further recall that the general form of $W(\varepsilon_\alpha,\varepsilon_\beta)$ is (cf. chapter 7)

$$W(\varepsilon_\alpha,\varepsilon_\beta) = K(\varepsilon_\alpha,\varepsilon_\beta)|L_{d\alpha}|^2|L_{\beta a}|^2\rho(\varepsilon_\alpha)\rho(\varepsilon_\beta)\cdot \quad (9.16)$$

$$\exp[-E_A(\varepsilon_\alpha,\varepsilon_\beta)/k_BT$$

where $K(\varepsilon_\alpha, \varepsilon_\beta)$ is a function which depends on the potential energy surfaces but varies relatively slowly with ε_α and ε_β. Eq.(9.15) can then be written as

$$W = K(\varepsilon_{\alpha^*}, \varepsilon_{\beta^*}) |L_{d\alpha^*}|^2 |L_{\beta^* a}|^2 \rho(\varepsilon_{\alpha^*}) (\varepsilon_{\beta^*}) I \qquad (9.17)$$

where

$$I = \int_{-\infty}^{\infty} d\varepsilon_\alpha \int_{-\infty}^{\infty} d\varepsilon_\beta [1-n(\varepsilon_\alpha)] n(\varepsilon_\beta) \exp[-E_A(\varepsilon_\alpha, \varepsilon_\beta)/k_B T \qquad (9.18)$$

and α^* and β^* are the values of α and β which contribute preferentially to the integrals. The remaining integrations are now relatively straightforward (112). For example, if we consider harmonic motion with no frequency shift, then $E_A(\varepsilon_\alpha, \varepsilon_\beta)$ is

$$E_A(\varepsilon_\alpha, \varepsilon_\beta) = [E_r^{mi} + \Delta E_F^{mi} + (\varepsilon_\alpha - \varepsilon_F)]^2 / 4E_r^{mi} \qquad (9.19)$$

$$E_A(\varepsilon_\alpha, \varepsilon_\beta) = [E_r^{fm} + \Delta E_F^{fm} - (\varepsilon_\beta - \varepsilon_F)]^2 / 4E_r^{fm} + \Delta E_F^{mi} + (\varepsilon_\alpha - \varepsilon_F) \qquad (9.20)$$

when the initial-intermediate and intermediate-final intersection points (for $\varepsilon_\alpha = \varepsilon_\beta = \varepsilon_F$), respectively, are the higher. ΔE_F is here the value of the difference between the minima of the appropriate potential energy surfaces when $\varepsilon_\alpha = \varepsilon_\beta = \varepsilon_F$. Integration of eq.(9.18) gives the corresponding values of I

$$I = N_\alpha^{mi} N_\beta^{fm} \exp[-(E_r^{mi} + \Delta E_F^{mi})^2 / 4E_r^{mi} k_B T] \qquad (9.21)$$

and

$$I = M_\alpha^{fm} M_\beta^{mi} \exp \left\{ -[E_r^{fm} + \Delta E_F^{fm})^2 / 4E_r^{fm} + \Delta E_F^{mi}]/k_B T \right\} \qquad (9.22)$$

The activation energies are seen to be those corresponding to ET to or from the Fermi level (cf. electrochemical processes in the 'normal' overvoltage region). The N's and M's are energy intervals, i.e. ρN and ρM are numbers of levels which contribute to W. For example, with reference to fig.(9.2) we see that when eq.(9.21) is valid, ρN_{α}^{mi} is a number of levels of the order $k_B T \rho$ around the Fermi level which contribute and in which $\varepsilon_{\alpha *}$ must be located. ρN_{β}^{fm} is similarly the number of levels between the minimum of the final state potential energy surface for $\varepsilon_{\beta} = \varepsilon_F$ and the particular value of ε_{β} which would cause the two intersection points to coincide. Many electronic levels, i.e. many more than $k_B T \rho$ will thus generally contribute, and the resulting equations therefore bear a certain resemblance to the expressions for second order processes in which high-frequency modes are involved (eq.(7.28)).

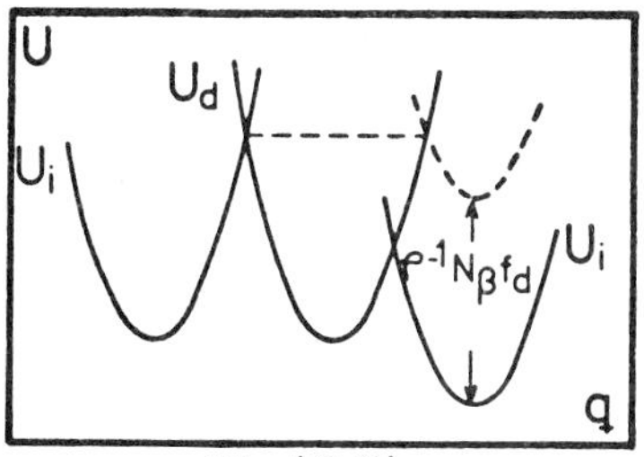

Fig (9.2)

We conclude that this formalism may illuminate certain important features of membrane electron transfer processes in a compact and transparent form. Moreover, several of the features predicted should be subject to experimental investigations. In particular, the free energy relationship of the process is expected to display a considerable structure from which important conclusions about the mechanism (i.e. second order ET via valence or conduction band, elastic and inelastic tunnelling etc.) might be drawn.

9.4 Conformational Dynamics

Conformational changes in biological processes are to a certain extent analogous to the solvent motion in solvents of low molecular weight. Conformational changes are thus a collective low-frequency motion, and can be viewed as an additional classical subsystem. The role of the dynamics of electron-conformational interactions in enzyme reactions was noted by Lumry and Biltonen (384). The matter was also discussed by Blumenfel'd (385) and Vol'kenstein (383) in the context of such biological phenomena as enzyme-substrate complex formation, and electron transport in membranes. Attempts towards a quantitative formulation are represented by the introduction of the conformon concept (371), analogous to the polaron concept, i.e. the conformon consists of a charge distribution together with the conformational changes induced in the surroundings, which in turn creates a conformational potential well.

The role of the conformational dynamics can, however, be given a more rigorous formulation in terms of the multiphonon rate theory of chemical processes. Biological macromolecules possess a large number of configurations stable to small atomic displacements. These quasiequilibrium states determine the secondary and tertiary structures and are commonly named conformational states. The conformations are determined by forces of different kinds, such as interaction energy of nonbonded atoms, torsional and deformational energy, electrostatic interaction energy between different segments, energy in hydrogen bonds, and interaction energy with the surrounding solvent. As a result of thermal motion the macromolecule may spontaneously be transformed from one conformational state to another. However, conformational transitions cannot straightaway be identified with a particular low-frequency mode as occasionally attempted (371). The nature of conformational transitions is qualitatively different from usual low-frequency modes and more comparable to a kind of chemical conversion.

Various conformational states usually have different energies. However, the specific functions of biological macromolecules are related to the fact that the energy of one particular conformation is lower than that of all others (e.g. the helix conformation). The conformational states may also be distorted by external forces such as pressure or electric fields. If the pertubation is sufficiently small, the distortion only induces a shift in the equilibrium positions of the atoms but not a transition to a new conformational state which would require a more drastic reorganization of the secondary and tertiary bond systems. Chemical (ET) processes may therefore induce both distortions in a particular conformational state and shifts in the relative energies of different conformational states, and these two classes are conveniently named conformationally adiabatic and nonadiabatic, respectively.

We incorporate the conformational dynamics by means of a method analogous to the one previously invoked for the influence of a solvent (110). Conformational coordinates thus span a many-dimensional potential energy surface on which each minimum corresponds to a quasiequilibrium conformational state and is determined by a large number of atomic coordinates (fig.(9.3)). If the chemical reaction does not induce changes in the conformational states, this means that the interaction between the electronic and conformational subsystems is weak and can be ignored in the first approximation (fig.(9.3a)). In such cases the reaction centre is only coupled to intramolecular and solvent modes and the theoretical formalism is identical to the one described in previous chapters. If the centre is also coupled to conformational modes the potential energy surfaces in the initial and final states are located as in fig.(9.3b). The electron transfer is then preceded by a complicated motion on the initial state potential energy surface from ξ_{io} to ξ^*. After the electronic transition in the intersection region the system now finds itself in a conformational state which strongly differs from absolute equilibrium, towards which the system approaches subsequently ($\xi^* \rightarrow \xi_{fo}$).

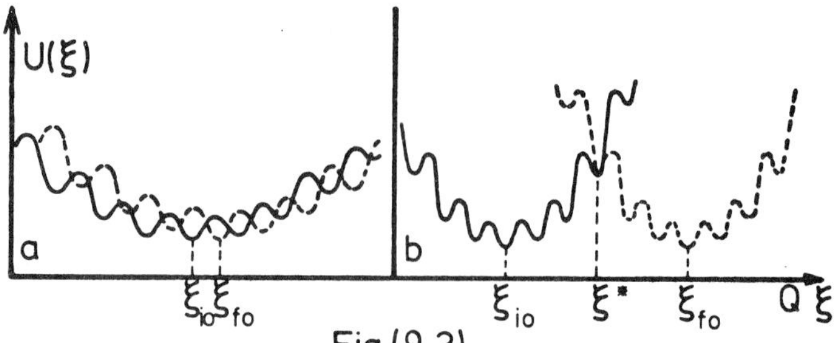

Fig (9.3)

Potential energy surfaces spanned by a conformational coordinate ξ. a: weak coupling. b: strong coupling.

It is now reasonable to assume that the coupling to the conformational subsystem is determined by a relatively small number of averaged field parameters, Π_n. These parameters are equivalent to the polarization components P_ν which we introduced in chapter 2, where ν referred to different kinds of molecular motion. For a linear medium (in the sense discussed in chapter 2) Π_n is then related to a set of normal coordinates, $\xi_{\rho n}$, by the equation

$$\Pi_n = \sum_\rho A_{\rho n} \xi_{\rho n} \tag{9.23}$$

where n refers to a given class of conformational motion. ρ is the different coupled corformational modes inside the class n, and $A_{\rho n}$ is a set of coefficients. The following 'effective' conformational Hamiltonians can then be constructed (cf. chapter 2)

$$H_i^{con} = T + \frac{1}{2} \sum_{n,\rho} \hbar\omega_{\rho n} (\xi_{\rho n} - \xi^i_{\rho no})^2 + \mathcal{U}_{io} \tag{9.24}$$

$$H_f^{con} = T + \frac{1}{2}\sum_{n,\rho} \hbar\omega_{\rho n}(\xi_{\rho n} - \xi^f_{\rho no})^2 + \mathcal{U}_{fo} \qquad (9.25)$$

where T is the nuclear kinetic energy, the M$\mathcal{U}$'s the electronic energies including the conformational equilibrium energies, and $\xi^i_{\rho no}$ and $\xi^f_{\rho no}$ the equilibrium values of in the initial and final states, respectively. As for low-molecular weight solvents the frequencies and oscillator strengths are related to the linear response functions of the conformational subsystem, $G_n(\omega_{\rho n})$. Introducing the Hamiltonians for the intramolecular and solvent subsystems as well, we can write the total potential energy surfaces for the macromolecular system in the initial and final states in the form

$$U_i = \frac{1}{2}\sum_{n,\rho} \hbar\omega_{\rho n}(\xi_{\rho n} - \xi^i_{\rho no})^2 + \frac{1}{2}\sum_{\vec{k},\nu} \hbar\omega_{\vec{k}\nu}(q_{\vec{k}\nu} - q_{\vec{k}\nu o})^2 + U_i^{ac}(\vec{Q}) + U_{io} \qquad (9.26)$$

$$U_f = \frac{1}{2}\sum_{n,\rho} \hbar\omega_{\rho n}(\xi_{\rho n} - \xi^f_{\rho no})^2 + \frac{1}{2}\sum_{\vec{k},\nu} \hbar\omega_{\vec{k}\nu}(q_{\vec{k}\nu} - q_{\vec{k}\nu o})^2 + U_f^{ac}(\vec{Q}) + U_{fo} \qquad (9.27)$$

where U^{ac} and $\vec{Q}$ refer to the potential energies and normal coordinates of the intramolecular modes around the active centre, and the U 's now include both the conformational and solvent equilibrium energies.

The dynamics of the conformational subsystem is now incorporated in such a way that the formalism becomes identical to the one developed previously for reactions in low-molecular weight solvents. Since conformational fluctuations are furthermore classical for all practical purposes ($\omega_{\rho n}$ varies between 10^{-1} and 10^{8} s^{-1} (385)), the transition probability can also be written straightaway in the familiar form

$$W = \varkappa \frac{\omega_{eff}}{2\pi} Z^{-1}_{int} \sum_{v,w} \exp(-\varepsilon_v/k_B T) S_{v,w} \times \qquad (9.28)$$

$$\exp[-(E_r^{con} + E_r^{sol} + \varepsilon_w - \varepsilon_v + \Delta E)^2 / 4(E_r^{con} + E_r^{sol}) k_B T]$$

where

$$E_r^{con} = \frac{1}{2} \sum_{n,\rho} \hbar \omega_{\rho n} (\xi^f_{\rho no} - \xi^i_{\rho no})^2 \tag{9.29}$$

is the conformational reorganization energy, and the transmission coefficient

$$\varkappa = 2\pi |V|^2 [\pi/\hbar^2 \omega^2_{eff} (E_r^{con} + E_r^{sol})]^{\frac{1}{2}} \tag{9.30}$$

The effective frequency, ω_{eff}, is

$$\omega_{eff} = [\frac{1}{2} \sum_{n,\rho} \hbar(\omega_{\rho n})^3 (\xi^f_{\rho no} - \xi^i_{\rho no})^2 \tag{9.31}$$

$$+ \frac{1}{2} \sum_{\vec{k},\nu} \hbar(\omega_{\vec{k}\nu})^3 (q^f_{\vec{k}\nu o} - q^i_{\vec{k}\nu o})^2] / (E_r^{con} + E_r^{sol})$$

E_r^{sol} is the solvent reorganization energy, and the other symbols have all been defined previously. These equations have the following implications:

(1) The coupling to the conformational subsystem is reflected in E_r^{con} and ΔE. In conformationally nonadiabatic processes this coupling is strong, while it is weak in conformationally adiabatic processes. This means that conformationally nonadiabatic processes are generally expected to be slower than conformationally adiabatic processes.

(2) The effective frequency, ω_{eff} , depends on the frequencies associated with transitions between conformational states, provided that (cf.eq.(5.47))

$$\frac{1}{2}\sum_{n,\rho} \hbar(\omega_{\rho n})^3(\xi^f_{\rho no}-\xi^i_{\rho no})^2 > \frac{1}{2}\sum_{\vec{k},\nu} \hbar(\omega_{\vec{k}\nu})^3(q^f_{\vec{k}\nu o}-q^i_{\vec{k}\nu o})^2 \quad (9.32)$$

i.e. for conformationally nonadiabatic processes. However, since $\varkappa$ is inversely proportional to ω_{eff} , the important conclusion can be drawn that the rate of electronically adiabatic processes does not depend on the relaxation time for the conformational modes. On the other hand, sufficiently exothermic (electronically) adiabatic processes, for which $\varkappa S_{v,w} \rightarrow 1$, have rates directly determined by ω_{eff} . This conclusion obviously modifies the suggestion of Blumenfel'd that the elementary act of enzyme reactions should be identified with the conformational relaxation (385).

(3) Even though eq.(9.29) formally contains a weighted sum of Arrhenius factors, the temperature dependence of W may be more involved. This is because the response functions may depend on the temperature in a complicated fashion. This effect is particularly strong in certain regions corresponding to the occurrence of such phenomena as denaturation, phase transitions, or other major reorganization of the secondary and tertiary structures.

If the linear response approximation is inadequate, the rate probability can still be reformulated in terms of the semiclassical theory discussed in chapter 5. However, we shall rather conclude the present chapter by noting the special importance of higher order processes through low-lying intermediate states in relation to conformational relaxation in biological systems. The characteristic classical frequencies in low-molecular weight solvents are $10^{11}-10^{13}$ s^{-1} . Conclusions about such reaction paths therefore have to rest either on indirect evidence, such as free energy relationships, or on fast measuring techniques such as picosecond absorption spectroscopy. Conformational relaxation times are much longer ($1-10^{-8}$ s), and direct detection of the 'quantum dynamic' processes with respect to these nuclear subsystems is therefore expected to be more feasible.

As we have noted above, the elementary biological ET processes e.g. in the photosynthetic and mitochondrial membranes, are arranged in the order of monotonically decreasing redox potentials and rate constants. This is also illustrated by the following scheme which shows the series of ultimate ET steps in the mitochondrial chain with the figures referring to the half-times (ms) of the processes (364)

$$O_2 \xleftarrow{0.4} a_3 \xleftarrow{0.51} a \xleftarrow{2.5} c \xleftarrow{5} c_1 \xleftarrow{80} b \xleftarrow{300} fp \xleftarrow{500} fp \xleftarrow{750} DPN \tag{9.33}$$

At various links in the chain the endothermic steps are coupled to other, exothermic processes, such as the conversion of ADP to ATP in which way the liberated energy is recuperated and stored. The processes are represented by a corresponding set of potential energy surfaces spanned by conformational and molecular coordinates. The measured ET processes are now much faster than the characteristic conformational relaxation times determined for isolated members of the ET chain (388,389). A representation by potential energy surfaces such as figs.(7.1) and (7.2) must therefore involve the additional restriction that the processes cannot be accompanied by conformational relaxation in the intermediate states. In other words, when an electronic transition has occurred at a particular intersection region between two zero order potential energy surfaces, the system also reaches the intersection region corresponding to a subsequent step before dissipating its potential energy.

When chains of biological ET steps are viewed as such 'quantum dynamic' processes, the formalism of higher order processes outlined in chapter 7, is available. In particular, this theory provides a convenient rationalization of two important concepts in the general descriptions of enzyme and phosporylation processes (385,394), namely specificity and control, and energy recuperation.

If we consider again fig.(7.2) we recognize that for the many-dimensional representation of real systems the trajectory along which the system arrives to the final state, is determined not only by the energy of the highest saddle point of all the intersecting potential energy surfaces but also by their relative positions. This may be taken as a quantitative formulation of the selectivity of enzyme catalysis and other biological processes. Thus, if fig.(7.2) represents two possible reaction paths, and if the saddle points along the two paths have the same energy, the system follows preferentially the route of smallest curvature. For this route, and provided that the system has gained enough energy to reach the highest saddle point, there is a high probability that it will reach the second saddle point after having passed the first one, i.e. before dissipating the energy liberated by the first step. If the surfaces are located unfavourably (fig.(7.2b)), more energy is required than corresponding to the highest saddle point, or the trajectory becomes more involved, in both cases giving a longer reaction time in the intermediate state.

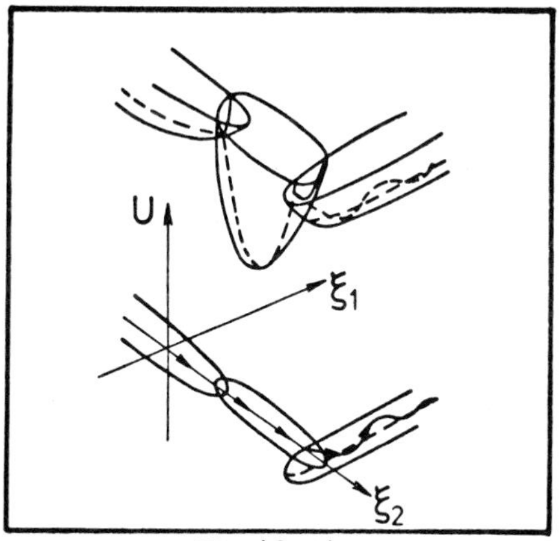

Fig (9.4)

Fig.(9.4) shows potential energy surfaces for a three-level reaction located favourably for energy recuperation. The first step is exothermic, corresponding to the formation of a chemical bond or to one of the exothermic ET steps in photosynthesis or respiration. The second step is endothermic corresponding to the rupture of some other chemical bond, or to the primary step in the synthesis of 'energy-rich' compounds (the conversion of ADP to ATP). If the potential energy surfaces were located unfavourably relative to eachother, the energy liberated in the first step would be dissipated. On the other hand, a favourable position, such as the one shown in the figure, may ensure that the system reaches the second saddle point with practically no loss of energy, and the second, endothermic step occurs by means of energy recuperated from the first transition.

We can conclude that the concepts of 'effective Hamiltonian' and quantum dynamic processes provide a useful theoretical framework, which can account for a number of important features of biological processes including in particular the role of conformational fluctuations, as well as the effects (1)-(6) listed in section 9.1.

Appendix 1

A1.1 Derivation of the Sum Rules(eq.(2.49)).

In the limit of an infinite continuous isotropic medium the Fourier component of the polarization correlation function, $\Pi(\vec{k},\omega)$, is

$$\Pi(\vec{k},\omega) = \int_{-\infty}^{\infty} d\tau\, e^{i\omega\tau} \int_{-\infty}^{\infty} d\vec{\rho}\, e^{-ik\cdot\rho} \langle \vec{P}(\vec{\rho},\tau)\vec{P}(0,0)\rangle \tag{A.1}$$

where $\tau = t-t'$, and $\vec{\rho} = \vec{r}-\vec{r}'$. We assume furthermore that the medium is characterized by different polarization branches, ν , and that there is no correlation between different branches. Eq.(A.1) can then be rewritten

$$\Pi(\vec{k},\omega) = \sum_{\nu} \int_{-\infty}^{\infty} d\tau\, e^{i\omega\tau} \langle P_{\vec{k}\nu}(\tau) P_{\vec{k}\nu}(0)\rangle \tag{A.2}$$

In order to calculate the statistical average $\langle P_{\vec{k}\nu}(\tau)P_{\vec{k}\nu}(0)\rangle$ it is convenient to transform the polarization components to second quantized form,

$$q_{\vec{k}\nu} = 2^{-\frac{1}{2}}(a^{+}_{\vec{k}\nu} + a_{\vec{k}\nu}) \tag{A.3}$$

$$P_{\vec{k}\nu} = 2^{-\frac{1}{2}} i(a^{+}_{\vec{k}\nu} - a_{\vec{k}\nu}) \tag{A.4}$$

where $a^{+}_{\vec{k}\nu}$ and $a_{\vec{k}\nu}$ are the creation and annihilation operators, respectively, for the elementary excitation of wave number k. Together with eqs.(2.36) and (2.37)

$$\langle P_{\vec{k}\nu}(\tau) P_{\vec{k}\nu}(0)\rangle = \hbar\omega_{\vec{k}\nu}\, c_{k\nu} \langle a^{+}_{\vec{k}\nu}(\tau) a_{\vec{k}\nu}(0)\rangle \tag{A.5}$$

Since

$$a^{+}_{\vec{k}\nu}(\tau) = e^{i\varepsilon_{\vec{k}\nu}\tau/\hbar}\, a^{+}_{\vec{k}\nu}(0); \quad a_{\vec{k}\nu}(\tau) = e^{i\varepsilon_{\vec{k}\nu}\tau/\hbar}\, a_{\vec{k}\nu}(0) \qquad (A.6)$$

where $\varepsilon_{\vec{k}\nu}$ is the energy of the appropriate excitation, and

$$\langle a^{+}_{\vec{k}\nu} a_{\vec{k}\nu} \rangle = \langle a_{\vec{k}\nu} a^{+}_{\vec{k}\nu} \rangle - 1 = (e^{\hbar\omega_{\vec{k}\nu}/k_B T} - 1)^{-1} \qquad (A.7)$$

$$\langle a^{+}_{\vec{k}\nu} a^{+}_{\vec{k}\nu} \rangle = \langle a_{\vec{k}\nu} a_{\vec{k}\nu} \rangle = 0 \qquad (A.8)$$

eq.(A.2.2) can be transformed into

$$\Pi(\vec{k},\omega) = \frac{1}{4}\sum_{\nu} \hbar\omega_{\vec{k}\nu}\, c_{k\nu}\,(e^{\hbar\omega_{\vec{k}\nu}/k_B T} - 1)^{-1}[\delta(\omega - \omega_{\vec{k}\nu}) + \delta(\omega + \omega_{\vec{k}\nu})] =$$

$$\frac{1}{4}\sum_{\nu} \hbar\omega c_{k\nu}\,(e^{\hbar\omega/k_B T} - 1)^{-1}[\delta(\omega - \omega_{\vec{k}\nu}) - \delta(\omega + \omega_{\vec{k}\nu})] \qquad (A.9)$$

This gives for the space correlation function

$$\Pi(\vec{k}) = \frac{1}{2\pi}\int_{-\infty}^{\infty} (\vec{k},\omega)\,d\omega = \frac{1}{8\pi}\hbar\omega_{\vec{k}\nu}\, c_{k\nu}\, \mathrm{cth}\frac{\hbar\omega_{\vec{k}\nu}}{2k_B T} \qquad (A.10)$$

We shall subsequently make use of the fluctuation dissipation theorem of linear response theory. This theorem is a relation between the general response function and the correlation function of the response in time and space. It is expressed as

$$G(k,\omega) = -\frac{1}{\hbar}\int_{-\infty}^{\infty}(e^{\hbar u/k_BT}-1)S(\vec{k},u)\frac{du}{\omega-u+i\gamma}\quad \gamma \to 0+ \qquad (A.11)$$

where

$$S(\vec{k},\omega) = \Pi(\vec{k},\omega)/2\pi\hbar \qquad (A.12)$$

$G(k,\omega)$ can be separated in real and imaginary components

$$G(k,\omega) = \frac{1}{\hbar}\,P\int_{-\infty}^{\infty}(e^{\hbar u/k_BT}-1)S(\vec{k},u)\frac{du}{u-\omega} + \qquad (A.13)$$

$$+\ i\,\frac{\pi}{h}(e^{\hbar\omega/k_BT}-1)S(\vec{k},\omega)$$

where P stands for the principal part. Thus,

$$\mathrm{Im}G(k,\omega) = \mathrm{Im}\,\varepsilon(k,\omega)/4\pi\,|\varepsilon(k,\omega)|^2 = \frac{\pi}{h}(e^{\hbar\omega/k_BT}-1)S(\vec{k},\omega) \qquad (A.14)$$

When eq.(A.9) is inserted into eq.(A.13), eq.(2.54) is straightaway obtained.

A1.2 Derivation of Eq.(2.56).

Inserting eq.(A.9) into eq.(A.11) and integrating with respect to u gives

$$G(k,\omega) = \sum_{\nu} \frac{\omega^{2}_{\vec{k}\nu}\, c_{k\nu}}{4\pi} \frac{1}{\omega^{2}_{\vec{k}\nu} - \omega^{2} - i\gamma \frac{\omega}{|\omega|}} \qquad (A.15)$$

For well separated absorption bands the absorption displays maxima at characteristic frequencies, $\omega^{l}_{\vec{k}\nu}$, whereas Im$\varepsilon(k,\omega) \approx 0$ for $\omega^{l-1}_{\vec{k}\nu} < \omega < \omega^{l}_{\vec{k}\nu}$. For a given nondissipative zone which separates the l'th and the (l+1)'th absorption band, the inequalities $\omega^{l-1}_{\vec{k}\nu} \ll \omega \ll \omega^{l}_{\vec{k}\nu}$ are valid. Eq.(A.14) can then be written approximately

$$G(k,\omega) \approx \frac{1}{4}\,[1 - \varepsilon_{l}(k)^{-1}] \approx \sum_{i=1}^{L} \frac{c^{i}_{k\nu}}{4\pi} - \sum_{i=1}^{l-1} \frac{\omega^{l2}_{\vec{k}\nu}}{\omega^{2}} \frac{c^{i}_{k\nu}}{4\pi} \qquad (A.16)$$

where L is the number of absorption bands. Since, finally, $\omega^{l}_{\vec{k}\nu} \ll \omega$ at the low-frequency side of the l'th absorption band, the second term on the right-hand side of eq.(A.15) can be neglected to give

$$\sum_{\nu \in l} c^{l}_{k} \approx \varepsilon_{l+1}(k)^{-1} - \varepsilon_{l}(k)^{-1} \qquad (A.17)$$

where $\varepsilon_{l}(k)$ is the (real) static dielectric constant in the nondissipative region which separates the l'th and the (l+1)'th absorption bands.

Index

A

B

C

D

E

F

G

H

I

K

L

M

N

O

P

Q

R

S

T

Index

B

C

D

E

F

G

H

J

K

L

M

N

O

P

R

S

T

V

W

Z

REFERENCES

1) Partington, J.R.: A History of Chemistry, London: McMillan 1964, Vol. 4.

2) Jortner, J.: Pure and Appl. Chem. 24, 165 (1970).

3) Freed, K.F.: Topics in Applied Physics 15, 23 (1976).

4) Diestler, D.J.: Topics in Applied Physics 15,169 (1976).

5) Fong, F.: Theory of Molecular Relaxation. New York: Wiley 1975.

6) Anderson, P.W., Halparin, B.I. Varma, C.M.: Phil. Mag. 25, 1 (1972).

7) Jäckle, J.: Z. Phys. 257, 212 (1972).

8) Phillips, W.A.: J. Low Temp. Phys. 7, 331 (1972).

9) Harmony, M.D.: Chem. Soc. Rev. 1, 211 (1972).

10) Narayanamurti, V.,Pohl, R.O.: Rev. Mod. Phys. 42, 201 (1970).

11) Brønsted, J.M., Pedersen, K.J.: Z. Phys. Chem. 108, 185 (1924).

12) Gurney, R.W.: Proc. Roy. Soc. A134, 137 (1931).

13) Taube, H.: Electron Transfer Reactions in Solution. New York: Academic Press 1970.

14) Halpern, J.: Quart. Rev. 15, 207 (1961).

15) Sutin, N.: Ann. Rev. Phys. Chem. 17, 119 (1966).

16) Reynolds, W.L., Lumry, R.W.: Mechanisms of Electron Transfer. New York: The Ronald Press 1966.

17) Taube, H., Gould, E.S.: Acc. Chem. Res. 2, 321 (1969).

18) Dogonadze, R.R., Ulstrup, J., Kharkats, Yu.I.: J.C.S. Faraday Trans. II 68, 744 (1972).

19) Bell, R.P.: The Proton in Chemistry, 2nd. ed. London: Chapman and Hall 1973.

20) Søndergård, N.C., Sørensen, P.E., Ulstrup, J.: Acta Chem. Scand. A23, 709 (1975).

21) Kramers, H.A.: Physica 1, 182 (1934).

22) Anderson, P.W.: Phys. Rev. 79, 350 (1950).
23) Zener, C.: Phys. Rev. 82, 403 (1951).
24) Taube, H., Myers, H.: J.Amer.Chem.Soc. 76, 2103 (1953).
25) Levich, V.G.: Adv. Electrochem. Electrochem. Eng. 4, 249 (1966).
26) Dogonadze, R.R. and Kuznetsov, A.M.: J. Electroanal. Chem. 65, 545 (1975).
27) Amis, E.S.: Solvent Effects on Reaction Rates and Mechanisms. New York and London: Academic Press 1966.
28) Waisman, E., Worry, G., Marcus, R.A.: J. Electroanal. Chem. 82, 9 (1977).
29) Dogonadze, R.R.: Theory of Molecular Electrode Kinetics. In: Reactions of Molecules at Electrodes. Hush, N.S. (ed.). New York: Wiley 1971, pp. 135-227.
30) Marcus, R.A.: Ann. Rev. Phys. Chem. 15, 155 (1964).
31) Earley, J.E.: Progr. Inorg. Chem. 13, 243 (1968).
32) Basolo, F., Pearson, R.G.: Mechanisms of Inorganic Reactions, 2nd. ed. New York: Wiley 1967.
33) Dogonadze, R.R., Kuznetsov, A.M.: Physical Chemistry. Kinetics. Moscow: VINITI 1973.
34) Dogonadze, R.R., Kuznetsov, A.M.: Progr. Surf. Sci. 6, 1 (1975).
35) Matthews, D.B., Bockris, J.O'M.: Mod. Asp. Electrochemistry 6, 242 (1971).
36) Appleby, A.J., Bockris, J.O'M, Sen, R.K., Conway, B.E.: MTP Int. Rev. Sci. Tech. Electrochemistry. London: Butterworth 1972.
37) Conway, B.E.: Can. J. Chem. 37, 178 (1959).
38) Conway, B.E., Salomon,M.E.: J. Phys. Chem. 68, 2009 (1964). J. Chem. Phys. 41, 3169 (1964).
39) a. Bockris, J.O'M.,Matthews, D.R.: Proc. Roy. Soc.A232, 479 (1966); b. J. Chem. Phys. 44, 298 (1966).

40) Christov, S.G.: Z. Elektrochem. 62, 567 (1958). 64, 840 (1960) Z. Physik. Chem. 214, 40 (1960). Electrochim. Acta 4, 194, 306 (1961). J. Res. Inst. Catalysis, Hokkaido Univ. 16, 169 (1968).
41) Butler, J.A.V.: Proc. Roy. Soc. A157, 423 (1936).
42) Gerischer, H.: Z. Physik. Chem. (Frankfurt) 26, 223, 325 (1960).
43) Marcus, R.J., Zwolinskij, B.J., Eyring, H.: J. Phys. Chem. 58, 432 (1954).
44) Sacher, E., Laidler, K.J.: Trans. Faraday. Soc. 59, 396 (1963).
45) Glasstone, S., Laidler, K.J. Eyring, H.: The Theory of Rate Processes. New York, London and Tokyo: McGraw-Hill 1941.
46) Horiuti, J., Polanyi, M.: Acta Physicochim. URSS 2, 505 (1935).
47) Bell, R.P.: Proc. Roy. Soc. A154, 414 (1936).
48) Frumkin, A.N.: Z. Physik. Chem. (A) 160, 116 (1932).
49) a. Marcus, R.A.: J. Chem. Phys. 24, 966 (1956); b. 26, 867, 872 (1957); c. Disc. Faraday Soc. 29, 21 (1960); d. J. Phys. Chem. 67, 893 (1963); e. J. Chem. Phys. 43. 679 (1965).
50) Levich, V.G., Dogonadze, R.R.: Dokl. Akad. Nauk. SSSR. Ser. Fiz. Khim. 124, 123 (1959).
51) Levich, V.G., Dogonadze, R.R.: Coll. Czech. Chem. Comm. 26, 193 (1961).
52) Dogonadze, R.R.: Dokl. Akad. Nauk. SSSR, Ser. Fiz. Khim. 142, 1108 (1962).
53) Libby, W.F.: J. Phys. Chem. 56, 863 (1952).
54) Weiss, J.J.: Proc. Roy. Soc. A 222, 128 (1954).
55) Platzmann, R., Franck, J.: Z. Phys. 138, 411 (1954).
56) Fröhlich, H.: Theory of Dielectrics. Oxford: The Clarendon Press. 1949.

57) Pekar, S.I.: Untersuchungen über die Elektronentheorie der Kristalle. Berlin: Akademie-Verlag 1954.

58) Holstein, T: Ann. Phys. 8, 325 (1959).

59) Schmidt, P.P.: Specialist Periodical Report, Electrochemistry. London: The Chemical Society 1975, Vol.5, pp. 21-131.

60) a. Kharkats, Yu.I.: Elektrokhimiya 9, 881 (1973); b. 10, 612 (1974).

61) a. Dulz, G., Sutin, N.: Inorg. Chem. 2, 917 (1964); b. Diebler, H., Sutin, N.: J. Phys. Chem. 68, 174 (1964).

62) Sutin, N., Gordon, B.M.: J. Amer. Chem. Soc. 83., 70 (1961).

63) Miller, J.D., Prince, R.H.: J. Chem. Soc. 1370 (1966).

64) Vetter, K.J.: Electrochemical Kinetics. New York, London: Academic Press 1967.

65) Scherer, G., Willig, F.: J. Electroanal. Chem. 85, 77 (1977).

66) Eigen, M.: Angew. Chem. 75, 489 (1963).

67) Kresge, A.J.: Chem. Soc. Rev. 2, 475 (1973).

68) Holzwarth, J., Strohmaier, L., Gerischer, H.: Ber. Bunsenges. Physik.Chem. 76, 1048 (1972).

69) Suga, K., Aoyagui, S.: Bull. Chem. Soc. Japan 46, 755 (1973).

70) Suga, K., Mizota, H., Kanzaki, Y., Aoyagui, S.: J. Electroanal. Chem. 41, 313 (1973).

71) Meisel, D.: Chem. Phys. Lett. 34, 263 (1975).

72) Saveant, J.-M., Tessier, D.: J. Phys. Chem. 81, 2192 (1977).

73) Bindra, P., Brown, A.P., Fleischmann, M., Pletcher, D.: J. Electroanal. Chem. 58, 39 (1975).

74) Candlin, J.P., Halpern, J., Trimm, D.L.: J. Amer. Chem. Soc. 86, 1013 (1964).

75) Dorfman, L.M., Brandon, J.R.: J. Chem. Phys. 53, 3849 (1970).
76) Campion, R.J., Purdie, N., Sutin, N.: Inorg. Chem. 3, 1091 (1964).
77) Wilkins, R.G., Yelin, R.E., Inorg. Chem. 12, 2667 (1968).
78) Pladziewicz, J.R., Espenson, J.H.: J. Phys. Chem. 75, 3381 (1971).
79) Winograd, N., Kuwana, T.: J. Amer. Chem. Soc. 93, 4343 (1971).
80) Hodges, H.L., Holwerda, R.A., Gray, H.B.: J. Amer. Chem. Soc. 96, 3132 (1974).
81) a. McArdle, J.V., Gray, H.B., Creutz, C., Sutin, N.: J. Amer. Chem. Soc. 96, 5737 (1974); b. Holwerda, R.A., Wherland, S., Gray, H.B.: Ann. Rev. Biophys. Bioeng. 5, 363 (1976).
82) Hush, N.S.: J. Chem. Phys. 28, 962 (1958).
83) Dogonadze, R.R., Chizmadzhev, Yu.A.: Dokl. Akad. Nauk. SSSR, Ser. Fiz. Khim. 144, 1077 (1962). 145, 848 (1962). 150, 333 (1963).
84) Dogonadze, R.R., Kuznetsov, A.M., Chizmadzhev, Yu.A.: Zhur. Fiz. Khim. 38, 1195 (1964).
85) Dogonadze, R.R., Kuznetsov, A.M.: Elektrokhimiya 7, 763 (1971).
86) Dogonadze, R.R., Kornyshev, A.A., Kuznetsov, A.M.: Teor. Mat. Fiz. 15, 127 (1973).
87) Vorotyntsev, M.A., Dogonadze, R.R., Kuznetsov, A.M.: Dokl. Akad. Nauk. SSSR, Ser. Fiz. Khim. 195, 1135 (1970).
88) Dogonadze, R.R., Kuznetsov, A.M., Vorotyntsev, M.A., Zakaraya, M.G.: J. Electroanal. Chem. 75, 315 (1977).
89) Dogonadze, R.R., Kuznetsov, A.M.: Elektrokhimiya 2, 1324 (1967).
90) Vorotyntsev, M.A., Dogonadze, R.R., Kuznetsov, A.M.: Phys. Stat. Sol. 54, 125, 425 (1972).

91) Kestner, N.R., Logan, J., Jortner, J.: J. Phys. Chem. 78, 2148 (1974).
92) Schmidt, P.P.: J.C.S. Faraday Trans II 69, 1104, 1123 (1973).
93) Dogonadze, R.R., Kuznetsov, A.M.: Teor. Eksp. Khim. 6, 298 (1970).
94) Dogonadze, R.R., Urushadze, Z.D.: J. Electroanal. Chem. 32, 235 (1971).
95) Levich, V.G., Dogonadze, R.R., Kuznetsov, A.M.: Electrochim. Acta 13, 1025 (1968).
96) Levich, V.G., Dogonadze, R.R., German, E.D., Kuznetsov, A.M., Kharkats, Yu.I.: Electrochim. Acta. 15, 353 (1970).
97) Vorotyntsev, M.A., Dogonadze, R.R., Kuznetsov, A.M.: Dokl. Akad. Nauk. SSSR, Ser. Fiz. Khim. 209, 1135 (1973).
98) German, E.D., Dogonadze, R.R.: J. Res. Inst. Catalysis Hokkaido Univ. 20, 34 (1972).
99) German, E.D., Dogonadze, R.R.: Int. J. Chem. Kin. 6, 457, 467 (1974).
100) Van Duyne, R.P., Fischer, S.F.: Chem. Phys. 5, 183 (1974).
101) Ulstrup, J., Jortner, J.: J. Chem. Phys. 63, 4358 (1975).
102) Schmickler, W.: J.C.S. Faraday Trans. II 72, 307 (1976)
103) Efrima, S., Bixon, M.: Chem. Phys. 13, 447 (1976).
104) Dogonadze, R.R., Kuznetsov, A.M., Vorotyntsev, M.A.: Z. Phys. Chem. NF 100, 1 (1976).
105) Volkenstein, M.V., Dogonadze, R.R., Madumarov, A.K., Kharkats, Yu.I.: Dokl. Akad. Nauk. SSSR, Ser. Fiz. Khim. 199, 124 (1971).
106) a. Dogonadze, R.R., Ulstrup, J., Kharkats, Yu.I.: J.C.S. Faraday Trans. II 70, 64 (1974); b. J. Electroanal. Chem. 39, 47 (1972).

107) Kharkats, Yu.I., Madumarov, A.K., Vorotyntsev, M.A.: J.C.S. Faraday Trans. II 70, 1578 (1974).

108) Jortner, J.: J. Chem. Phys. 64, 4860 (1976).

109) Kuznetsov, A.M., Søndergård, N.C., Ulstrup, J.: Chem. Phys. 29, 383 (1978).

110) Dogonadze, R.R., Kuznetsov, A.M., Ulstrup, J.: J.Theor. Biol. 69, 239 (1977).

111) Dogonadze, R.R., Kuznetsov, A.M., Zakaraya. M.G., Ulstrup, J.: In: Proc. Int. Symp. on Low-Temperature Tunnelling in Biological Systems. New York: Academic Press, in press.

112) Dogonadze, R.R., Ulstrup, J., Kharkats, Yu.I.: J.Theor. Biol. 40, 259, 279 (1973).

113) Goldberger, M.L., Watson, K.M.: Collision Theory. New York: Wiley 1964.

114) Huang, K., Rhys, A.: Proc. Roy. Soc. A204, 406 (1950).

115) Lax, M.: J. Chem. Phys. 20, 1752 (1952).

116) O'Rourke, R.C.: Phys. Rev. 91, 265 (1953).

117) Markham, J.J.: Rev. Mod. Phys. 31, 956 (1959).

118) Rebane, K.K.: Impurity Spectra of Solids. New York: Plenum (1970).

119) Kubo, R., Toyozawa, Y.: Progr. Theor. Phys. 13, 160 (1955).

120) a. Krivoglaz, M.A.: Zhur. Eksp. Teor. Fiz. 25, 191 (1935); b. Rickayzen, G.: Proc. Roy. Soc. A241, 480 (1957).

121) Sturge, M.D.: Phys. Rev. B8, 6 (1973).

122) a. Förster, T.: Disc. Faraday Soc. 27, 7 (1959); b. Dexter, D.L.: J. Chem. Phys. 21, 836 (1953).

123) Arnett, E.M.: Acc. Chem. Res. 6, 404 (1973).

124) Muirhead-Gould, J.S., Laidler, K.J.: Trans. Faraday Soc. 63, 955 (1967).

125) Clementi, E., Popkic, H.: J. Chem. Phys. 57, 1077 (1972).

126) Kistenmacher, H. Popkic, H.: J. Chem. Phys. 58, 1689 (1973); 61, 799 (1974).
127) Watts, R.O., Clementi, E., Fromm, J.: J. Chem. Phys. 61, 2550 (1974).
128) Flynn, C.P.: Point Defects and Diffusion. Oxford: Clarendon 1972.
129) a. Schockley, W., Bardeen, J.: Phys. Rev. 77, 407 (1950); 80 72 (1950); c. Kittel, C.: Quantum Theory of Solids. New York: Wiley 1963.
130) Kharkats, Yu.I., Kornyshev, A.A., Vorotyntsev, M.A.: J.C.S. Faraday Trans. II 72, 361 (1976).
131) Zubarev, D.N.: Nonequilibrium Statistical Thermodynamics. New York: Plenum 1974.
132) a. Saxton, R.: Proc. Roy. Soc. A213, 473 (1952); b. Hasted, J.B.: Specialist Periodical Report. Dielectric and Related Molecular Processes. London: The Chemical Society 1972, Vol.1, pp. 121-162.
133) Eisenberg, D., Kauzmann, W.: The Structure and Properties of Water. Oxford: Clarendon 1969.
134) Dogonadze, R.R., Kornyshev, A.A.: J.C.S. Faraday Trans. II 70 1121 (1974).
135) Holub, K., Kornyshev, A.A.: Z. Naturforsch. 31a, 1601 (1976).
136) Dogonadze, R.R., Kuznetsov, A.M., Marsagishvili, T.A.: Phys. Stat. Sol., submitted.
137) Schmidt, P.P., McKinley, J.M.: J.C.S. Faraday Trans. II 72, 143 (1976).
138) Schmidt, P.P.: J.C.S. Faraday Trans II 72, 1048 (1976).
139) Appel, J.: Solid State Physics 21, 133 (1968).
140) Copeland, D.A., Kestner, N.R., Jortner, J.: J. Chem. Phys. 53, 1189 (1970).
141) Schiff, L.I.: Quantum Mechanics, 3rd ed. Tokyo: McGraw-Hill 1968.

142) Landau, L.D., Lifshitz, E.M.: Quantum Mechanics, 3rd ed. Oxford: Pergamon 1965.

143) Feynman, R.P., Hibbs, A.R.: Quantum Mechanics and Path Integrals. New York: McGraw-Hill 1965.

144) a. Spears, K.G., Rice, S.A.: J. Chem. Phys. 55, 5561 (1971); b. Abramson, A.S., Spears, K.G., Rice, S.A.: J. Chem. Phys. 56, 2291 (1972); c. Rice, S.A.: J. Chem. Phys. 61, 651 (1974).

145) Landau, L.D., Lifshitz, E.M.: Statistical Physics, 2nd ed. Oxford: Pergamon 1969.

146) Lin, S.H., Lee, S.T., Yoon, Y.H., Eyring, H.: Proc. Nat. Acad. Sci. 73, 2533 (1976).

147) Born, M., Huang, K.: Dynamical Theory of Crystal Lattices. Oxford: Clarendon 1954.

148) a. Robinson, G.W., Frosch, R.P.: J. Chem. Phys. 37, 1962 (1962); b. 38, 1187 (1963).

149) Bixon, M., Jortner, J.: J. Chem. Phys. 48, 715 (1968).

150) Dvali, V.G., Dogonadze, R.R.: Elektrokhimiya 12, 937 (1976). Original No.404-76 filed at VINITI 1976.

151) Schmidt, P.P.: J. Electroanal. Chem. 82, 29 (1977).

152) a. Schmickler, W.: Electrochim. Acta 21, 777 (1976); b. Ratner, M.A., Madhukar, A.: Chem. Phys. 30. 201 (1978).

153) a. Nitzan, A., Jortner, J.: J. Chem. Phys. 56, 3360 (1972); b. Freed, K.F., Lin, S.H.: Chem. Phys. 11, 409 (1975).

154) Morse, P.M., Feshbach, H.: Methods of Theoretical Physics. New York: McGraw-Hill 1953.

155) Gradshtein, I.S., Ryzhik, I.M.: Tables of Integrals, Sums, Series, and Products. Moscow: Nauka 1971.

156) De Bruijn, N.G.: Asymptotic Methods in Analysis. Amsterdam: North Holland 1958.

157) Dogonadze, R.R., Itskovitch, E.M., Kuznetsov, A.M., Vorotyntsev, M.A.: J. Phys. Chem. 79, 2827 (1975).
158) Ovchinnikov, A.A., Ovchinnikova, M.Ya.: Zh. Eksp. Teor. Fiz. 56, 1278 (1969).
159) Specialist Periodical Report. Dielectric and Related Processes. Davies, M. (ed). London: The Chemical Society, Vol.1 1972, Vol.2 1975.
160) Dolin, S.P., German, E.D., Dogonadze, R.R.: J.C.S. Faraday Trans. II 73, 648 (1977).
161) Taube, H.: Adv. Chem. Ser. 162, 127 (1977).
162) Chou, M., Creutz, C., Sutin, N.: J. Amer. Chem. Soc. 99, 5615 (1977).
163) Dogonadze, R.R., Kuznetsov, A.M., Ulstrup, J.: J. Electroanal. Chem. 79, 267 (1977).
164) a. Fischer, H., Tom, G.M., Taube, H.: J. Amer. Chem. Soc. 98, 5512 (1976); b. Rieder, K., Taube, H.: J. Amer. Chem. Soc. 99, 7891 (1977).
165) Schlag, E.W., Schneider, S., Fischer, S.F.: Ann. Rev. Phys. Chem. 22, 465 (1971).
166) Lin, S.H.: J. Chem. Phys. 44, 3759 (1966).
167) a. Englman, R., Jortner, J.: Mol. Phys. 18, 145 (1970); b. Freed, K.F., Jortner, J.: J. Chem. Phys. 52, 6272 (1970).
168) Lin, S.H., Eyring, H.: Proc. Nat. Acad. Sci. USA 69, 3192 (1972).
169) Byrne, J.P., McCoy, E.F., Ross, I.G.: Austr. J. Chem. 18, 1589 (1965).
170) a. Fong, F.K., Naberhuis, S.L., Miller, M.M.: J. Chem. Phys. 56, 4020 (1972); b. Fong, F.K., Wassam, W.A.: J. Chem. Phys. 58, 956 (1973).
171) a. Lauer, H.V., Fong, F.K.: J. Chem. Phys. 60, 274 (1974); b. Fong, F.K., Lauer, H.V., Chilver, C.R.: J. Chem. Phys. 63, 366 (1975).

172) Tsai, S.C., Robinson, G.W.: J. Chem. Phys. 49, 3184 (1968).

173) Siebrand, W.: J. Chem. Phys. 47, 2411 (1967).

174) a. Murata, S., Iwanaga, C., Toda, T., Kokubun, H.: Chem. Phys. Lett. 13, 101(1972); b. Chem. Phys. Lett. 15, 152 (1972).

175) a. Partlow, W.D., Moos, H.W.: Phys. Rev. 157, 252 (1967); b. Riseberg, L.A., Moos, H.W.: Phys. Rev. 174, 429 (1968).

176) Lin, S.H.: J. Chem. Phys. 56, 2648 (1972).

177) a. Kellogg, R.E., Wueth, N.C.: J. Chem. Phys. 45, 3156 (1966); b. Martin, T.E., Kalantar, A.H.: J. Chem. Phys. 48, 4996 (1968).

178) Lin, S.H., Bersohn, R.E.: J. Chem. Phys. 48, 2732 (1968).

179) Robin, M.B., Day, P.: Adv. Inorg. Chem. Radiochem. 10, 247 (1967).

180) a. Hush, N.S., Allen, G.C.: Progr. Inorg. Chem. 8, 357 (1967); b. Hush, N.S.: Progr. Inorg. Chem. 8, 391 (1967).

181) Hush, N.S.: Electrochim. Acta 13, 1005 (1968).

182) a. Callahan, R.W., Brown, G.M., Meyer, T.J.: Inorg. Chem. 14, 1443 (1975); b. Callahan, R.W., Meyer, T.J.: Chem. Phys. Lett. 39, 82 (1976); c. Powers, M.J., Callahan, R.W., Salmon, D.J., Meyer, T.J.: Inorg. Chem. 15, 1457 (1976); d. Callahan, R.W., Keene, F.R. Meyer, T.J., Salmon, D.J.: J. Amer. Chem. Soc. 99, 1064 (1977); e. Meyer, T.J.: Acc. Chem. Res. 11, 94 (1978).

183) Stynes, H.C., Ibers, J.A.: Inorg. Chem. 10. 2304 (1971).

184) a. Vogler, A., Kunkely, H.: Ber. Bunsenges. Phys. Chem. 79, 83 (1975); b. Brown, G.M., Meyer, T.J., Cowan, D.O., LeVanda, C., Kaufman, F., Roling, P.V., Rausch, M.D.: Inorg. Chem. 14, 506 (1975); c. Powers, M.J., Callahan, R.W., Salmon, D.J., Meyer, T.J.: Inorg. Chem. 15, 894 (1976).

185) Tom, G.M., Taube, H.: J. Amer. Chem. Soc. 97, 5310 (1975).

186) a. Creutz, C., Taube, H.: J. Amer. Chem. Soc. 91, 3988 (1969); b. J. Amer. Chem. Soc. 95, 1086 (1973).

187) a. Mayoh, B., Day, P.: J. Amer. Chem. Soc. 94, 2885 (1972); b. Inorg. Chem. 13, 2273 (1974).

188) Elias, J.H., Drago, R.S.: Inorg. Chem. 11, 415 (1972).

189) Beattie, J.K., Hush, N.S., Taylor, P.R.: Inorg. Chem.. 15, 992 (1976).

190) Beattie, J.K., Raston, C., White, A.: J.C.S. Dalton Trans. 1121 (1977).

191) Bunker, B.C., Drago, R.S., Hendrickson, D.N., Richman, R.N., Kessell, S.L.: J. Amer. Chem. Soc. 100, 3805 (1978).

192) Freed, K.F., Fong, F.K.: J. Chem. Phys. 63, 2890 (1975).

193) Efrima, S., Bixon, M.: Chem. Phys. 13, 447 (1976).

194) Siebrand, W.: J. Chem. Phys. 46, 440 (1967).

195) Søndergård, N.C., Ulstrup, J., Jortner, J.: Chem. Phys. 17, 417 (1976).

196) a. Brüniche-Olsen, N., Ulstrup, J.: J.C.S. Faraday Trans.II 74, 1690 (1978); b. J.C.S. Faraday Trans.I 75, 205 (1979).

197) Kharkats, Yu.I., Ulstrup, J.: J. Electroanal. Chem. 65, 555 (1975).

198) Fraser, P.A., Jarmain, W.R.: Proc. Phys. Soc. (London) A66, 1145 (1953).

199) Razem, D., Hamill, W.H., Funabashi, K.: J. Chem. Phys. 67, 5404 (1977).
200) Kawabata, K.: J. Chem. Phys. 55, 3672 (1971).
201) Rehm, D., Weller, A.: Israel J. Chem. 8. 259 (1970).
202) Scheerer, R., Grätzel, M.: J. Amer. Chem. Soc. 99, 865 (1977).
203) Afanasev, I.B., Prigoda, S.V., Mal'tseva, T.Yu., Samokhvalo, G.I.: Int. J. Chem. Kin. 6, 643 (1974).
204) Romashov, L.V., Kiryukhin, Yu.I., Bagdasar'yan, Kh.S.: Dokl. Akad. Nauk. SSSR, Ser. Khim. 230, 1145 (1976).
205) Schomburg, H., Staerk, H., Weller, A.: Chem. Phys. Lett. 22, 1 (1973).
206) Van Duyne, R.P., Fischer, S.F.: Chem. Phys. 26, 9 (1977).
207) Creutz, C., Sutin, N.: J. Amer. Chem. Soc. 99, 241 (1977).
208) Frank, A.J., Grätzel, M., Henglein, A., Janata, E.: Ber. Bunsenges. Phys. Chem. 80, 294, 547 (1976).
209) Berdnikov, V.M.: Zh. Fiz. Khim. 49, 2988 (1975).
210) Rentzepis, P.M., Jones, R.P., Jortner, J.: Chem. Phys. Lett. 15, 480 (1972).
211) Hart, E.J., Anbar, M.: The Hydrated Electron. New York: Wiley 1970.
212) Miller, J.R., Clifft, B.E., Hines, J.J., Runowski, R.F., Johnson, K.W.: J. Phys. Chem. 80, 457 (1976).
213) Miller, J.R.: J. Phys. Chem. 82, 767 (1978).
214) Zamaraev, K.I., Khairutdinov, R.F.: Chem. Phys. 4, 181 (1974).
215) Girina, E.L., Erskov, B.G., Pikaev, A.K.: Izv. Akad. Nauk. SSSR, Ser. Khim 1250 (1977).
216) Miller, J.R.: J. Phys. Chem. 79, 1050 (1975).
217) Zamaraev, K.I., Khairutdinov, R.F., Miller, J.R.: Chem. Phys. Lett. 57, 311 (1978).

218) Beitz, J.V., Miller, J.R.: In: Proceedings of the Conference on Low-Temperature Tunnelling in Biological Systems. Chance, B. (ed.). Philadelphia: Academic Press 1977, in press.

219) Zener, C.: Proc. Roy. Soc. A137, 696 (1932).

220) Coulson, C.A., Zalewski, K.: Proc. Roy. Soc. A268, 437 (1962).

221) Nikitin, E.E.: Theory of Nonadiabatic Reactions. In: Chemische Elementarprozesse. Hartmann, H. (ed.). Berlin: Springer-Verlag 1968, pp. 43-76.

222) Child, M.S.: Disc. Faraday Soc. 53, 18 (1972).

223) Vorotyntsev, M.A., Dogonadze, R.R., Kuznetsov, A.M.: Vestn. Mosk. Univ., Fiz. 28, 224 (1973).

224) German, E.D., Dvali, V.G., Dogonadze, R.R., Kuznetsov, A.M.: Elektrokhimiya 12, 667 (1976).

225) German, E.D.: Izv. Akad. Nauk SSSR, Ser. Khim. 10, 1269 (1974).

226) German, E.D., Dogonadze, R.R.: Izv. Akad. Nauk SSSR, Ser. Khim 2, 2155 (1973).

227) Clack, D.W., Farrinmod, M.S.: J. Chem. Soc. A299 (1971).

228) Barnet, M.T., Graven, B.M., Freeman, H.C., Kime, N.E., Ibers, J.A.: Chem. Comm. 307 (1966).

229) a. Sacconi, L., Sabatini, A., Gans, P.: Inorg. Chem. 3, 1772 (1964); b. Schimanouchi, T., Nakagawa, I.: Inorg. Chem. 3, 1805 (1964).

230) a. Sutin, N.: Ann. Rev. Nucl. Sci. 12, 285 (1962); b. Sykes, A.G.: Adv. Inorg. Chem. Radiochem. 10, 153 (1967).

231) Adams, D.M.: Metal-Ligand and Related Vibrations. London: E. Arnold 1967.

232) Nakagawa, I., Shimanouchi, T.: Spectrochim. Acta 20, 429 (1964).

233) a. Stranks, D.R.: Disc. Faraday Soc. 29, 73 (1960); b. Biradar, N.S., Stranks, D.R., Vaidya, M.S.: Trans. Faraday Soc. 58, 2421 (1962).

234) Stynes, H.C., Ibers, J.A.: Inorg. Chem. 10, 2304 (1971).

235) Allen, A.D., Senoff, C.V.: Can. J. Chem. 45, 1337 (1967).

236) Vargaftik, M.N., German, E.D., Dogonadze, R.R., Syrkin, Ya. K.: Dokl. Akad. Nauk SSSR, Ser. Fiz. Khim. 206, 370 (1972).

237) Langford, C.H., Gray, H.B.: Ligand Substitution Processes. New York: Benjamin 1965.

238) Bender, M.L.: Mechanisms of Homogeneous Catalysis from Protons to Proteins. New York: Wiley-Interscience 1971.

239) Narayanamurti, V., Pohl, R.O.: Rev. Mod. Phys. 42, 201 (1970).

240) Hofacker, G.L., Maréchal, Y., Ratner, M.A.: Dynamical Properties of Hydrogen Bonded Systems. In: The Hydrogen Bond. Recent Developments in Theory and Experiment. Schuster, P. (ed.). Amsterdam: North Holland 1975, pp. 295-357.

241) Stoneham, A.M.: Theory of Defects in Solids. Oxford: Clarendon 1975.

242) Flynn, C.P., Stoneham, A.M.: Phys. Rev. B 1, 3966 (1970).

243) Cleeton, C.E., Williams, N.H.: Phys. Rev. 45, 234 (1934).

244) Cotton, F.A., Wilkinson, G.: Advanced Inorganic Chemistry, 3rd ed. New York: Interscience 1972.

245) Hund, F.: Z. Phys. 43, 805 (1927).

246) Nordheim, L.: Z. Phys. 46, 833 (1927).

247) Gamov, G.: Z. Phys. 51, 204 (1928).

248) Bourgin, D.G.: Proc. Nat. Acad. Sci. USA 15, 357 (1929).

249) Roginskij, S.Z., Rozenkevitch, L.S.: Z. Phys. Chem. B 10, 47 (1930).
250) Wigner, E.P.: Z. Phys. Chem. B 19, 203 (1923).
251) Bell, R.P.: Proc. Roy. Soc. A 139, 466 (1933).
252) Caldin, E.F.: Chem. Rev. 69, 135 (1969).
253) Harmony, M.D.: Chem. Soc. Rev. 1, 211 (1972).
254) Kresge, A.J.: Chem. Soc. Rev. 2, 475 (1973).
255) Eigen, M.: Angew. Chem. Internat. Ed. 1, 3 (1964).
256) a. Bell, R.P., Millington, J.F., Pink, J.M.: Proc. Roy. Soc. A303, 1 (1968); b. Bell, R.P., Crtichlow, J.E.: Proc. Roy. Soc. A325, 35 (1971); c. Bell, R.P., Sørensen, P.E.: J.C.S.Perkin II 1740 (1972). d. Kjær, A.M., Sørensen, P.E., Ulstrup, J.: J.C.S. Perkin II 51 (1978).
257) Westheimer, F.H.: Chem. Rev. 61, 265 (1961).
258) a. Wang, J.-T., Williams, F.: J. Amer. Chem. Soc. 94, 2930 (1972); b. Campion, A., Williams, F.: J. Amer. Chem. Soc. 94, 7633 (1972).
259) a. Dogonadze, R.R., Krishtalik, L.I.: Usp. Khim. 44, 1387 (1975); b. Krishtalik, L.I.: Experimental Investigation of the Elementary Act of Electrode Processes. In.: Itogi Nauki. Elektrokhimiya, Vol. 12. Moscow: VINITI 1977, pp. 5-55.
260) Marcus, R.A.: J. Phys. Chem. 72, 891 (1968).
261) Johnston, H.S.: Adv. Chem. Phys. 3, 131 (1960).
262) Berry, R.S.: Rev. Mod. Phys. 32, 447 (1960).
263) Zimmermann, H.: Angew. Chem. 76, 1 (1964).
264) Somojai, R.L., Hornig, D.F.: J. Chem. Phys. 36, 1980 (1962).
265) Brickmann, J., Zimmermann, H.: Ber. Bunsenges. Phys. Chem. 70, 157 (1966).
266) Brickmann, J., Zimmermann, H.: J. Chem. Phys. 50, 1608 (1969).

267) Vorotyntsev, M.A., Dogonadze, R.R., Kuznetsov, A.M.: Dokl. Akad. Nauk SSSR, Ser. Fiz. Khim. 209, 1135 (1973).
268) Gelbart, W.M., Freed, K.F., Rice, S.A.: J. Chem. Phys. 52, 2460 (1970).
269) Conway, B.E., Bockris, J.O'M.: Can. J. Chem. 35, 1124 (1957).
270) Vorotyntsev, M.A., Chonishvili, G.M.: Dokl. Akad. Nauk SSSR, Ser. Fiz. Khim. 225, 116 (1975).
271) a. Gol'danskij, V.I., Frank.Kamenetskij, M.D., Barkalov, I.M.: Dokl. Akad. Nauk SSSR, Ser. Fiz. Khim. 211, 133 (1973); b. Science 182, 1344 (1973).
272) Gol'danskij, V.I.: Ann. Rev. Phys. Chem. 27, 85 (1976).
273) Austin, R.H., Beeson, K.W., Eisenstein, L., Frauenfelder, H., Gunsalus, I.C.: Biochemistry 14, 5355 (1975).
274) Alberding, N., Austin, R.H., Beeson, K.W., Chan, S.S., Eisenstein, L., Frauenfelder, H., Nordlund, T.M.: Science 192, 1002 (1976).
275) Bolton, W., Perutz, M.F.: Nature 228, 551 (1970).
276) Perutz, M.F., Heidner, E.J., Ladner, J.E., Beetlestone, J.G., Ho, C., Slade, E.F.: Biochemistry 13, 2187 (1974).
277) a. Watson, H.C.: Progr. Stereochemistry 4, 299 (1968); b. Nordwell, J.C., Nunes, A.C., Schoeborn, B.P.: Science 190, 568 (1975).
278) Hoard, J.L., Scheidt, W.R.: Proc. Nat. Acad. Sci. USA 70, 3919 (1973).
279) Peng, S.-M., Ibers, J.A.: J. Amer. Chem. Soc. 98, 8032 (1976).
280) Jortner, J., Ulstrup, J.: J. Amer. Chem. Soc., submitted.
281) Eaton, W.A., Hanson, L.K., Stephens, P.J., Sutherland, J.C., Dunn, J.B.R.: J. Amer. Chem. Soc. 100, 4991 (1978).

282) German, E.D., Dogonadze, R.R., Kuznetsov, A.M., Levich, V.G., Kharkats, Yu.I.: J. Res. Inst. Catalysis Hokkaido Univ. 19, 99 (1971).

283) German, E.D., Dogonadze, R.R., Kuznetsov, A.M., Levich, V.G., Kharkats, Yu.I.: Elektrokhimiya 6, 350 (1970).

284) German, E.D., Kharkats, Yu.I.: Izv. Akad. Nauk SSSR, Ser. Khim. 1029 (1972).

285) German, E.D.: Izv. Akad. Nauk SSSR, Ser. Khim. 959, 2802 (1977).

286) Bordwell, F.G., Boyle, Jr., W.J.: J. Amer. Chem. Soc. 97, 3447 (1976).

287) Dogonadze, R.R., Ulstrup, J., Kharkats, Yu.I.: Dokl. Akad. Nauk SSSR, Ser. Fiz. Khim. 207, 640 (1972).

288) Dogonadze, R.R., Kuznetsov, A.M.: Elektrokhimiya 13, 672 (1977).

289) a. Anderson, P.W.: Phys. Rev. 79, 350 (1950); b. 115, 2 (1959).

290) Yamashita, J., Kondo, J.: Phys. Rev. 109, 730 (1958).

291) Halpern, V.: Proc. Roy. Soc. A291, 113 (1966).

292) Jortner, J., Rice, S.A.: J. Chem. Phys. 44, 3364 (1966).

293) Jortner, J., Ben-Reuven, A.: Chem. Phys. Lett. 41, 401 (1976).

294) Albrecht, A.C.: J. Chem. Phys. 34, 1476 (1961).

295) Clark, R.J.H.: Adv. Infrared Raman Spectroscopy 1, 143 (1975).

296) George, P., Griffiths, J.S.: Electron Transfer and Enzyme Catalysis. In: The Enzymes. Myrbäck, K. (ed.). New York: Academic Press, Vol. 1, pp. 347-389.

297) Halpern, J., Orgel, L.E.: Disc. Faraday Soc. 29, 32 (1960).

298) Manning, P.V., Jarnagin, R.C., Silver, M.: J. Phys. Chem. 68, 265 (1964).

299) Schmidt, P.P.: Austr. J. Chem. 22, 673 (1969).
300) Ulstrup, J.: J. Electroanal. Chem. 79, 191 (1977).
301) a. Schmickler, W.: J. Electroanal. Chem. 82, 65 (1977); b. 83, 87 (1977).
302) a. Doblhofer, K., Ulstrup, J.: J. Phys. 38, C-5, 49 (1977); b. Doblhofer, K., Nölte, D., Ulstrup, J.: Ber. Bunsenges. Phys. Chem. 82, 403 (1978).
303) McConnell, H.A.: J. Chem. Phys. 35, 508 (1961).
304) a. Kharkats, Yu. I.: Elektrokhimiya 8, 1300 (1972); b. 9, 1860 (1973).
305) Kuznetsov, A.M., Kharkats, Yu. I.: Elektrokhimiya 12, 1277 (1976).
306) Laidler, K.J.: J. Chem. Phys. 10, 34, 43 (1942).
307) Taube, H.: Ber. Bunsenges. Phys. Chem. 76, 966 (1972).
308) a. Taube, H.: J. Amer. Chem. Soc. 77, 4481 (1955); b. Sebera, D.K., Taube, H.: J. Amer. Chem. Soc. 83, 1785 (1961); c. Gould, E.S., Taube, H.: J. Amer. Chem. Soc. 85, 3706 (1963); d. Gould, E.S.: J. Amer. Chem. Soc. 87, 4730 (1965).
309) Nordmeyer, F., Taube, H.: J. Amer. Chem. Soc. 90, 1162 (1968).
310) Zanella, A., Taube, H.: J. Amer. Chem. Soc. 94, 6403 (1972).
311) Gaunder, R.G., Taube, H.: Inorg. Chem. 9, 2627 (1970).
312) a. Robson, R., Taube, H.: J. Amer. Chem. Soc. 89, 6487 (1967); b. French, J. E., Taube, H.: J. Amer. Chem. Soc. 92, 6951 (1969); c. Norris, C., Nordmeyer, F.R.: J. Amer. Chem. Soc. 93, 4044 (1971); Hoffman, M.Z.: Simic, M.: J. Amer. Chem. Soc. 94, 1157 (1972).
313) Ulstrup, J.: Acta Chem. Scand. 27, 1067 (1973).
314) a. Ulstrup, J.: Acta Chem. Scand. 23, 3091 (1969); b. 25, 3397 (1971); c. Trans. Faraday Soc. 67, 2645 (1971).

315) a. Sutin, N., Foreman, A.: J. Amer. Chem. Soc. 93, 5274 (1971); b. Przystas, T.J., Sutin, N.: J. Amer. Chem. Soc. 95, 5545 (1973).

316) Campion, R.J., Deck, C.F., King, Jr., P., Wahl, A.C.: Inorg. Chem. 6, 672 (1967).

317) a. Gjertsen, L., Wahl, A.C.: J. Amer. Chem. Soc. 79, 1020 (1959); b. Meyers, O.E., Sheppard, J.E.: J. Amer. Chem. Soc. 83, 4730 (1961).

318) Holzwarth, J., Strohmaier, L.: Ber. Bunsendes. Phys. Chem. 77, 1145 (1973).

319) Peter, L.M., Dürr, W., Bindra, P., Gerischer, H.: J. Electroanal. Chem. 71, 31 (1976).

320) Zandstra, P.J., Weismann, S.I.: J. Amer. Chem. Soc. 84, 4408 (1962).

321) a. Shimada, K., Moshuk, G., Connor, H.D., Caluwe, P., Szwarc, M.: Chem. Phys. Lett. 14, 396 (1972); b. Connor, H.D., Shimada, K., Szwarc, M.: Macromolecules 5, 801 (1972); c. Chem. Phys. Lett. 14. 402 (1972).

322) a. Herriman, J.E., Maki, A.H.: J. Chem. Phys. 39, 778 (1963); Gerson, F., Martin, W.B.: J. Amer. Chem. Soc. 91, 1883 (1969).

323) a. Shimada, K., Szwarc, M.: Chem. Phys. Lett. 28, 540 (1974); b. 34, 503 (1975).

324) a. Wentworth, W.E., Lovelock, J.E., Chen, E.C.: J. Phys. Chem. 70, 445 (1966); b. Chen, E.C.M., Wentworth, W.E.: J. Chem. Phys. 63, 3183 (1975).

325) Dogonadze, R.R., Kornyshev, A.A., Kuznetsov, A.M., Marsagishvili, T.A.: J. Physique 38, C5-35 (1977).

326) Kliewer, K.L., Fuchs, R.: Adv. Chem. Phys. 27, 355 (1974).

327) Bockris, J.O'M., Devanathan, M.A., Müller, K.: Proc. Roy. Soc. A274, 55 (1963).

328) Damaskin, B.B., Frumkin, A.N.: Electrochim. Acta 19, 173 (1973).

329) Parsons, R.: J. Electroanal. Chem. 59, 279 (1975).

330) Dogonadze, R.R., Kuznetsov, A.M.: Elektrokhimiya 7, 172 (1971).

331) Vorotyntsev, M.A., Dogonadze, R.R., Kuznetsov, A.M.: Elektrokhimiya 7, 306 (1971).

332) a. Gerischer, H.: Adv. Electrochemistry Electrochem. Eng. 1, 139 (1961); b. In: Physical Chemistry. An Advanced Treatise. Eyring, H., Henderson, D., Jost, W. (eds.). New York: Academic Press 1970, Vol. 9B. pp. 461-540.

333) Dogonadze, R.R., Kuznetsov, A.M., Vorotyntsev, M.A.: J. Electroanal. Chem. 25, App. 17 (1970).

334) Dogonadze, R.R., Kuznetsov, A.M., Vorotyntsev, M.A.: Croat. Chim. Acta 44, 257 (1972).

335) Vorotyntsev, M.A., Dogonadze, R.R., Kuznetsov, A.M.: Fiz. Tverd. Tela 12, 1605 (1970).

336) a. Dogonadze, R.R., Chizmadzhev, Yu.A.: Dokl. Akad. Nauk SSSR, Ser. Fiz. Khim. 150, 333 (1963); b. Dogonadze, R.R., Kuznetsov, A.M., Chizmadzhev, Yu.A.: Zh. Fiz. Khim. 38, 1195 (1964).

337) Frumkin, A.N.: Z. Phys. Chem. (A) 164, 121 (1933).

338) Frumkin, A.N., Petrij, O.A., Nikolajeva-Fedorovitch, N.V.: Electrochim. Acta 8, 177 (1963).

339) Angell, D.H., Dickinson, I.: J. Electroanal. Chem. 35, 55 (1972).

340) a. Samec, Z., Weber, J.: J. Electroanal. Chem. 77, 163 (1977); b. Marecek, V., Samec, Z., Weber, J.: J. Electroanal. Chem. 94, 169 (1978).

341) Weaver, M.J., Anson, F.C.: J. Phys. Chem. 80, 1861 (1976).

342) Pleskov, Yu.V., Tyagai, V.A.: Dokl. Akad. Nauk SSSR, Ser. Fiz. Khim. 141, 1135 (1961).

343) Delahay, P.: Double Layer and Electrode Kinetics. New York: Interscience 1965.

344) Capon, A., Parsons, R.: J. Electroanal. Chem. 46, 215 (1973).

345) Vojnovic, M.V., Sepa, D.B.: J. Chem. Phys. 51, 5344 (1969).

346) a. Bockris, J.O'M., Mannan, R.J., Damjanovic, A.: J. Chem. Phys. 48, 1898 (1968); Galizzioli, D., Trassatti, S.: J. Electroanal. Chem. 44, 367 (1973).

347) Krishtalik, L.I.: Adv. Electrochemistry Electrochem. Eng. 7, 283 (1970).

348) Krishtalik, L.I.: Electrochim. Acta 13, 1045 (1968).

349) Rotenberg, Z.A., Lakomov, V.I., Pleskov, Yu.I.: Elektrokhimiya 6, 515 (1970).

350) Dogonadze, R.R., Kuznetsov, A.M., Ulstrup, J.: Electrochim. Acta 22, 967 (1977).

351) a. Schmickler, W., Ulstrup, J.: Chem. Phys. 19, 217 (1977); Schmickler, W.: Ber. Bunsenges. Phys. Chem. 82, 477 (1978).

352) Duke, C.B.: Tunnelling in Solids. New York, Academic Press 1969.

353) a. Schultze, J.W., Vetter, K.J.: Electrochim. Acta 18, 889 (1973); Ber. Bunsenges. Phys. Chem. 77, 945 (1973); c. Kohl, P., Schultze, J.W.: Ber. Bunsenges. Phys. Chem. 77, 953 (1973).

354) Many, A., Goldstein, Y., Grover, N.B.: Semiconductor Surfaces. Amsterdam: North Holland 1965.

355) a. Parker, G.M., Mead, C.A.: Appl. Phys. Lett. 14, 21 (1969); b. Gundlach, K.H.: Festkörperprobleme 11, 237 (1971).

356) Schultze, J.W., Stimming, U.: Z. Phys. Chem. NF 98, 285 (1975).
357) Lambe, J., Jaklevic, R.C.: Phys. Rev. 165, 821 (1968).
358) Coleman, R.V., Clark, J.M., Korman, C.S.: Inelastic Tunnelling Spectroscopy with Application to Biology and Surface Physics. In: Inelastic Tunnelling Spectroscopy. Wolfram, T. (ed.). Berlin: Springer-Verlag 1978, pp. 34-61.
359) Dogonadze, R.R., Kuznetsov, A.M.: Izv. Akad. Nauk SSSR, Ser. Khim. 1085 (1964).
360) Overhof, H: Festkörperprobleme 16, 239 (1976).
361) Eichhorn, G.I. (ed.): Inorganic Biochemistry. Amsterdam: Elsevier 1973.
362) a. DeVault, D., Chance, B.: Biophys. J. 6, 825 (1966); b. DeVault, D., Parkes, J.H., Chance, B.: Nature 215, 642 (1967); c. Gutman, F.: Nature 219, 1359 (1967).
363) a. Dutton, P.L., Kaufman, K.J., Chance, B., Rentzepis, P.M.: Biophys. J. 23, 207 (1978).
364) Chance, B., DeVault, D., Legallais, V., Meta, L., Yonetani, T.: Kinetics of Electron Transfer Reactions in Biological Systems. In: Fast Reactions and Primary Processes in Chemical Kinetics. Claesson, S. (ed.). New York: Interscience 1967, pp. 437-468.
365) Blow, D.H.: Acc. Chem. Res. 9, 145 (1976).
366) Peters, K., Applebury, M.L., Rentzepis, P.M.: Proc.Nat. Acad. Sci. USA 74, 3119 (1977).
367) a. Wang, J.H.: Science 161, 328 (1968); b. Proc. Nat. Acad. Sci. USA 66, 874 (1970).
368) a. Jordan, P.: Naturwiss. 26, 693 (1938); Szent-Györgyi, A.: Science 93, 609 (1941); Nature 148, 157 (1941).
369) Gutman, F., Lyons, L.E.: Organic Semiconductors. New York: Wiley 1967.

370) a. Winfield, M.E.: J. Mol. Biol. 12, 600 (1965); b. Tanako, T., Kallai, O.B., Swanson, R., Dickerson, R.E.: J. Biol. Chem. 248, 5234 (1973).

371) a. Cope, F.W., Straub, K.D.: Bull. Math. Biophys. 31, 761 (1969); Kemeny, G., Rosenberg, B.: J. Chem. Phys. 53, 3549 (1970). c. Vol'kenstein, M.V.: J. Theor.Biol. 34, 193 (1972); d. Kemeny, G., Goklany, I.M.: J.Theor. Biol. 40, 107 (1973).

372) Grigorov, L.N.: Izv. Akad. Nauk SSSR, Ser. Biol. 3, 447 (1969).

373) a. Grigorov, L.N., Chernavskij, D.S.: J. Theor. Biol. 39, 1 (1973).

374) Vol'kenstein, M.V., Dogonadze, R.R., Madumarov, A.K., Urushadze, Z.D., Kharkats, Yu.I.: Molekulyarnaya Biologiya 6, 431 (1972).

375) Hopfield, J.J.: Proc. Nat. Aca. Sci. USA 71, 3640 (1974).

376) Dogonadze, R.R., Kharkats, Yu.I., Ulstrup, J.: Chem. Phys. Lett. 37, 360 (1976).

377) Koshland, Jr., D.E.: J. Theor. Biol. 2, 75 (1962).

378) a. Ferreira, R.: J. Theor. Biol. 39, 665 (1973); b. Damjanovich, S., Somogyi, B.: J. Theor. Biol. 41, 567 (1973).

379) Storm, D.R., Koshland, Jr., D.E.: Proc. Nat. Acad. Sci. USA 66, 445 (1970).

380) Perutz, M.F.: Proc. Roy. Soc. B167, 448 (1967).

381) Vallee, B.L., Williams, R.J.P.: Proc. Nat. Acad. Sci. USA 59, 498 (1968).

382) Krishtalik, L.I.: Molekulyarnaya Biologiya 8, 91 (1974).

383) Vol'kenstein, M.V.: Izv. Akad. Nauk. SSSR, Ser. Biol. 6, 805 (1971).

384) Lumry, R., Biltonen, R.: Thermodynamic and Kinetic Aspects of Protein Conformations in Relation to Physiological Function. In: Structure and Stability of Biological Macromolecules. New York: Dekker 1969, pp.65-212.

385) a. Blumenfel'd, L.A.: Biofizika 16, 724, 954 (1971); b. J. Theor. Biol. 58, 269 (1976).

386) a. McElroy, J.D., Feher, G., Mauzerall, D.C.: Biochim. Biophys. Acta 333, 261 (1974); b. Dutton, P.L., Kihara, T., McCray, J.A., Thornber, J.P.: Biochim. Biophys. Acta 226, 81 (1971); c. Hales, B.J.: Biophys. J. 16, 471 (1976); d. Kihara, T., McCray, J.A.: Biochim. Biophys. Acta 292, 297 (1973); e. Clayton, R.K., Yau, H.F.: Biophys. J. 12, 867 (1972); f. Hsi, E.S.P., Bolton, J.R.: Biochim. Biophys. Acta 347, 126 (1976); g. Romijn, J.C., Amesz, J.: Biochim. Biophys. Acta 423, 164 (1976).

387) Dutton, P.L., Prince, R.G., Tiede, D.M., Petty, K.M., Kaufman, K.J., Netzel, T.L., Rentzepis, P.M.: Brookhaven Symp. Biol. 28, 213 (1976).

388) Sutin, N.: Adv. Chem. Ser. 162, 154 (1977).

389) Pecht, I., Faraggi, M.: Proc. Nat. Acad. Sci. USA 69, 902 (1972).

390) Mandel, N., Mandel, G., Trus, B.L., Rosenberg, J., Carlson, G., Dickerson, R.E.: J. Biol. Chem. 252, 4619 (1977).

391) Jacks, C.A., Bennett, L.E., Raymond, W.N., Lovenberg, W.: Proc. Nat. Acad. Sci. USA 71, 1118 (1974).

392) Rawlings, J., Wherland, S., Gray, H.B.: J. Amer. Chem. Soc. 98, 2177 (1976).

393) Tien, H.T.: Bilayer Lipid Membranes. New York: Dekker 1974.

394) Boyer, P.D., Chance, B., Ernster, L., Mitchell, P., Racker, E., Slater, E.C.: Ann.Rev.Biochem. 46, 955 (1977).

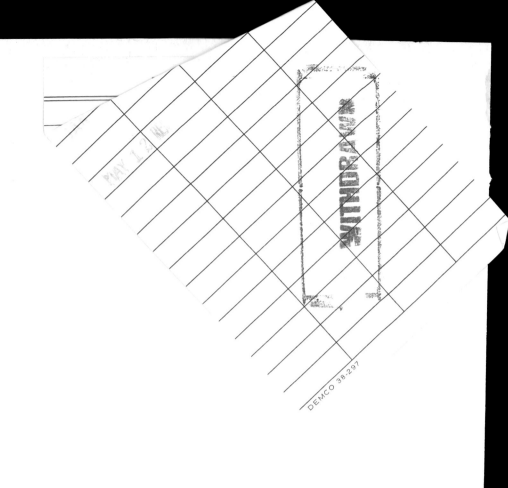